Microstructure and Mechanical Behaviour of Structural Materials

Microstructure and Mechanical Behaviour of Structural Materials

Guest Editors

Saif Haider Kayani
Byung-Joo Kim

MDPI

Basel • Beijing • Wuhan • Barcelona • Belgrade • Novi Sad • Cluj • Manchester

Guest Editors
Saif Haider Kayani
Herbert Gleiter Institute of Nanoscience
Nanjing University of Science and Technology
Nanjing
China

Byung-Joo Kim
Advanced Production Technology Center
Research Institute of Medium and Small Shipbuilding
Busan
Republic of Korea

Editorial Office
MDPI AG
Grosspeteranlage 5
4052 Basel, Switzerland

This is a reprint of the Special Issue, published open access by the journal *Crystals* (ISSN 2073-4352), freely accessible at: https://www.mdpi.com/journal/crystals/special_issues/64ZFT5G999.

For citation purposes, cite each article independently as indicated on the article page online and as indicated below:

Lastname, A.A.; Lastname, B.B. Article Title. *Journal Name* **Year**, *Volume Number*, Page Range.

ISBN 978-3-7258-6027-2 (Hbk)
ISBN 978-3-7258-6028-9 (PDF)
https://doi.org/10.3390/books978-3-7258-6028-9

Cover image courtesy of Saif Haider Kayani

Contents

About the Editors

Saif Haider Kayani

Saif Haider Kayani is a Postdoctoral Fellow at the School of Materials Science and Engineering, Herbert Gleiter Institute of Nanoscience, Nanjing University of Science and Technology, Nanjing, China. He holds a Ph.D. in Advanced Materials Engineering from the University of Science and Technology (UST), Republic of Korea. Dr. Kayani's research focuses on the microstructure, mechanical properties, and corrosion of metallic materials, with an emphasis on their interrelationships, explored through cutting-edge characterization techniques such as electron microscopy and atom probe tomography. His work has garnered multiple accolades, including the *Outstanding Contribution in Materials Photography* award from the Korea Institute of Metals and Materials. In 2025, he was honored with a prestigious academic award from the Korea Institute of Materials Science for his significant contributions to the field of metallic materials. Dr. Kayani's current research is dedicated to the development of sustainable metallic alloys using post-consumer scrap. His innovative work has resulted in numerous patents and publications in leading peer-reviewed journals. Additionally, he actively contributes to the academic community as a peer reviewer for top journals in metallurgy.

Byung-Joo Kim

Byung-Joo Kim is a Senior Researcher at the Enterprise Collaboration Division of the Research Institute of Medium and Small Shipbuilding (RIMS) in the Republic of Korea. He earned his doctoral degree in Materials Science and Engineering from Pusan National University, Republic of Korea. He has contributed significantly to numerous research projects and is currently focused on advancing welding processes and developing non-ferrous materials used in shipbuilding. His research has led to several patents and publications in leading peer-reviewed journals. Acknowledging his achievements, he was honored with presentation awards at the Global Conference on Innovation Materials 2023 and the 74th World Foundry Congress 2024. His research interests include metallic and composite materials, sustainable practices, welding innovation, and the study of deformation mechanisms.

crystals

MDPI

Editorial

Microstructure and Mechanical Behavior of Structural Materials

Saif Haider Kayani [1,*] and Byung-Joo Kim [2,*]

[1] Lightweight Materials Research Division, Korea Institute of Materials Science, Changwon 51508, Republic of Korea
[2] Research Institute of Medium and Small Shipbuilding, Busan 46757, Republic of Korea
* Correspondence: saifkayani@kims.re.kr (S.H.K.); bjkim@rims.re.kr (B.-J.K.)

Metallic materials continue to attract significant interest for structural applications globally, ranging from high-performance sectors like aerospace and automotive industries to everyday household items. The mechanical performance of these structural metallic materials is predominantly governed by their microstructure, which directly influences their mechanical properties. Consequently, understanding the microstructure–property relationship is crucial for predicting and optimizing the performance of structural components in various applications.

This Special Issue, titled "Microstructure and Mechanical Behavior of Structural Materials", presents a collection of 13 original research articles and 2 comprehensive review papers, each contributing to the understanding of diverse structural materials and their microstructure–property correlations. Several studies focus on Fe-based alloys, exploring the microstructural and mechanical behavior of both low-carbon and stainless steels. For instance, Shekhawat et al. [1] examined the influence of initial grain orientation on grain rotation in low-carbon steel, revealing that grains aligned with the γ-fiber (ND//<111>) orientation remained stable, while grains in other orientations experienced notable rotation. Pompeo et al. [2] investigated 17-4PH stainless steel fabricated via bound metal deposition, demonstrating that a 45° infill angle reduced δ-ferrite content and resulted in enhanced hardness.

This Special Issue further includes findings on lightweight (Ti- and Al-based) alloys highlighting how processing techniques and microstructural evolution influence the mechanical properties. Kim et al. [3] investigated the intermetallic compound layer thickness and morphology in Al-Si-Mg/STS420 joints, demonstrating a significant effect of reaction time and temperature on IMC layer thickness and hardness gradients. Hulka et al. [4] examined Ti-20Zr alloy as a potential biomaterial, revealing a lamellar α and β structure post-heat treatment and air cooling. The softer α phase, approximately 30% softer than the β phase, along with a stable corrosion-resistant oxide layer formed in simulated body fluid, promotes bone integration, supporting its biocompatibility. Krawczyk et al. [5] showed that aging Ti-3Al-8V-6Cr-4Zr-4Mo alloys at 550 °C significantly enhances texture strength, whereas aging at lower temperatures leads to reduced texture intensity and lower mechanical performance. Kayani et al. [6] analyzed the impact of powder size on pore density and IMC formation in porous TiAl alloys, concluding that finer powders (325 mesh) increase porosity and facilitate faster diffusion during sintering, facilitating the formation of IMCs, especially at lower temperatures. Rawles et al. [7] optimized the rolling process of Mg-Al and Ti-6Al-4V alloys, achieving significant improvements in strength and hardness after cold working. Korchef and Souid [8] demonstrated that optimized equal-channel angular pressing processing refines the grain size in Al alloys, increasing hardness and yield strength. Kušter et al. [9] studied polymer composites containing Al-Cu-Fe quasicrystals and found that adding more than 20% of quasicrystals improved impact strength and wear resistance, while an excess of 30% compromised mechanical performance. Newport et al. [10] focused on the post-processing heat treatment of Al-Mg-Sc composites produced via direct energy deposition (DED), showing that a 300 °C heat treatment forms $Al_3(Sc,Zr)$ phases, optimizing tensile strength without significant grain coarsening.

Several studies in this Special Issue provide valuable insights into the fracture and damage mechanisms in specialized materials. Deng et al. [11] investigate the three-point bending fracture process of Yellow River ice, emphasizing the role of grain size and crack depth ratio in influencing fracture resistance. Wang et al. [12] employ multiscale simulations to study the spallation characteristics of Ta under high-strain-rate impacts, revealing the evolution of damage mechanisms and the effects of strain rate on spallation strength. Liu et al. [13] examine the impact of secondary orientation on the deformation behavior of Ni-based single-crystal superalloys at room temperature (RT) and 850 °C. Their findings indicate that the secondary orientation has a negligible influence on mechanical properties at RT but plays a significant role at elevated temperatures, contributing to the enhanced ductility of superalloys.

Lastly, two review articles provide additional insights into the relationship between microstructure evolution and mechanical properties. Zarinejad et al. [14] focus on additive manufacturing techniques for W alloys, addressing persistent issues like thermal stress-induced cracking and oxide formation. The review discusses innovative methods, including laser-based DED and binder jetting, that enhance W-based material performance. Thool et al. [15] reviewed the microstructural and textural evolution in HCP metals such as Zr, Mg, and Ti, emphasizing the impact of twinning and dislocation movements on anisotropic behavior. Both studies offer a comprehensive view of emerging trends and future directions in advanced structural materials research. Overall, the studies in this Special Issue provide valuable insights into how microstructural modifications impact the mechanical properties of structural materials, supporting advancements in material design and process optimization.

Author Contributions: Conceptualization, S.H.K. and B.-J.K.; methodology, S.H.K. and B.-J.K.; validation, S.H.K. and B.-J.K.; resources, S.H.K. and B.-J.K.; data curation, S.H.K. and B.-J.K.; writing—original draft preparation, S.H.K. and B.-J.K.; writing—review and editing, S.H.K. and B.-J.K. All authors have read and agreed to the published version of the manuscript.

Conflicts of Interest: The authors declare no conflict of interest.

References

1. Shekhawat, S.; Khatirkar, R.; Demiral, M. Lattice Rotation Dependence on Microstructural Characteristics in a Low Carbon Steel. *Crystals* **2024**, *14*, 186. [CrossRef]
2. Di Pompeo, V.; Santecchia, E.; Santoni, A.; Sleem, K.; Cabibbo, M.; Spigarelli, S. Microstructure and Defect Analysis of 17-4PH Stainless Steel Fabricated by the Bound Metal Deposition Additive Manufacturing Technology. *Crystals* **2023**, *13*, 1312. [CrossRef]
3. Kim, B.-J.; Lim, H.-Y.; Kayani, S.H.; Lee, Y.-S.; Kim, S.-H.; Cha, J.-H. Formation and Evolution of Interfacial Structure in Al–Si–Mg/Stainless Steel Bimetals during Hot-Dipping Process. *Crystals* **2024**, *14*, 387. [CrossRef]
4. Hulka, I.; Uțu, I.-D.; Brito-Garcia, S.; Verdu-Vazquez, A.; Mirza-Rosca, J.C. Electrochemical Study and Mechanical Properties of Ti-Zr Alloy for Biomedical Applications. *Crystals* **2024**, *14*, 493. [CrossRef]
5. Krawczyk, J.; Frocisz, Ł.; Goły, M.; Tomasik, S.; Śleboda, T. The Analysis of Changes in the Crystal Structure of Near-Beta Titanium Alloy in the Solution-Treated and Aged Conditions after Static Tensile Testing. *Crystals* **2023**, *13*, 1223. [CrossRef]
6. Kayani, S.H.; Ajmal, H.M.S.; Kim, B.-J.; Park, N.-K.; Euh, K. Influence of Powder Size on Pore Characteristics and Intermetallic Phase Kinetics in Porous Ti-Al Alloys. *Crystals* **2024**, *14*, 559. [CrossRef]
7. Rawles, J.; Fialkova, S.; Hubbard, K.; Xu, Z.; Hale, C.; Sankar, J. Optimizing the Rolling Process of Lightweight Materials. *Crystals* **2024**, *14*, 582. [CrossRef]
8. Korchef, A.; Souid, I. Grain Refinement and Strengthening of an Aluminum Alloy Subjected to Severe Plastic Deformation through Equal-Channel Angular Pressing. *Crystals* **2023**, *13*, 1160. [CrossRef]
9. Kušter, M.; Samardžija, Z.; Komelj, M.; Huskić, M.; Bek, M.; Pierson, G.; Kouitat-Njiwa, R.; Dubois, J.-M.; Šturm, S. Effect of Al-Cu-Fe Quasicrystal Particles on the Reinforcement of a Polymer–Matrix Composite: From Surface to Mechanical Properties. *Crystals* **2024**, *14*, 216. [CrossRef]
10. Boillat-Newport, R.; Isanaka, S.P.; Liou, F. Heat Treatment Post-Processing for the Improved Mechanical Properties of Scalmalloy® Processed via Directed Energy Deposition. *Crystals* **2024**, *14*, 688. [CrossRef]
11. Deng, Y.; Wang, J.; Meng, Y.; Zhu, Y. Simulation Analysis of Three-Point Bending Fracture Process of Yellow River Ice. *Crystals* **2024**, *14*, 729. [CrossRef]

12. Wang, Y.; Guo, Y.; Li, S.; Li, Y.; Yang, M.; Shi, W. Multiscale Simulation Study on the Spallation Characteristics of Ductile Metal Ta under High Strain Rate Impact. *Crystals* **2024**, *14*, 833. [CrossRef]
13. Liu, S.; Zong, C.; Ma, G.; Zhao, Y.; Huang, J.; Guo, Y.; Chen, X. Effects of Temperature and Secondary Orientations on the Deformation Behavior of Single-Crystal Superalloys. *Crystals* **2024**, *14*, 996. [CrossRef]
14. Zarinejad, M.; Tong, Y.; Salehi, M.; Mu, C.; Wang, N.; Xu, Y.; Rimaz, S.; Tian, L.; Kuah, K.X.; Chen, X. Advancements and Perspectives in Additive Manufacturing of Tungsten Alloys and Composites: Challenges and Solutions. *Crystals* **2024**, *14*, 665. [CrossRef]
15. Thool, K.; Yazar, K.U.; Kavimani, V.; Gupta, A.; Choi, S.-H. Microstructural and Textural Evolution in Hexagonal Close-Packed Metals: The Case of Zirconium, Magnesium, and Titanium. *Crystals* **2024**, *14*, 727. [CrossRef]

crystals

MDPI

Article

Lattice Rotation Dependence on Microstructural Characteristics in a Low Carbon Steel

Satish K. Shekhawat [1], Rajesh K. Khatirkar [2,*] and Murat Demiral [3,*]

[1] Powergrid Corporation of India Limited, Technology Development Department, PARTeC, Panchgaon, Manesar, Gurgaon 122413, Haryana, India; satishshekhawat@powergrid.in
[2] Department of Metallurgical and Materials Engineering, Visvesvaraya National Institute of Technology (VNIT), South Ambazari Road, Nagpur 440010, Maharashtra, India
[3] College of Engineering and Technology, American University of the Middle East, Egaila 54200, Kuwait
* Correspondence: rajeshk@mme.vnit.ac.in (R.K.K.); murat.demiral@aum.edu.kw (M.D.)

Abstract: The lattice rotation behavior of low-carbon (LC) steel subjected to tensile deformation was studied by electron backscatter diffraction (EBSD). The EBSD scans of the same region were taken before and after tensile deformation. The rotation of the grains was found to depend on a number of factors like the initial orientation, the size of the grains, the number of neighboring grains and the region of the grain. The region near the grain boundaries was found to have significant deviation from that of the interior of the grain. The lattice rotations were also simulated using DAMASK software. The simulations gave information about the state of stress for each grain and the strain gradients developed during the deformation. The orientation dependence of misorientation and geometrically necessary dislocations (GNDs) was also studied. It was found that the misorientations changed more gradually in α-fiber grains than in γ-fiber grains.

Keywords: EBSD; grain rotation; DAMASK; GNDs

1. Introduction

Low-carbon steel is the perfect material for structural elements and welded constructions in a variety of industries, including construction and automobile manufacturing, because of its exceptional weldability, ductility and toughness. Its adaptability to applications needing a blend of strength and flexibility is highlighted by its resistance to deformation and capacity to absorb impact energy. Additionally, because of its lower cost in comparison to alloy and high-carbon steels, it is the material of choice for applications where high strength is not the only need, which helps to streamline manufacturing processes.

The mechanical properties of alloys, notably low-carbon steel, are intricately tied to several factors, with the initial grain size and texture and the size and distribution of second-phase particles playing pivotal roles. The evolution of the macroscopic texture in low-carbon steel is particularly influential, significantly impacting the engineering properties, a phenomenon directly linked to the crystal plasticity behavior of individual grains [1]. The development of deformation texture hinges on the activities of slip within specific slip systems. This induces a gradual rotation of grains from their initial orientation, allowing them to remain cohesive with neighboring grains and delaying fracture. The consequence of such slip activities in different regions of grains, coupled with the associated lattice rotation, leads to a substantial alteration in the shape of grains. This intricate interplay of factors underscores the complex nature of how mechanical properties are determined in low-carbon steel, highlighting the importance of understanding crystal plasticity and texture evolution for effective material design and engineering applications.

A few studies published in the literature examined the lattice rotation behavior of alloys including low-carbon steels. Zaeferer et al. [2] studied the low-alloyed TRIP steel's microstructure, transformation processes and the relationship between its microstructure and mechanical qualities. The formation of the microstructure occurred when γ-grains expanded during inter-critical annealing and then shrank during the subsequent cooling process, all without the nucleation of new α-grains. Using a combination of optical microscopy (OM), scanning electron microscopy (SEM), electron backscattered diffraction (EBSD) and X-ray diffraction (XRD), Dutta et al. [3] examined the deformation microstructure beneath the area of ultrasonic treatment in the deformed samples. A considerable decrease in both the dislocation density and the fraction of low-angle grain boundaries together with preferred grain rotation was noticed. Recovery was correlated with a notable decrease in sub-grain development during deformation due to the ultrasound's softening impact. Nesterova [4] investigated the inter-relation between the microstructure and texture evolutions and their joint influence on the mechanical behavior. It was demonstrated that grains with well-developed pre-strain dislocation barriers and involvement with specific orientation groups exhibited a high localization of plastic flow inside microbands that followed an orthogonal strain-path mutation. Dhinwal et al. [5] studied the processing conditions' effects on texture and microstructure evolution during the multi-pass asymmetric rolling of extra-low-carbon steel. The applied thickness reduction during each pass, the sheet's overall thickness reduction and the degree of imposed asymmetry all had an impact on the development of both texture and microstructure. Gupta et al. [6] studied the relationship between strain path and microstructure with texture development in BCC Ti-15V-3Cr-3Sn-3Al alloy. It was observed that the texture growth was more strongly reliant on the strain path than the original microstructure during cold rolling. Mahadule et al. [7] analyzed the texture development during cold rolling in a Ti-15V-3Cr-3Sn-3Al alloy via experiments and numerical modeling based on crystal plasticity. It was observed that γ-(normal direction, ND//<111>) and α-(rolling direction, RD//<110>) fibers strengthened with cold rolling. Moura [8] studied the effect of crystallographic texture and the strain hardening behavior of UNS S32304 lean duplex stainless steel under a hot tensile test. Because of the austenite to ferrite phase transformation, the strain rate had an impact on the strain hardening behavior of duplex stainless steel. With an increase in strain rate, the transition strain between deformation phases grew. Nagarajan et al. [9] used in situ EBSD examination and crystal plasticity modeling to try and understand the impact of the coordinated deformation of polycrystalline Al-Mg alloy on its lattice rotation behavior during tensile deformation. It was observed that more rotations of grains were possible than the number of slip systems that were triggered during deformation. A potential shortest path rotated directly to the nearby stable end orientation when the misorientations shared by the immediate neighbors were more than 30°. Liu et al. [10] investigated the deformation mechanism in terms of slip behavior, the evolution of a grain boundary and the development of geometrically necessary dislocations (GNDs) for ferrite in a low-carbon Al-killed steel. While the microstructural unit size limited the dislocation's mobility, small grains were more likely to collect GNDs. Grain orientation had less of an impact on all grains when their Schmid Factor (SF) was high. An in situ EBSD analysis of nickel-based superalloy deformation behavior during uniaxial tensile tests was conducted by Zhang et al. [11]. As strain increased, the average misorientation angle rose and subsequently fell, and this was related to the number of dislocations. A number of coordinated strengthening mechanisms worked together during the deformation process to boost the alloy's strength. Chen and Wang [12] conducted a comparable work in which they examined 409 L ferritic stainless steel in situ using EBSD during tensile testing. The tensile axis of the rolling direction specimens rotated in the direction of <101> during tensile deformation, which was the stable end orientation of body-centered cubic (BCC) metals. However, because less advantageous slip systems were in place in transverse-direction specimens, the rotation of the tensile axis towards <101> was constrained. In both specimens, the most advantageous slip system was {110}<111>. Mehmet Sahin Atas et al. [13–15] studied model Ni-15Al, Ni-Al-Y and Ni-Al-Nb superalloys and showed that aging time significantly influenced the size, shape and type of precipitates in these systems. They also studied these systems ex situ.

Our main goal in this work is to use quasi in situ electron backscatter diffraction (EBSD) methods to examine the lattice rotation behavior of low-carbon steel under tension. Our analysis encompasses an examination of active slip systems, lattice rotation rates and rotation paths, with a particular focus on understanding the influence of misorientation and grain boundary segments shared with immediate neighbors. Additionally, we delve into the nuances of the kinematics associated with local lattice reorientation induced by second-phase particles, slip plane kinking and the inherent deformation heterogeneity across neighboring grains. In tandem with experimental findings, the study employs DAMASK software for the simulation of lattice rotations. This computational approach not only validates experimental observations but also provides a deeper understanding of the stress states within grains and the evolving strain gradient during deformation. The integration of experimental and simulated results offers a holistic perspective on the mechanical intricacies at the microstructural level.

2. Materials and Methods

2.1. Material and Processing

The chemical composition of the low carbon used in the present investigation was determined by optical emission spectroscopy (OES) and is given in Table 1. The initial sheet was 0.5 mm thick.

Table 1. Chemical composition (in weight% alloying elements) of the low-carbon steel.

	C	Mn	Al	Si	P	Cr	Fe
LC steel	0.008	0.33	0.22	0.50	0.05	0.011	Balance

In order to investigate the effect of deformation on grain rotation, tensile tests were carried out. The tensile specimen was prepared along the rolling direction (RD) as per ASTM E8 [16]. The gauge section of the tensile specimen was metallographically polished followed by electropolishing, and finally, sub-micron colloidal silica [17–19] was used to carry out the EBSD investigations before and after deformation. A square region was considered, and to identify same region before and after deformation, one microindentation was placed at each corner. Tensile tests were carried out on an INSTRON^{TM} 3345 machine at a cross head speed of 1 mm/min. The experiments were repeated at strains of 2%, 3%, 5% and 7%, and the same area underwent EBSD microstructural analysis. To monitor the rotation and deformation behavior of the same grain (or region), interrupted tests were required.

2.2. Microstructure Characterization

EBSD was used to characterize the microstructure. The initial sample (before deformation) was electropolished with an electrolyte consisting of a mixture of ethanol and perchloric acid at a ratio of 80:20 (by volume) with Struer's Letropol-5^{TM} electro-polisher for 30 s at 18 V DC and 0 °C. EBSD measurements were conducted on an FEI Quanta 200HV scanning electron microscope (SEM) equipped with an EDAX, USA, EBSD system. The step size used was 1 μm to cover a large area. To prevent any negative impact on the quality of the data, the camera settings were maintained throughout the scanning process for every sample. EBSD data were post processed with TSL 7.2 software [20], and all the data points with questionable accuracy, i.e., confidence index CI) < 0.1 [21], were neglected during the analysis. Kernel average misorientation [22], the density of geometrically necessary dislocation [23] and grain orientations were tracked during deformation.

2.3. Crystal Plasticity (CP) Tensile Simulations

CP tensile simulations were carried out using the open source code Düsseldorf Advanced MAterial Simulation Kit (DAMASK) [24] with the integrated spectral solver. The orientation input files were generated from the experimental EBSD data of the initial undeformed sample. In the next section, the constitutive equations used in the simulations are summarized.

The deformation gradient splits into a plastic and elastic component through multiplication as follows:

$$F = F^e\,F^p, \tag{1}$$

where F^e denotes the rotation and elastic distortion of the crystal lattice and F^p represents the inelastic shear deformation on the slip planes.

The following generalized Hook's Law is used to calculate the material's stress response:

$$S = C : E, \tag{2}$$

with S being the Second Piolo–Kirchoff stress, C the forth-order anstropic elastic tensor and E the Green–Lagrange strain.

The shear contributions of the slip systems α make up the plastic velocity gradient as follows:

$$L^p = \sum_{\alpha} \dot{\gamma}^{\alpha} m^{\alpha} \otimes n^{\alpha}, \tag{3}$$

where $\dot{\gamma}^{\alpha}$ is the slip rate of the slip system, m^{α} the slip direction and n^{α} the normal to the slip plane which forms the Schmidt tensor.

The shear rates of the slip systems α are calculated in relation to the selected material model as follows:

$$\dot{\gamma}^{\alpha} = \dot{\gamma}_0^{\alpha} \left| \frac{\tau^{\alpha}}{g^{\alpha}} \right|^{n} \mathrm{sgn}(\tau^{\alpha}), \tag{4}$$

with $\dot{\gamma}_0^{\alpha}$ being the reference slip rate, τ^{α} determining the resolved shear stress, n denoting the rate sensitivity and g^{α} being the slip resistance of the respective slip system.

After that, the slip resistance itself asymptotically changes from an initial value of g_0^{α} to a saturation value of g_{sat}^{α} using the following equation:

$$\dot{g}^{\alpha} = h_0 \left| \dot{\gamma}^{\beta} \right| \left| 1 - \frac{g^{\beta}}{g_{sat}^{\beta}} \right|^{\alpha} sgn\left(1 - \frac{g^{\beta}}{g_{sat}^{\beta}} \right) h^{\alpha\beta}. \tag{5}$$

Here, h_0 and α represent the self-hardening coefficient and the hardening exponent, respectively, and $h^{\alpha\beta}$ is components of the slip interaction matrix with a value of 1.0 and 1.4 for coplanar and non-coplanar slip systems, respectively.

The constants used in CP tensile simulations using DAMASK are given in Table 2, taken from Refs. [25,26]. The simulations were carried out with a step size of 0.0001 considering two slip systems ({110}<111>+{112}<111>). The tensile boundary conditions were defined using the strain matrix [1 0 0, 0 −0.5 0, 0 0 −0.5].

Table 2. Material parameters of low-carbon steel used in CP simulations.

Parameter	Value
C_{11}	233.3 GPa
C_{12}	135.5 GPa
C_{44}	118.0 GPa
$\dot{\gamma}_0^{\alpha}$	0.001 1/s
h_0	1000 MPa
n	20
α	2

Table 2. *Cont.*

Parameter	Value
g_0^{α}	95 MPa {110} 97 MPa {112}
g_{sat}^{α}	222.0 MPa {110} 412.7 MPa {112}

3. Results and Discussion

Figure 1 shows the true stress–true strain curve and $\varphi_2 = 45°$ constant section of the orientation distribution function (ODF) [27] of the LC steel used in the present investigation. The stress–strain curve shows that the material work hardened with an increase in deformation. The ductility was around 20%. The ODF showed a pronounced γ-fiber (ND//<111>) texture with significant Goss ({110}<100>). An exact cube ({100}<100>) was not present, but near-cube components were present. The spread around the cube was ~10–15° [27].

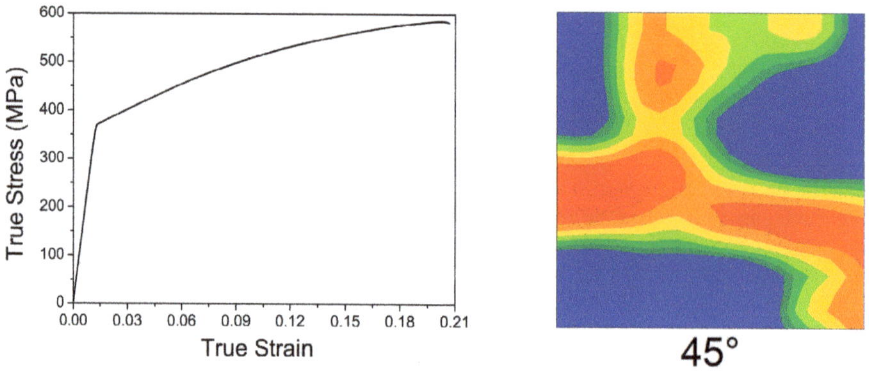

(**a**) (**b**)

Figure 1. (**a**) True stress–true strain curve for the LC steel used in the present investigation. The tests were interrupted at a strain of 0.03, 0.05 and 0.05 for microstructural characterization. (**b**) $\varphi_2 = 45°$ constant section of the orientation distribution function (ODF) for the initial material.

Figure 2 shows the distribution of KAM with the increase in tensile strain. As expected, KAM increased with an increase in deformation. The increase in green color signifies an increase in the magnitude of KAM. KAM is expected to scale with GNDs, i.e., excess dislocations [28]. The excess dislocations are expected to create large misorientations in the vicinity. Statistically stored dislocations (SSDs) are redundant and do not give rise to an increase in misorientations. It can be clearly seen that the near-boundary zone had higher GNDs than the interior of the grains. These were visible at small deformations, usually less than 10%. After 10% deformation, the dislocation density became too large to be clearly resolved and, hence, the increase in misorientation saturated. Figure 3 shows the overall GND distribution for the entire scanned area. The GND number fraction increased to a higher value with the increase in deformation, i.e., the strain. Both distributions of the KAM map and GND indicated the increase in dislocations with an increase in the deformation [29].

The developments in misorientations are also expected to be orientation-sensitive and depend on a number of factors including the grain size, the number of nearest neighbors and their misorientations, etc. Therefore, it is very difficult to quantify this phenomenon. However, it has also been reported that the near-boundary gradient zone also plays an important role. The DAMASK simulation results (Figure 4) indicated an increase in the stress level with an increase in strain. There was also a significant change in the overall orientation distribution. However, it was clear that some grains rotated more than others after the deformation. The IPF map, i.e., the bottom row of Figure 4, shows that some grains completely reoriented after deformation, even if the deformation was very small. In order to gain more clarity on this, 15 grains were studied in the present work. These grains were selected randomly to understand this phenomenon.

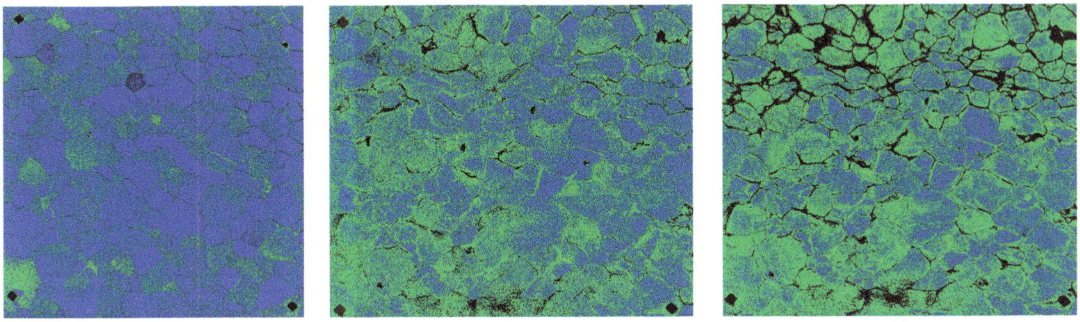

(**a**) (**b**) (**c**)

Figure 2. The distribution of Kernel average misorientation (KAM) of the undeformed (**a**), 3% deformed (**b**) and 7% deformed (**c**) sheet. 0° 5° is the color scale.

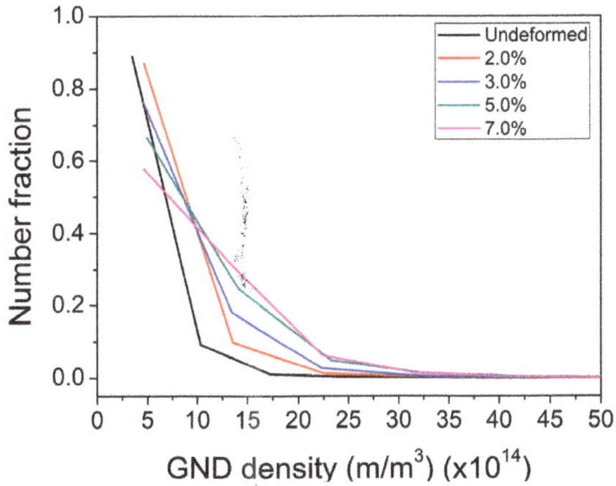

Figure 3. Geometrically necessary dislocation density (GND) distribution for undeformed and deformed samples.

As seen in Figure 5, 15 grains were chosen at random to observe grain rotation during deformation. The grain identification is marked for the undeformed IPF map as numbers from 1 to 15 (see Figure 5a). It can be seen from Figure 5b–d that there was a significant change in the color of the grains. This change implies that during deformation, the grains rotated and generated orientation gradients. After around 7% strain, the larger grains such as the ones numbered 1, 12 and 13 led to the development of significant spread and also their gradients. This gradient was more apparent near the grain boundaries. It has been reported [30] that this orientation gradient near grain boundaries, often termed as the near-boundary gradient zone (NBGZ), is responsible for the changes in texture during deformation. The cube grains (4, 7, 8 and 9) were found to be relatively more stable than the others, i.e., other orientations. Figure 6 shows the IPF for the undeformed and deformed conditions for these 15 grains.

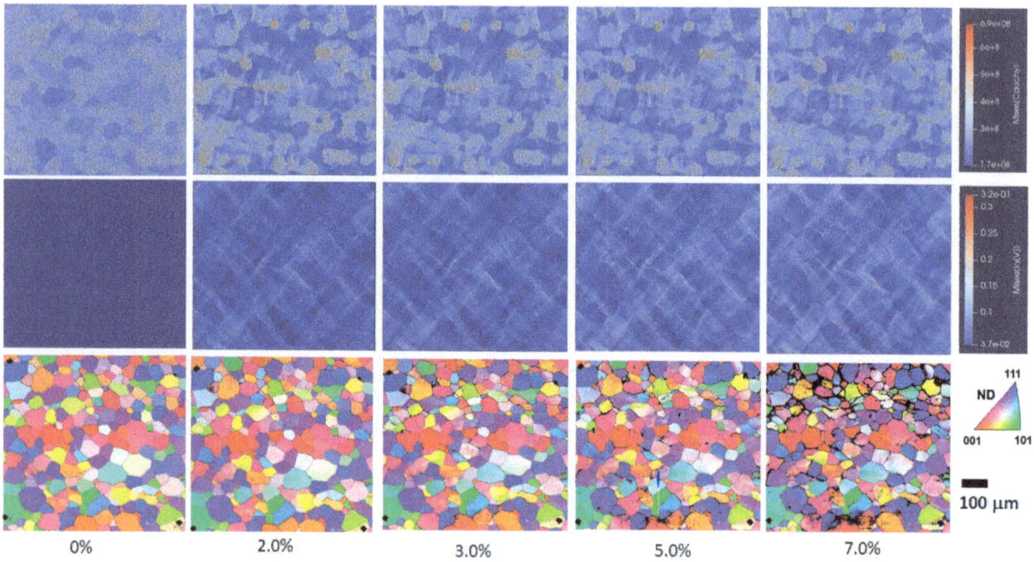

Figure 4. The von-Mises stress (top row), von-Mises strain (middle row) and inverse pole figure (IPF) map (bottom row) after DAMASK simulations.

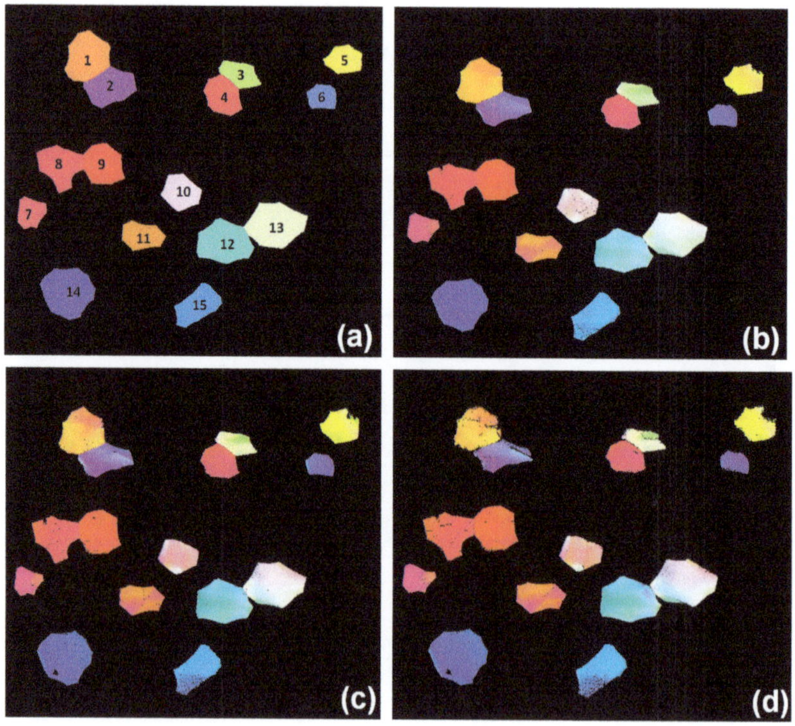

Figure 5. ND IPF map of the 15 grains tracked during deformation. (**a**) Undeformed, (**b**) after 3% strain, (**c**) after 5% strain and (**d**) after 7% strain.

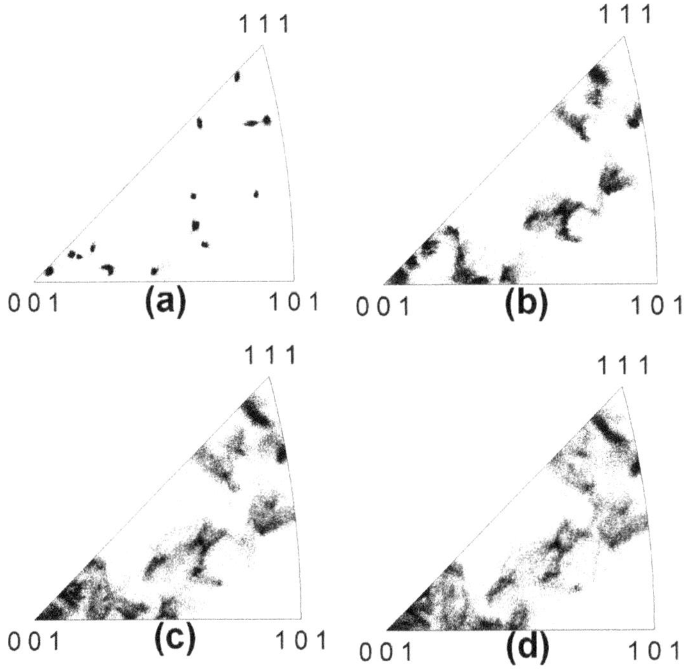

Figure 6. ND IPF of the 15 grains showing rotation and orientation spread during deformation. (**a**) Undeformed, (**b**) after 3% strain, (**c**) after 5% strain and (**d**) after 7% strain.

γ-fiber (ND//<111>) grains also developed gradients, but these grains did not move away significantly from the ND//<111> orientation. Grains which had orientations other than cube and γ-fiber showed significant rotations. This is very clearly visible in the IPF of Figure 6. Even after 7% strain, the IPF changed to a very large extent, and, hence, after 10% deformation, BCC materials showed a change in crystallographic texture independent of the mode of deformation (tensile, compression or cold rolling). Table 3 gives numerical values of the orientation, grain size, GOS and number of neighbors for these 15 grains. The grain size and initial orientation seemed to be the most important parameters governing the grain rotation and hence texture change during the deformation.

It is always believed that the rotation of grains occurs from unstable to stable end orientations during deformation. Inagaki [31–34] has shown that in polycrystalline iron, grain boundary constraints play an important role in the formation of cold deformation texture. In poly-crystals, crystal rotation occurs during rolling along different paths from that observed in single crystals [35]. It was found that crystal rotations occur along two paths (Path A: {001}<100> to {001}<110> to {112}<110> to {223}<110> and Path B: {110}<001> to {554}<225> to {111}<112> to {111}<110> to {223}<110>). The stable end orientation of cold rolling textures in poly-crystals was therefore identified as {223}<110>. It was also suggested that {111}<112> orientations are unstable and rotate towards {111}<110> [31]. A strong γ -fiber (ND//<111>) texture is developed if the initial orientation (before cold deformation) is {111}<112>. If two {111}<uvw>-type grains are neighbors, their grain boundary regions rotate readily into common stable end orientation. It has been reported (using rate-dependent full constraint and relaxed constraint theory) that the principal component inherited from the recrystallized hot band, i.e., {100}<110>, is a fairly stable end orientation and, hence, its intensity increases with cold rolling [36]. There is no concrete evidence of the exact mechanism of rotation of grains since it is a very complex phenomenon. However, it is clear that initial orientations represent one of the most important parameters during deformation. To this end, if anyone wishes to customize the final crystallographic texture of a deformed or annealed material, the materials' initial crystallographic textures need to be customized.

Table 3. Orientation, grain size, number of neighbors and grain orientation spread (GOS) for the tracked grains.

Grain_ID	Phi1 (°)	Phi (°)	Phi2 (°)	Grain Size (μm)	# Neighbors	Grain Orientation Spread (GOS, °)			
						Undeformed	3% Strain	5% Strain	7% Strain
1	79.1	13.8	280.3	164.6	8	0.4	1.9	3.1	3.5
2	208.9	39.6	137.3	151.8	7	0.4	2.5	4.2	4.7
3	127.3	31	258.3	112.8	6	0.3	1.4	2.3	2.9
4	103.7	9.4	242	134.7	8	0.3	1.1	1.6	1.9
5	343.9	21.9	4.7	115.6	7	0.4	1.6	2.1	2.7
6	220	46.1	125	99.4	6	0.5	1.5	2	2.5
7	167.9	12.2	208.7	108.7	7	0.3	1.5	2.5	2.9
8	326.6	8.4	55.5	149.3	7	0.3	0.9	1.6	1.9
9	117.8	3.5	236.4	145.6	8	0.5	2.2	3.7	4.2
10	112.8	32	207.2	129.8	6	0.2	1.9	2.9	3.3
11	179.6	14.1	187	124.9	7	0.3	2.3	3.9	4.6
12	16.9	41.5	339.3	174.2	8	0.2	1.7	2.7	3.4
13	149.9	30.4	198.4	180.8	9	0.4	2	2.9	3.6
14	14.4	48.7	315.8	181	9	0.3	1.8	2.9	3.5
15	219.8	51.7	148.1	139.8	7	0.3	1.2	1.9	2.2

4. Conclusions

The following conclusions can be drawn from the present work:

- The rotation of grains towards stable end orientation depends largely on the initial orientation of the grains. Rotations are larger if the initial orientations deviate significantly from the stable end orientation.
- Large grains (in size) tend to develop large in-grain misorientations and lattice rotations along with NBGZ. They have a higher tendency to fragment than reorient.
- The mode of deformation and number of neighbors can be significant for the reorientation of grains. However, in the present investigation, the initial orientation of the grain was found to be more significant.

Author Contributions: S.K.S.—conceptualization, data curation, methodology, formal analysis, writing—original draft. R.K.K.—conceptualization, data curation, methodology, formal analysis, writing—review and editing, supervision. M.D.—conceptualization, data curation, methodology, formal analysis, writing—review and editing, supervision. All authors have read and agreed to the published version of the manuscript.

Funding: The APC was funded by the American University of the Middle East, Kuwait.

Data Availability Statement: The raw/processed data required to reproduce these findings cannot be shared at this time as the data also form a part of an ongoing study.

Acknowledgments: The authors would like to acknowledge the use of National Facility for Texture and the OIM—A DST-IRPHA project for EBSD measurements.

Conflicts of Interest: The authors declare no conflicts of interest.

References

1. Skrotzki, W.; Pukenas, A.; Odor, E.; Joni, B.; Ungar, T.; Völker, B.; Hohenwarter, A.; Pippan, R.; George, E.P. Microstructure, Texture, and Strength Development during High-Pressure Torsion of CrMnFeCoNi High-Entropy Alloy. *Crystals* **2020**, *10*, 336. [CrossRef]
2. Zaefferer, S.; Ohlert, J.; Bleck, W. A study of microstructure, transformation mechanisms and correlation between microstructure and mechanical properties of a low alloyed TRIP steel. *Acta Mater.* **2004**, *52*, 2765–2778. [CrossRef]

3. Dutta, R.K.; Petrov, R.H.; Delhez, R.; Hermans, M.J.M.; Richardson, I.M.; Böttger, A.J. The effect of tensile deformation by in situ ultrasonic treatment on the microstructure of low-carbon steel. *Acta Mater.* **2013**, *61*, 1592–1602. [CrossRef]
4. Nesterova, E.V.; Bacroix, B.; Teodosiu, C. Microstructure and texture evolution under strain-path changes in low-carbon interstitial-free steel. *Met. Mater. Trans. A* **2001**, *32*, 2527–2538. [CrossRef]
5. Dhinwal, S.S.; Toth, L.S.; Hodgson, P.D.; Haldar, A. Effects of Processing Conditions on Texture and Microstructure Evolution in Extra-Low Carbon Steel during Multi-Pass Asymmetric Rolling. *Materials* **2018**, *11*, 1327. [CrossRef]
6. Gupta, A.; Khatirkar, R.K.; Kumar, A.; Thool, K.; Bibhanshu, N.; Suwas, S. Microstructure and texture development in Ti-15V-3Cr-3Sn-3Al alloy—Possible role of strain path. *Mater. Charact.* **2019**, *156*, 109884. [CrossRef]
7. Mahadule, D.; Demiral, M.; Mulki, H.; Khatirkar, R.K. Experiments and Crystal Plasticity Finite Element Simulations of Texture Development during Cold Rolling in a Ti-15V-3Cr-3Sn-3Al Alloy. *Crystals* **2023**, *13*, 137. [CrossRef]
8. de Moura, A.N.; Neto, C.A.R.; Castro, N.A.; Vieira, E.A.; Orlando, M.T.D.A. Microstructure, crystallographic texture and strain hardening behavior in hot tensile tests of UNS S32304 Lean Duplex stainless steel. *J. Mater. Res. Technol.* **2021**, *12*, 1065–1079. [CrossRef]
9. Nagarajan, S.; Jain, R.; Gurao, N.P. Microstructural characteristics governing the lattice rotation in Al-Mg alloy using in-situ EBSD. *Mater. Charact.* **2021**, *180*, 111405. [CrossRef]
10. Liu, M.; Liu, Y.; Li, H. Deformation mechanism of ferrite in a low carbon Al-killed steel: Slip behavior, grain boundary evolution and GND development. *Mater. Sci. Eng. A* **2022**, *842*, 143093. [CrossRef]
11. Zhang, C.; Ya, R.; Sun, M.; Ma, R.; Cui, J.; Li, Z.; Tian, W. In-situ EBSD study of deformation behavior of nickel-based superalloys during uniaxial tensile tests. *Mater. Today Commun.* **2023**, *35*, 105522. [CrossRef]
12. Chen, J.; Wang, J. In-situ EBSD study of 409 L ferritic stainless steel during tensile testing. *Mater. Res. Express* **2022**, *9*, 116508. [CrossRef]
13. Atas, M.S.; Yildirim, M. Temporal evolution, coarsening behavior and oxidation resistance of Ni-15Al superalloy. *J. Alloys Compd.* **2019**, *9*, 151784. [CrossRef]
14. Atas, M.S.; Yildirim, M. Structural properties and cyclic oxidation behavior of Ni-Al-Y superalloys. *Kokove Mater.* **2022**, *60*, 281–292.
15. Atas, M.S.; Yildirim, M. Morphological developments, coarsening and oxidation behavior of Ni-Al-Nb superalloys. *J. Mater. Eng. Perform.* **2020**, *29*, 4421–4434. [CrossRef]
16. *E8-04*; Standard Test Method for Tensile Testing of Metallic Materials. ASTM: West Conshohocken, PA, USA, 2004.
17. Shekhawat, S.K.; Basavaraj, V.P.; Hiwarkar, V.; Chakrabarty, R.; Nemade, J.; Guruprasad, P.J.; Suresh, K.G.; Doherty, R.D.; Samajdar, I. Direct Experimental Observations on Concurrent Microstructure and Magnetic Property Developments in Non-Grain Oriented Electrical Steel. *Metall. Mater. Trans. A* **2014**, *45*, 3695–3698. [CrossRef]
18. Khatirkar, R.; Basavaraj, V.; Shekhawat, S.K.; Haldar, A.; Samajdar, I. Orientation Dependent Recovery in Interstitial Free Steel. *ISIJ Intl.* **2012**, *52*, 84–893. [CrossRef]
19. Khatirkar, R.; Krishna, K.V.M.; Kestens, L.A.I.; Petrov, R.; Pant, P.; Samajdar, I. Strain localization in ultra low carbon steel: Exploring the role of dislocations. *ISIJ Int.* **2011**, *51*, 849–856. [CrossRef]
20. OIM. *Analysis Version 7.2. User Manual*; TexSEM Laboratories Inc.: Draper, UT, USA, 2013.
21. Nowell, M.M.; Wright, S.I. Orientation effects on indexing of electron backscattered diffraction patterns. *Ultramicroscopy* **1984**, *103*, 41–58. [CrossRef] [PubMed]
22. Wright, S.I.; Nowell, M.M.; Field, D.P. A review of strain analysis using electron backscatter diffraction. *Microsc. Microanal.* **2011**, *17*, 316–329. [CrossRef] [PubMed]
23. Kundu, A.; Field, D.P. Geometrically Necessary Dislocation Density Evolution in Interstitial Free Steel at Small Plastic Strains. *Met. Mater Trans A* **2018**, *49*, 3274–3282. [CrossRef]
24. Roters, F.; Eisenlohr, P.; Kords, C.; Tjahjanto, D.D.; Diehl, M.; Raabe, D. DAMASK: The Düsseldorf advanced material simulation kit for studying crystal plasticity using an FE based or a spectral numerical solver. *Procedia IUTAM* **2012**, *3*, 3–10. [CrossRef]
25. Tasan, C.C.; Diehl, M.; Yan, D.; Zambaldi, C.; Shanthraj, P.; Roters, F. Integrated experimental-simulation analysis of stress and strain partitioning in multiphase alloys. *Acta Mater.* **2014**, *81*, 386–400. [CrossRef]
26. Deeparekha, N.; Gupta, A.; Demiral, M.; Khatirkar, R.K. Cold rolling of an interstitial free (IF) steel—Experiments and simulations. *Mech. Mater.* **2020**, *148*, 103420. [CrossRef]
27. Verlinden, B.; Driver, J.; Samajdar, I.; Doherty, R.D. *Thermo-Mechanical Processing of Metallic Materials*; Pergamon Materials, Series; Cahn, R.W., Ed.; Elsevier: Amsterdam, The Netherlands, 2007.
28. Field, D.P.; Trivedi, P.B.; Wright, S.I.; Kumar, M. Analysis of local orientation gradients in deformed single crystals. *Ultramicroscopy* **2005**, *103*, 33–39. [CrossRef]
29. Humphreys, F.J.; Hatherly, M. *Recrystallization and Related Annealing Phenomena*; Elsevier: Oxford, UK, 2004.
30. Mishra, S.; Pant, P.; Narasimhan, K.; Rollett, A.; Samajdar, I. On the widths of orientation gradient zones adjacent to grain boundaries. *Scr. Mater.* **2009**, *61*, 273–276. [CrossRef]
31. Inagaki, H. Fundamental Aspect of Texture Formation in Low Carbon Steel. *ISIJ Int.* **1994**, *34*, 313–321. [CrossRef]
32. Inagaki, H.; Suda, T. The developments of rolling textures in low-carbon steels. *Texture* **1972**, *1*, 129–140. [CrossRef]
33. Inagaki, H. Formation of {111} recrystallization texture in polycrystalline iron. *Trans. Iron Steel Inst. Jpn.* **1984**, *24*, 266–274. [CrossRef]

34. Inagaki, H. Stable End Orientations in the Rolling Textures of the Polycrystalline Iron. *Int. J. Mater. Res.* **1987**, *78*, 431–439. [CrossRef]
35. Demiral, M.; Roy, A.; El Sayed, T.; Silberschmidt, V.V. Influence of strain gradients on lattice rotation in nano-indentation experiments: A numerical study. *Mater. Sci. Eng. A* **2014**, *608*, 73–81. [CrossRef]
36. Khatirkar, R.K. Strain Localizations in Low Carbon Steels. Ph.D. Thesis, IIT Bombay, Mumbai, India, 2012.

crystals

MDPI

Article

Microstructure and Defect Analysis of 17-4PH Stainless Steel Fabricated by the Bound Metal Deposition Additive Manufacturing Technology

Valerio Di Pompeo, Eleonora Santecchia *, Alberto Santoni, Kamal Sleem, Marcello Cabibbo and Stefano Spigarelli

Department of Industrial Engineering and Mathematical Sciences (DIISM), Polytechnic University of Marche, Via Brecce Bianche 12, 60131 Ancona, Italy; v.dipompeo@pm.univpm.it (V.D.P.); a.santoni@pm.univpm.it (A.S.); k.sleem@pm.univpm.it (K.S.); m.cabibbo@staff.univpm.it (M.C.); s.spigarelli@staff.univpm.it (S.S.)

* Correspondence: e.santecchia@staff.univpm.it

Abstract: Metal additive manufacturing (AM) technologies can be classified according to the physical process involving the raw material as fusion-based and solid-state processes. The latter includes sintering-based technologies, which are aligned with conventional fabrication techniques, such as metal injection molding (MIM), and take advantage of the freeform fabrication of the initial green part. In the present work, 17-4PH stainless steel samples were fabricated by material extrusion, or rather bound metal deposition (BMD), a solid-state AM technology. The powder-based raw material was characterized together with samples fabricated using different angular infill strategies. By coupling different characterization technologies, it was possible to identify and classify major properties and defects of the raw material and the fabricated samples. In addition, microstructural modifications were found to be linked with the mesostructural defects typical of the BMD solid-state additive manufacturing technology applied to metals.

Keywords: solid-state additive manufacturing; material extrusion; metals and alloys; defects analysis; microstructural characterization

1. Introduction

Additive manufacturing (AM) is a class of freeform fabrication technologies based on layer-wise material addition [1]. The ASTM F2792 [2] regulation defines AM as "a process of joining materials to make objects from 3D model data, usually layer upon layer, as opposed to subtractive manufacturing methodologies" and the key of additive manufacturing and its layer-based approach is that material can be placed only where it is needed in a component. This allows for the redefining of the shape of a part by applying topology optimization and following the Design for Additive Manufacturing (DfAM) principles [3–5]. This will have a crucial impact not only on the performance of the final part but also on the cost and the sustainability of AM adoption by also reducing the manufacturing waste [3,5–8].

Among a large number of available metal AM technologies, beam-based ones rely on the complete melting of the feedstock [4–6], which can be in the form of micrometric powder or wire, using different heat sources. The most popular fusion-based metal AM technologies are laser beam (LB) and electron beam (EB) powder bed fusion (PBF), or rather LB-PBF and EB-PBF. The former stands out for its high printing resolution, while the latter has a higher production speed [9–11].

On the other hand, solid-state AM technologies rely on the use of mechanical deformation or sintering to achieve a strong metallurgical bonding between subsequent layers of material [12]. Among the sintering-based ones, it is worth mentioning that material extrusion and binder jetting have similar features since both rely on a multistep approach based on the use of an organic binder to 3D print the desired geometry [13].

One of the latest additions to the solid-state AM technology is Bound Metal Deposition (BMD), a material extrusion process provided by the Desktop Metal (DM) Studio System, which is characterized by three fabrication steps: 3D printing, debinding, and sintering. Typically, further postprocessing actions are performed to increase the mechanical properties and the surface finishing of the fabricated part [8,10,14]. The computer-aided design model of the object must be subjected to slicing, as all the 3D-printing processes, while the DM Fabricate software allow for the definition of the BMD processing parameters, such as part orientation, number of perimeter-wall layers, number of bottom and top layers, infill pattern, and infill-pattern density. As clarified by Abe et al. [11] compared to as-designed dimensions, the software increases the as-sliced dimensions by approximately 18.6%, 18.6%, and 19.5% in the x, y, and z directions, respectively, to account for the shrinkage taking place during the final sintering step. The BMD feedstock comprises composite rods made up of a thermoplastic media (wax and polymer binder) filled with metal powder of the chosen alloy [15]. These rods are extruded by a heated nozzle having a diameter of 250 or 400 μm. Supports are fabricated using the same material but with a different printing strategy and they are separated from the surface of the object by a ceramic-interface layer deposited by an additional print head. This ceramic interface facilitates the separation between the part and the supports during the furnace sintering. During 3D printing, the alloy and the interface material are extruded onto the build plate, shaping the final 3D part (the so-called "green part") layer by layer, according to the original and sliced geometry. Once the component is printed, the green part is placed in a tank and immersed in a proprietary DM organic solvent to dissolve most of the binder. The part resulting from the debinding process is called "the brown part". The final step is a thermal sintering which removes the remaining binder and densifies the part by applying a thermal treatment that promotes the necking formation and the following interparticle adhesion. Postprocessing steps include the easy removal of the raft and supports, followed by machining and heat treatments (if needed) which could affect the inner porosity and/or modify the microstructure and the mechanical performance of the part [8,16]. BMD process steps are comparable to metal injection molding (MIM) ones; although both use a polymer-wax binder within green state parts and take advantage of debinding and sintering steps, the main difference between the two technologies is how the green state part is obtained. Whereas BMD performs 3D printing via material extrusion, MIM injects the build material into a mold. Therefore, BMD parts will benefit from a larger design freedom but will be characterized by a higher roughness compared to MIM parts [14,17,18].

A chromium–nickel–copper precipitation-hardened steel, 17-4PH is used in a wide range of industrial applications, including those characterized by mildly corrosive environments and high-strength requirements [11,14,19–21]; 17-4PH has a martensitic microstructure in the annealed condition and is further strengthened by a low-temperature heat treatment, which results in the precipitation of a copper-rich phase in the alloy. Precipitation-hardening steels are difficult to manufacture with conventional fabrication technologies and, therefore, AM has the potential to be disruptive for this class of stainless steels [22].

Currently, several 17-4PH-based filaments are available on the market and can be processed by material extrusion-based 3D printers [23–25], while the novelty of the BMD process stands in the use of rigid rods instead of filaments, which guarantees a higher metal-to-binder ratio. Pellegrini et al. [23] investigated the effect of two aging treatments applied on 17-4PH samples fabricated by three material extrusion technologies, such as BMD, fused filament fabrication (FFF), and atomic diffusion additive manufacturing (ADAM). Akessa [26], Bjørheim, et al. [10] and Gonzalez-Gutierrez et al. [24] fabricated specimens for tensile tests, the first two with the ADAM technology and the latter with a highly filled polymeric filament of 17-4PH; the authors attributed the high standard deviation (up to 38%) of the measured tensile data to the large number of defects and to the lack of cohesion between the layers and deposited tracks. Galati et al. [25] mainly evaluated the density, accuracy, and surface roughness of cubic and complex reference samples, fabricated by the ADAM technology. Kedziora et al. [27] performed a comparison between 316L and 17-4PH stainless steels fabricated by fused filament fabrication (FFF) and selective laser melting (SLM), in order to validate the use of FFF for the fabrication of structural parts. For this purpose, the paper focused on tensile, fatigue, and impact mechanical properties as well as the surface roughness and hardness of the fabricated samples, showing the detrimental impact of internal defects on the randomness of the results.

Most papers addressing material extrusion of metals are focused on the density and mechanical performance of the fabricated samples; therefore, the current literature lacks a study focused on the microstructural implications of process parameter variations and the related defects formation.

The goal of the present study is to give an overview of the solid-state AM technology called Bound Metal Deposition (BMD), from feedstock characterization to the final as-sintered structures fabricated with two different infill deposition strategies, in order to identify the main defects, define their nature, and classify them accordingly.

2. Materials and Methods

The feedstock material used in the BMD process is stainless steel 17-4PH micrometric powder, embedded in a polymeric binder to form composite rods. The chemical composition (wt.%) of the alloy provided by Desktop Metal [28] is reported in Table 1.

Table 1. Chemical composition (wt.%) of the 17-4PH stainless steel feedstock.

Fe	Cr	Cu	Ni	Nb + Ta	Mn	C
Bal.	15.5–17.5	3–5	3–5	0.29	<1.00	<0.07

The investigated sample was fabricated in the form of a hollow square for two main reasons: (i) to maximize the success rate of solvent debinding, thermal debinding, and sintering and (ii) to simplify the fixing and cutting phases. The schematic of the 3D-printed sample is reported in Figure 1a and the indicated dimensions are as follows: h = 50 mm, s = 5 mm, a = 50 mm, b = 50 mm.

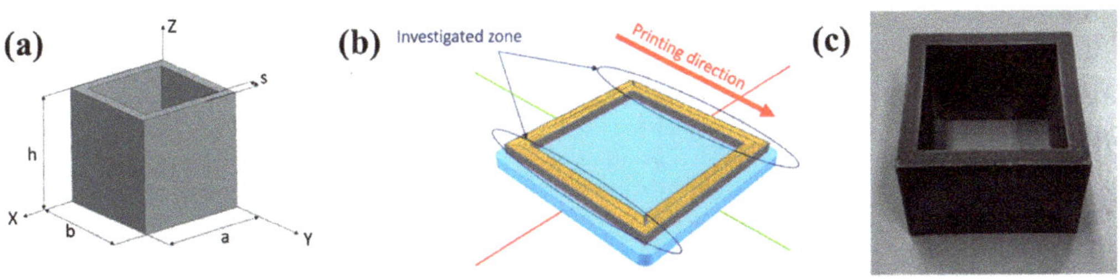

Figure 1. Details of the sample used in the current study: (**a**) Schematic design of the 3D-printed stainless steel sample used in the present work; (**b**) Section of the sample in the build preparation software printing software, (**c**) Picture of one of the fabricated samples. Investigated areas are arrowed in blue, while the printing direction is indicated in red.

Figure 1b shows a section for the printed sample in the Fabricate software for build preparation. It is worth noting that the printing direction indicates the direction in which the printing head is moving along the single deposited layer; this direction is fixed during the 3D-printing process.

The BMD process was performed with Desktop Metal Studio System equipment, with the combination of process parameters reported in Table 2.

Table 2. List of the process parameters used for the 3D-printing process.

Process Parameter	Value
Nozzle diameter [mm]	400
Printing speed [mm/s]	30
Infill strategy	100%
Overlap	0%
Layer height [mm]	0.15

Two different sets of two hollow square samples were fabricated by maintaining the same processing parameters and using a different deposition strategy in terms of infill direction with respect to the 3D-printed perimeter wall (Figure 2):

- BMD_90 sample, with a 90° deposition strategy;
- BMD_45 sample, with a 45° deposition strategy.

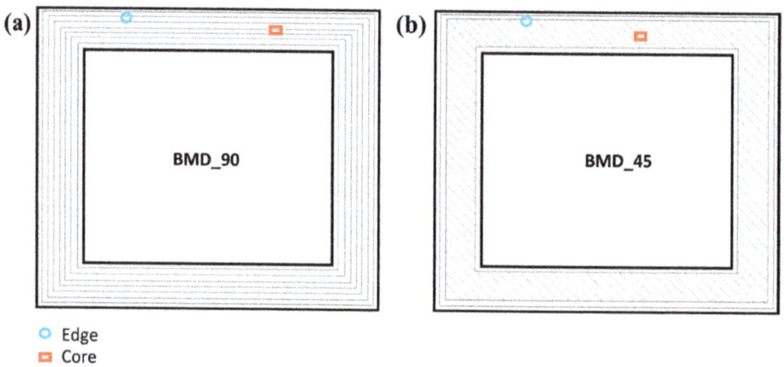

Figure 2. Schematics of the XY cross-section of the characterized samples: (**a**) BMD_90, (**b**) BMD_45. Microstructural investigations were performed on the edge (blue dot) and on the core (red rectangle) of the samples.

As per the BMD process specifications, the samples were fabricated on top of a raft and a ceramic interface layer, which are used to hold the part during the printing process and the final sintering step. After printing, the sample (green part) was subjected to the debinding operation for a time duration of about 30 h. To eliminate a large part of the polymeric binder, the debinding process was performed with the Desktop Metal equipment where the parts are immersed in a 300 × 300 × 300 mm^3 volume of solvent. After this step, the 3D-printed part becomes very fragile (brown part) since the metal-powder particles are still separated and kept together by the residual polymeric binder. The final step of the Desktop Metal process is the sintering of the brown part. The process was performed in a furnace under an inert gas atmosphere (Ar + 3%H_2) and consisted of two phases, an initial thermal debinding during which the residual binder is dissolved by heat, followed by a thermal cycle (sintering), during which metal particles are finally sintered at a temperature up to 1350 °C. The total sintering time was 35 h; the first 5 h are dedicated to thermal debinding, followed by 10 h for sintering, and the remaining time for cooling.

Specimens were cut from the original part in section XY (plane parallel to the build plate) and YZ (along the growth direction) and embedded in a phenolic resin for the metallographic preparation. Samples were then ground and polished on a Struers Tegramin-20 automatic grinding machine. Mirror-polished surfaces were used to investigate the pore formation and to quantify their dimensions and distribution. For this purpose, 20 optical micrographs were collected and all the defects were characterized using the software ImageJ (version 1.53 v21) [29]. To reveal the microstructure, samples were subjected to etching by immersion in a solution of 10 mL HNO_3 + 15 mL HCl + 12 mL acetic acid for

1 min and then washed with ethylic alcohol. A Leica DMi8 optical microscope was used to acquire micrographs on the mirror-polished and etched surfaces of all the samples. Phase quantification was performed on 5 micrographs acquired in a position close to the edge of the samples and 5 micrographs acquired on the core of the samples; quantification was performed according to the ASTM E562 standard [30] for the volume fraction manual point count, using the software ImageJ [29].

X-ray diffraction (XRD) patterns were acquired using a Bruker D8 Advance diffractometer operating at V = 40 kV and I = 40 mA, with a Bragg–Brentano geometry and Cu-Kα radiation. Measurements were performed on the YZ plane direction of both samples, using an angular range 2θ = 20–80° with a step size of 2θ = 0.02° and a dwell time of 2 s. Lattice parameters were extracted by Rietveld refinement performed using the Profex software (version 5.2.2) [31].

In order to fully characterize the features of the 17-4PH feedstock (composite rods) and as-built BMD samples, field-emission gun scanning electron microscopy (FEGSEM) observations were performed using a Zeiss Supra 40 microscope, equipped with a Bruker Quantax Z200 microanalysis to perform energy dispersive spectroscopy (EDS) investigations. EDS was used to analyze the chemical composition and to characterize inclusions and other defects. Backscattered electrons signal (BSE) was used for all the SEM investigations. Postacquisition statistical analysis on the SEM micrographs was performed using the ImageJ software [29]. Particle size distribution (PSD), cumulative fraction, and other representative granulometric quantities (i.e., percentiles D10, D50 D90) related to the metal-powder fraction of the rods were calculated from five SEM micrographs.

Hardness Rockwell C measurements were performed with an Ernst Hardness tester, using a load of 150 kgf. Five indentations were performed along the XY plane and five along the growth direction (YZ plane).

The punctual characterization of the mechanical response of the 3D-printed steel was performed by Vickers microhardness measurements using a REMET HX-1000 tester with a load of 300 gf and a dwell time of 10 s. Microhardness was tested by performing three systematic indentations every 2 mm 13 times in the XY plane (Figure 3) and three indentations every 2 mm 22 times in the YZ plane, in order to cover all the available space throughout the sample.

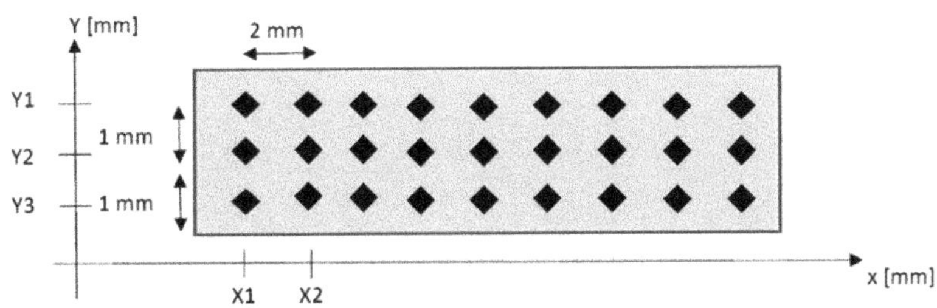

Figure 3. Schematic representation of the microhardness test plan indentations of the present work (number of indentations in the schematic is not representative).

3. Results and Discussion

3.1. Feedstock Characterization

The SEM analysis of the starting 17-4PH composite rods revealed the presence of mainly circular 17-4PH particles (Figure 4a), with the calculated particle size distribution and cumulative fraction reported in Figure 4b. From the data reported in Figure 4, representative percentiles have been calculated corresponding to the 10%, 50%, and 90% of the obtained particles' diameters (D10, D50, and D90, respectively); values are reported in Table 3.

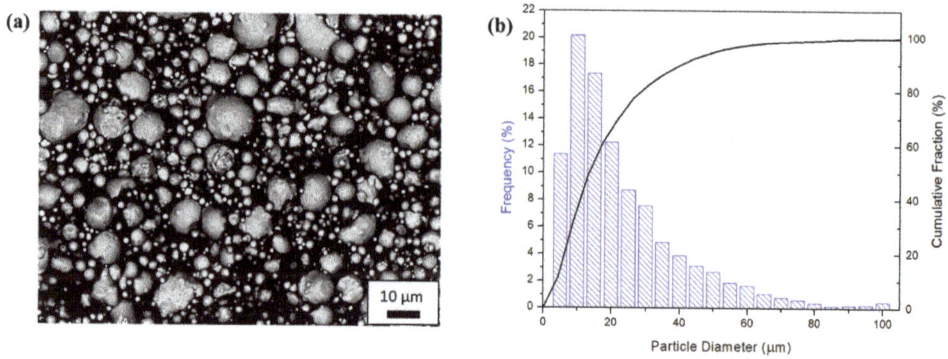

Figure 4. The 17-4PH composite rod analysis: (**a**) SEM-BSE micrograph and (**b**) particle size distribution and cumulative fraction.

Table 3. Representative percentiles calculated from SEM micrographs of the composite rod (AVG stands for average and SD for standard deviation).

	D10	D50	D90
AVG	4.5	15.5	45.0
SD	0.5	0.5	2.0

In principle, based on the paper by Hausnerova et al. [32], the same properties of the MIM powder feedstock can apply for the BMD composite rods, including shape, size distribution, loading, and packing density, which are properties also shared with the powder bed fusion feedstock. In particular, powder particle size plays a crucial role in realizing a dense sintered part with limited shrinkage, either with MIM or any other binder jetting or material extrusion-based (i.e., BMD) technology [33,34]. The particle population in Figure 4 comprises a large amount (more than half) of very fine particles, showing a diameter lower than 20 μm. The width distribution or distribution slope parameter, Sw, can be calculated using the following equation (Equation (1)) [35]:

$$Sw = \frac{2.56}{\log\left(\frac{d_{90}}{d_{10}}\right)} \tag{1}$$

Using the percentile values calculated on multiple SEM micrographs and reported in Table 3, the Sw value of the present distribution is 5.5 and clearly indicates a very wide distribution, as can be evinced also by the SEM-BSE micrograph reported in Figure 4. This wide dimensional range of the powder particles has no influence on the 3D-printing step of the BMD technology, being that the printing conditions are influenced only by the properties of the polymeric binder material. However, it is during the final sintering phase that the particle size dimension (PSD) becomes crucial because a dimensional range as wide as the one reported in Table 3 will result in the formation of pores due to the stacking of powder particles one on top of the other. While this phenomenon is mitigated during the MIM process owing to the pressure applied to the feedstock by the molds, the BMD freeform fabrication technology leads to residual porosities related to the particle size dimension of the metallic fraction.

While the investigation of the particle size distribution and the sphericity of the powder particles was performed with scanning electron microscopy, the energy dispersive spectroscopy (EDS) detector was used to investigate peculiarities in the chemical composition of the feedstock. Figure 5 shows the EDS elemental map acquired on the area indicated by the BSE micrograph and highlights the presence of contaminant particles with high concentrations of silicon, oxygen, aluminum, and manganese as well. It is worth mentioning that the contrast of the EDS map micrograph was intentionally enhanced in order to highlight the presence of a contaminant particle that was not visible with the standard contrast-balanced BSE micrograph.

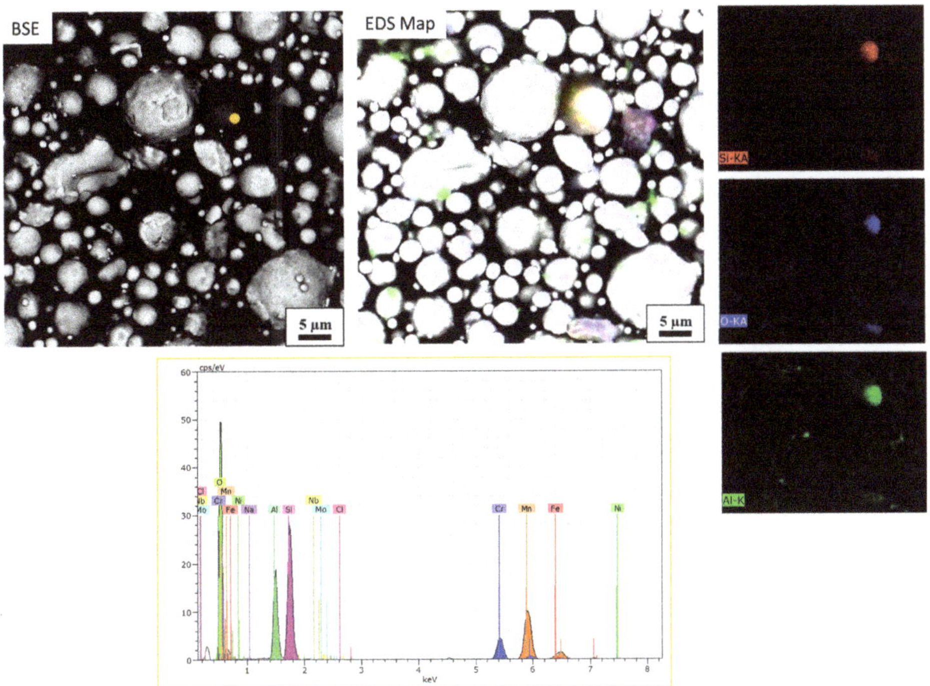

Figure 5. BSE micrograph and EDS elemental map of silicon and oxygen in the 17-4PH composite feedstock. The reported EDS spectrum was acquired in the orange dot position. Different colors of the EDS map correspond to Silicon (red), Oxygen (blue), and Aluminum (green).

While other papers in the literature report on silicon oxide formation on the surface of atomized metal-powder particles and on sintered stainless steels [36–38], the results of this investigation show that these oxides are in the form of free particles with the same granulometric size of the base alloy and totally different composition, owing also to the presence of high percentage of aluminum and total absence of Fe (confirmed by the peaks deconvolution reported in the spectrum of Figure 5). This suggests that, while the low atomization temperature and slow cooling rates can play a major role in the oxides formation, it is also reasonable to consider them as particles with a totally different nature inside the feedstock.

3.2. Analysis and Classification of Defects

The mirror-polished surfaces of the 17-4PH BMD_90 and BMD_45 samples allowed for the characterization of the structural integrity of the part and the eventual presence of pores within the structure. Figure 6 shows the optical microscopy results of the YZ section investigation and the SEM-BSE details of the extrusion defects.

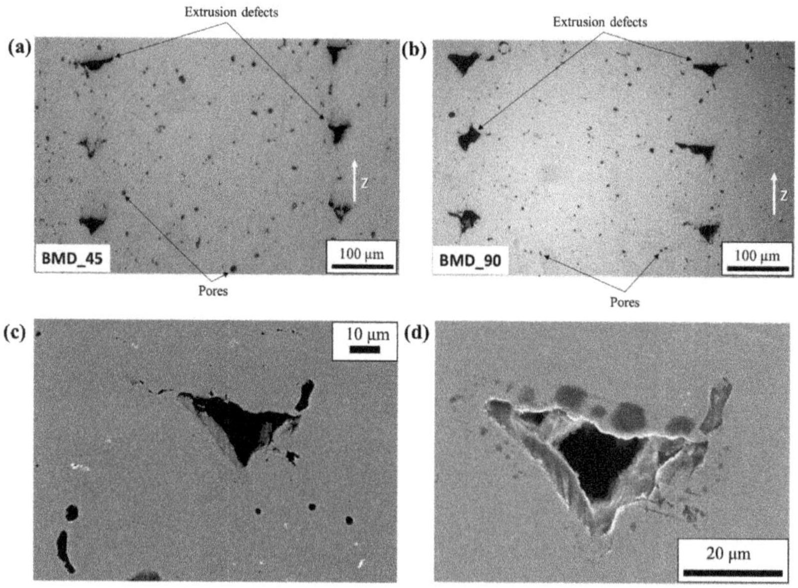

Figure 6. Optical micrographs of the YZ samples: (**a**) BMD_45, (**b**) BMD_90, and SEM-BSE micrographs of the mirror-polished YZ surface (BMD_90): (**c**) low magnification image with low brightness–high contrast balance to highlight the brightest inclusions, (**d**) high magnification image with high brightness–mid contrast balance to highlight the darker areas surrounding the extrusion-related pore.

Figure 6 highlights the presence of pores with a particular geometrical shape which can be ascribed to the space left by the extrusion of the composite material during the 3D-printing process, beyond the specific deposition strategy. It is worth mentioning that owing to the different deposition angles, pores related to the YZ plane of the BMD_45 sample (Figure 6a) appear smaller than for the YZ BMD_90 sample, thus increasing the density in the realized section perpendicular to the square. Figure 6 shows the structure of the BMD_45 and BMD_90 samples along the growth direction (YZ plane). The orientation of the pores is directly connected with the printing strategy and is enhanced by the binder dissolution and evaporation during the debinding and sintering phases, respectively. This suggests that the final thermal process is not able to limit the presence of such large defects and that a low bond between subsequent layers can be expected. The presence of the triangular-like defects highlights how the deposition of tracks and the following squash of the melted composite material by the extrusion nozzle play a major role in defining the shape of the pores. These findings are in line with the literature concerning the material extrusion of metals where, despite a different form of the starting feedstock (composite wire instead of composite rods), large geometrically shaped voids are found in the final part, confirming the link between the defect and the 3D-printing process itself [10,26,39].

Figure 6c,d show SEM-BSE micrographs acquired at different magnifications and using different levels of brightness–contrast. It is worth noting that while Figure 6c allows for highlighting the presence of very bright inclusions, the high brightness option used for Figure 6d enables the detection of darker areas concentrated around the edge of the extrusion-related pore corresponding to high concentrations of silicon oxides as the ones that will be discussed later in the paper.

The highly bright inclusions shown in Figure 6c were characterized by energy dispersive spectroscopy and the results showed the high presence of Nb and C (~40 ± 10 wt.% Nb, 10 ± 2 wt.% C, calculated by averaging the EDS values acquired from every bright region), suggesting that these inclusions correspond to niobium carbides (NbC). The presence of these carbides can be expected since the temperatures reached during the sintering phase and the corresponding dwell times are similar to the conditions at which solution annealing is typically performed on 17-4PH stainless steel (also known as "condition A") [26,40]. Compared to the recently published results by Akessa et al. [26], niobium carbides in the sample fabricated by BMD are located in different positions across the analyzed surface, while in the samples obtained by using the ADAM technology, carbides are found on the edges of the pores [26].

The optical micrographs of the BMD_45 and BMD_90 XY mirror-polished samples are shown in Figure 7.

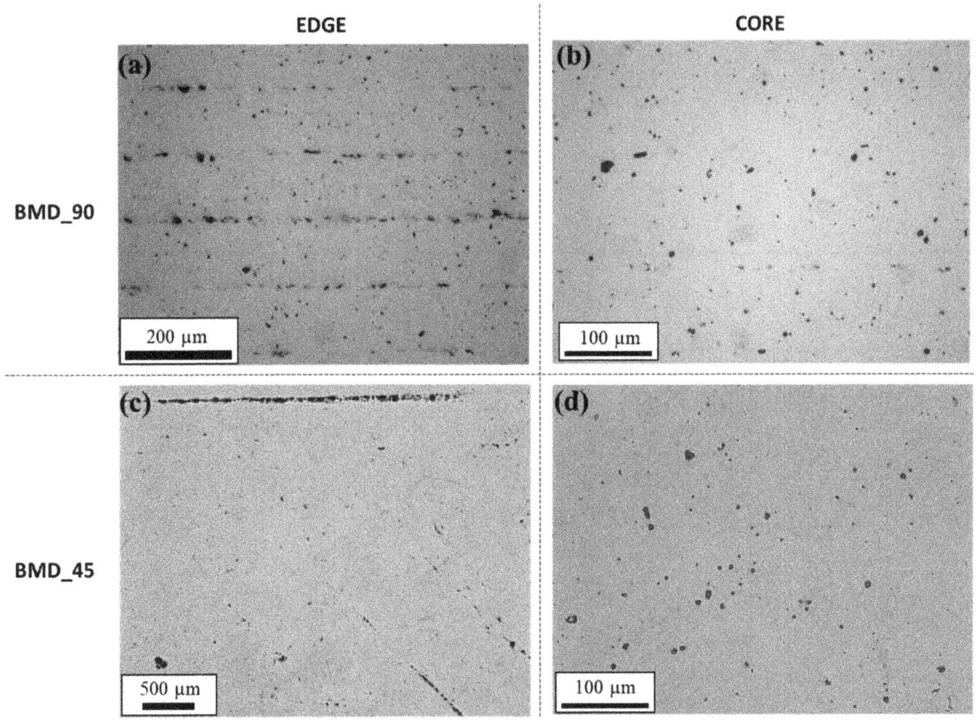

Figure 7. Optical micrographs of the XY samples: (**a**) BMD_90 close to the edge, (**b**) BMD_90 core of the sample, (**c**) BMD_45 close to the edge, and (**d**) BMD_45 core of the sample.

The planar arrangement of the deposited tracks (Figure 7) close to the edge of the fabricated square (Figure 7a,c) and in the central area (Figure 7b,d) for both the BMD_45 and BMD_90 samples is quite peculiar. A remarkable decohesion between different tracks is clearly visible in the areas closer to the outer edge of both samples (Figure 7a,c); the structure of both samples densifies with limited large defects formed next to the samples' core. Edge micrographs have a different scale bar to highlight the impact of the infill deposition strategy on the evolution of the defects.

The structure of the samples in the plane perpendicular to the growth direction shows large defects appearing as void lines close to the edges of the samples and aligned with the infill strategy (Figure 7a,c), while a higher density is reached in the core of the samples (Figure 7b,d). This effect is indeed independent of the infill deposition angle and is likely to be due to the combination of printing parameters such as the deposition strategy, overlap, and density profile chosen for the fabrication. While the literature shows continuous void lines [10,26,39], Figure 7a highlights discontinuous void lines preferentially located close to the outer wall of the 3D sample and suggests a better overall performance. During the sintering process, the evaporated residual binder increases the content of the pores since this thermal process takes place without the application of pressure and there is no further infiltration with additional material. Therefore, while the thermal cycle is able to densify the part and completely dissolve the polymeric binder, a residual porosity content should always be expected. Furthermore, as outlined above, the wide range of powder dimensions increases the inherent porosity and results in the formation of the round-shaped pores visible in the optical micrographs of Figures 6 and 7

Scanning electron microscopy investigations, together with element microanalysis, have been also performed on the polished surfaces of both sets of samples and the findings referring to the BMD_90 sample (YZ plane) are reported in Figure 8. It is worth mentioning that similar results were obtained in BMD_90 and BMD_45 samples.

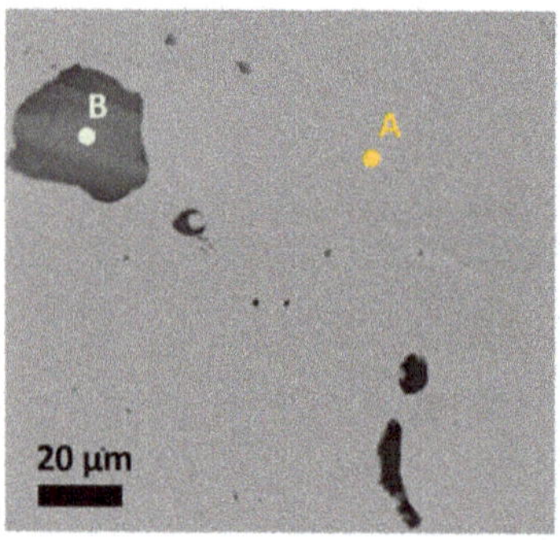

	Fe	Cr	Si	O
A	72.5	15.1	0.44	0.07
B	0.76	4.40	16.77	47.25

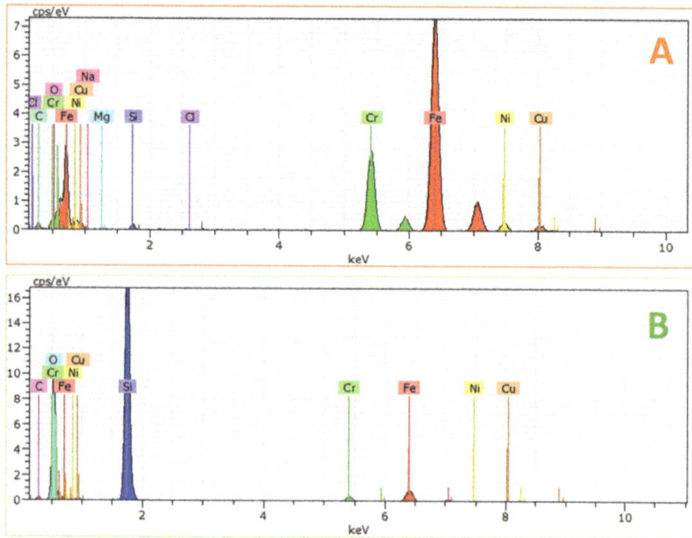

Figure 8. Energy dispersive spectroscopy (EDS) spectra taken in positions A and B of the SEM micrograph on the left (BMD_90 sample). Quantitative (wt.%) amounts of the main elements varying between point A and point B are also reported.

The BSE signal allowed for the highlighting of differences in the chemical composition of the final part (Figure 8) and the use of the energy dispersive microanalysis performed on specific positions (being A the matrix and B the discontinuity) allows identifying the presence of an inclusion rich in silicon and oxygen. It resulted that particles with a high amount of O, Si, Al, and Mn were found inside the metal-powder feedstock (Figure 5) and these are likely to be the cause of the inclusions reported here in Figure 8, with an associated redistribution of alloying elements during surface and volumetric diffusion occurring in the sintering furnace. These oxide particles are also responsible for the darker BSE contrast areas in Figure 6d and highlight how their concentration is particularly remarkable near extrusion-related pores.

Investigations allowed for the isolation of different defects and to correlate their nature with different production steps as (i) *extrusion-related defects*, related to large pores having remarkable dimensions (i.e., 20 microns) and a particular geometrical arrangement (from triangular to star-shaped in the growth direction) and (ii) *feedstock-related defects*, which include all the discontinuities inside the final part depending on the feedstock only. With reference to the latter, an additional classification of the feedstock's defects can be suggested: (i) *chemistry-related defects*, that is defects and inclusions connected with the chemical composition and peculiar metallurgical behavior of the alloy and (ii) *geometry-related defects*, such as rounded pores connected with the sintering of the metal particles. In this latter case, the surfaces of the particles impinge on one another and cause gaseous inclusions to form in the remaining space after the binding removal. These small, rounded pores are typical of AM sintering-based technologies, as well as MIM, and are linked with the formation of necks between metal particles during furnace sintering [18,23–27,35]. Decohesion of the deposited tracks, as highlighted in Figure 7, can be classified as an additional extrusion-related defect, being due to the particular BMD 3D-printing process. It is worth pointing out that chemistry-related defects are due to the presence of peculiar particles in the feedstock rods and to the formation of silicon oxides due to the high percentage of Si in the chemical composition to prevent the formation of other detrimental oxides [36].

3.3. Microstructural Characterization

Mirror-polished samples were subjected to etching to reveal the microstructure and the results are shown by the optical micrographs reported in Figure 9.

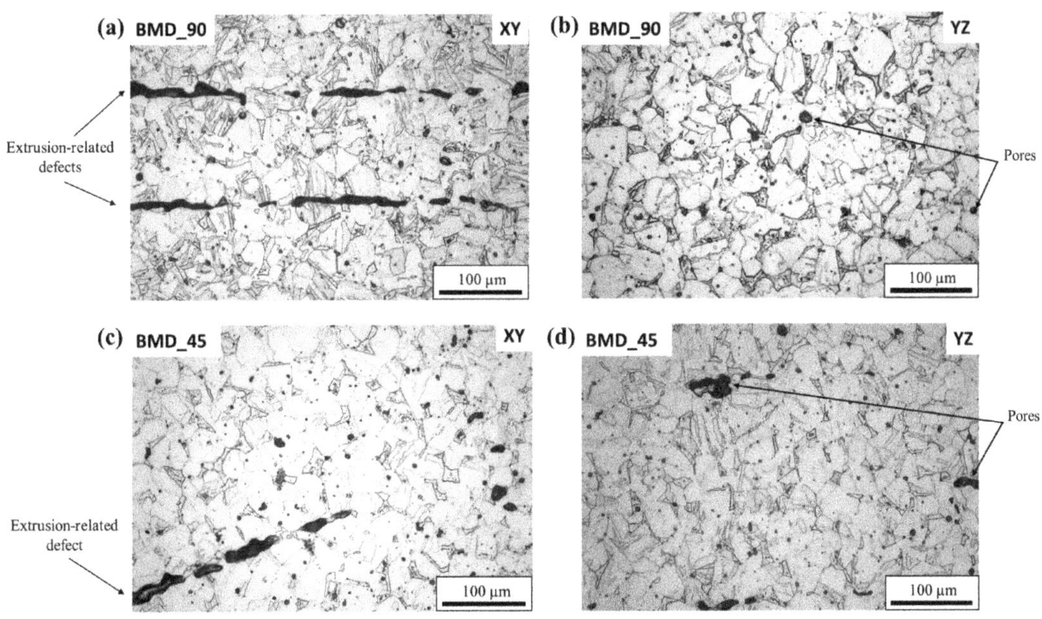

Figure 9. Optical micrographs of the etched investigated samples: (**a**) BMD_90 XY, (**b**) BMD_90 YZ, (**c**), BMD_45 XY, and (**d**) BMD_45 YZ.

The microstructure of all the samples revealed by optical microscopy (Figure 9) is a combination of ferrite and martensite; no remarkable differences can be highlighted between the two sets of samples.

Concerning the microstructure, Figure 9 shows the presence of ferrite and martensite for both the BMD_45 and BMD_90 samples; the phase distribution is consistent with the "condition A" status, that is, after a solubilization process [26,40,41]. In this respect, quantification of the δ-ferrite phase according to the ASTM E562 standard was performed for both samples in two different regions, close to the edge and in the core of the sample. In particular, the edges of both the BMD_45 and BMD_90 samples are characterized by higher δ-ferrite percentages, a trend which is more pronounced for the former sample, whose core shows the lowest δ-ferrite amount. As reported in the literature by Wu et al. [42] carbon diffusion is the driving mechanism of δ-ferrite formation in 17-4PH fabricated by injection molding. Diffusion is determined by the sintering temperature and dwell time and can lead to a remarkable variation of the martensite–ferrite contents in martensitic stainless steels [42,43]. Sintering parameters cannot be varied in the BMD system used in the present study and this suggests that any microstructural modification between edge and core must be related to the structural defects of the samples. The large linear pores (Figure 7) corresponding to tracks' decohesion occurring at the edges of the analyzed samples enhanced carbon diffusion and, therefore, the formation of a larger amount of δ-ferrite compared to the core of the samples. Therefore, the results of this quantification reported in Figure 10 highlight that the phase distribution is determined by the mesostructure of the sample.

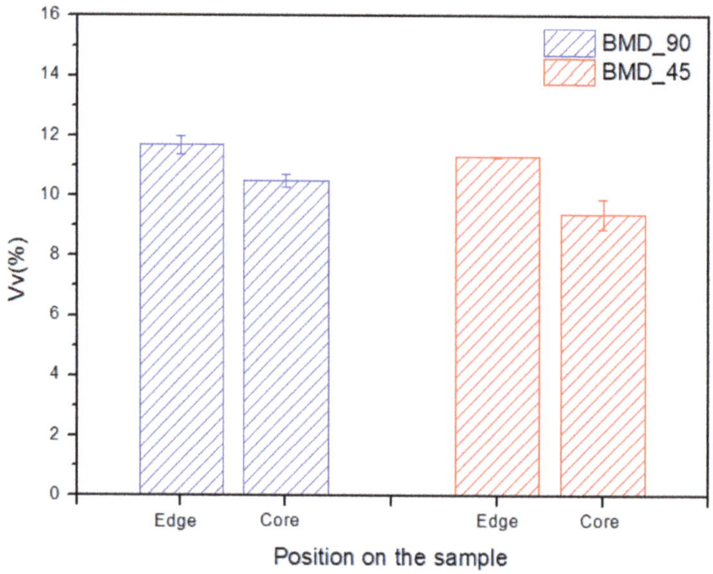

Figure 10. The δ-ferrite volume fraction resulting from phase quantification (ASTM E562) on the edge and on the core of both the BMD_90 (blue bars) and BMD_45 (red bars) samples. The error bar of the BMD_45 edge condition is very thin and overlaps with the contour of the bar itself.

This result can be considered in good agreement with the current literature about binder jetting of 17-4PH [43] and highlights that phase composition can be tailored with material extrusion by changing the extrusion pattern. This is of particular importance since a δ-ferrite content below 10% allows for the mitigation of potentially detrimental effects on the mechanical properties of the final parts [18,44].

The volume fraction of δ-ferrite was quantified from optical micrographs taken on the edge and in the core (see Figure 2) of both the BMD_45 and BMD_90 samples; the corresponding quantitative analyses are reported in Figure 10.

Calculated values of the δ-ferrite phase show a variation between 9% and 12%, with a significantly lower value recorded on the cores of both samples (Figure 10), a trend more pronounced for the BMD_45 sample. In the latter (Figure 10, red bars) the different deposition strategies of the filling pattern are shown to have a remarkable influence not only on the pore distribution (Figure 7) but also on the phase evolution during sintering.

To further investigate the phase distribution and crystallinity of the samples, X-ray diffraction was performed and the results are shown in Figure 11.

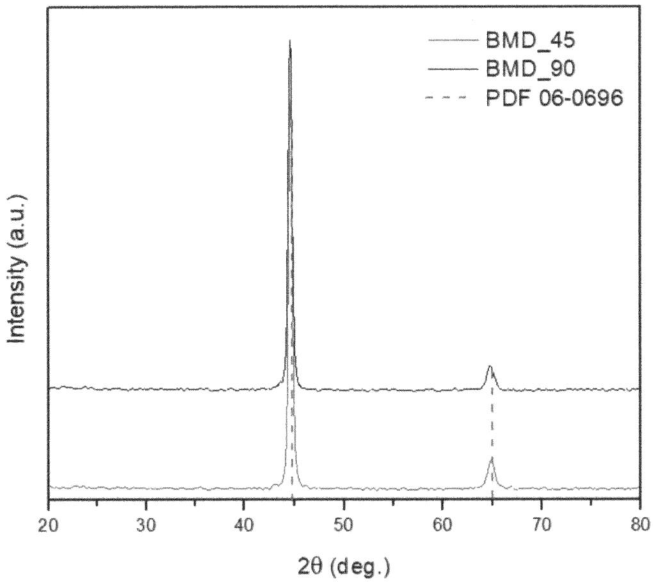

Figure 11. X-ray diffraction patterns of the two samples. Grey dashed lines are the peak positions identified by the PDF card 06-0696 (bcc ferrite).

The X-ray diffraction patterns of the BMD_45 (black line) and BMD_90 (blue line) samples reported in Figure 11 were acquired along the growth direction (YZ plane). The gray dashed lines identify the δ-ferrite peak positions, corresponding to the PDF card 06-0696. No peaks related to the fcc γ-iron phase are visible in the acquired patterns [26].

It is known that the lattice parameters of the bct martensite crystal vary with carbon content and temperature. In particular, in the case of low-carbon steels (<0.6 wt% C), the tetragonality of the martensite crystal is almost nonexistent [45–47]. For this reason, while the presence of both δ-ferrite and martensite has been clearly detected with metallography (Figure 9), X-ray diffraction patterns show only two clear peaks related to the bcc phase [48,49]. Rietveld refinement was performed on both the diffraction patterns of Figure 11 and the comparison between experimental values of the bcc and bct lattice parameters are reported in Table 4, together with the comparison with the nominal lattice parameter value of the ferrite PDF card 06-9606. Rietveld refinement performed on the X-ray diffraction patterns of both samples allowed for the calculation of the lattice parameters of both phases (Table 4) and to compare them with the reference PDF cards (06-0696 PDF card for ferrite and 44-1290 PDF card for martensite).

Table 4. Experimental lattice parameters were calculated from peak position in the XRD patterns by Rietveld refinement. First row shows the nominal lattice parameter value from the ferrite 06-0696 PDF card and martensite 44-1290 PDF card.

Sample	δ-Ferrite (nm)	α-Martensite (nm)
Reference	a = 0.2866 (PDF 06-0696)	a = 0.2859 (PDF 44-1290) c = 0.2937
BMD_45	a = 0.2863 ± 0.0001	a = 0.2851 c = 0.2878 ± 0.0001
BMD_90	a = 0.2863 ± 0.0001	a = 0.2851 c = 0.2877 ± 0.0001

The values of Table 4 show how the a and c parameters of the tetragonal body-centered lattice structure of martensite are strongly deformed, making this crystalline structure closer to a body-centered one, leading to the diffraction peaks visible in Figure 11. While the obtained ferrite/martensite peaks overlap is typical of 17-4PH stainless steel and similar diffraction patterns can be found in the literature [48,49], the current results are in contrast with the literature about ADAM processing of the same alloy [26], since no peaks related to retained austenite are visible.

3.4. Mechanical Characterization

Figure 12a shows the microhardness (μHV, 300 gf) profiles for the BMD_90 sample along the growth direction (YZ plane) and remarkable differences can be highlighted between this plane and the XY base plate one (Figure 12b). Along the growth direction, the average microhardness value varies between 289 and 303 HV, without following a particular trend. On the other hand, the average microhardness value on the XY plane varies between 252 and 262 HV. It should be mentioned that some spikes in the μHV average values characterized by a related large standard deviation can be noted in both directions; these can be ascribed to the presence of a large number of pores with different dimensions, ranging between ~5 and ~60 μm, as well as to the presence of inclusions having a chemical composition which differs from the 17-4PH one.

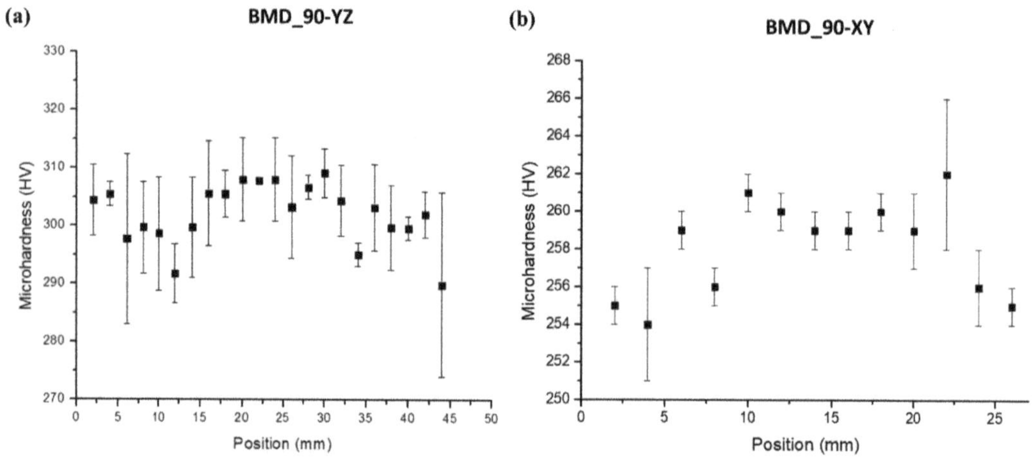

Figure 12. HV microhardness profiles (300 gf) of the BMD_90 sample collected: (**a**) on the growth direction (plane YZ) and (**b**) on the XY plane.

Figure 13 shows the microhardness profiles of the BMD_45 sample in the two considered planes. Along the growth direction, the average microhardness value varies between 279 and 297 HV, on the XY plane values vary between 283 and 293 HV.

Microhardness values of the BMD_45 samples show a trend towards lower average values from the bottom to the top of the YZ sample (Figure 12a), while in the XY direction (Figure 13b) values are quite in line with each other and there is no remarkable trend to be pointed out.

Microhardness results acquired on the XY and YZ planes are reported in Figure 12 for the BMD_90 and in Figure 13 for the BMD_45 one. There is an interesting trend followed by the BMD_45 sample towards lower average values from the bottom to the top of the YZ sample (Figure 13a), while in the XY direction (Figure 13b), values are quite in line with each other. By comparing Figures 12b and 13b, it is clear that the infill deposition strategy impacts also the microhardness values, together with the phase composition. The microhardness values obtained with the 45° (BMD_45) infill deposition strategy are remarkably higher compared to the 90° (BMD_90) strategy. Since all the samples were 3D printed and postprocessed in the same batch and with the same processing conditions, the microhardness results suggest that extrusion-related defects and decohesion of deposited tracks and layers play a major role in determining the local mechanical properties.

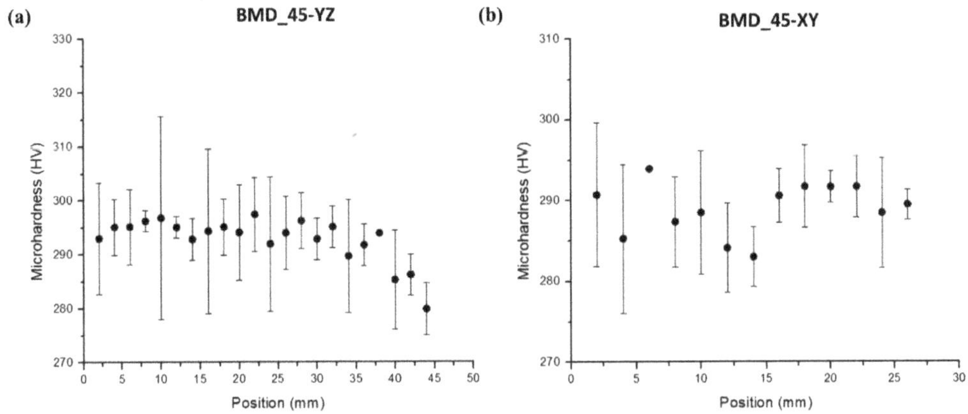

Figure 13. HV microhardness profiles (300 gf) of the BMD_45 sample collected: (**a**) and on the growth direction (plane YZ) and (**b**) on the XY plane.

Figure 13a suggests a major role for the sintering process, with a pronounced relaxation of the microstructure compared to the bottom side, where the locking effect of the raft and the ceramic interphase layer is consistent. In the XY plane of the BMD_90 sample (Figure 12b), microhardness values seem to be influenced more by the process parameter than by the sintering process. This explains the μHV trend towards higher values moving from the outer area; that is, the contour or the outer border of the sample to the core. While in the former (contour), the printing parameters are optimized to improve the surface finishing, close to the core the printing parameters are optimized to enhance the mechanical properties. In this regard, to improve the surface finishing, the printer extrudes less material on the edge than it does in the inner zone and the applied forces on the layer during the deposition phase are low. This behavior is typical of metal AM processes in general, where the hatching strategy influences the mechanical properties regardless of the nature of the process and the specific alloy, as demonstrated by Saboori et al. [50]. The shown anisotropy is in line with the typical behavior of material extrusion AM parts, suggesting how the combination of the extrusion technology for the green part production and the debinding and sintering operations to get the brown and the final 100% metallic part, are essential and represent an added value peculiar of the BMD technology.

A different mechanical behavior was observed during the Rockwell hardness (HRC) analysis because the measurement was less affected by the printing deposition strategy. Figure 14 summarizes and compares the hardness of three different fabrication processes applied to the 17-4PH stainless steel. Rockwell hardness results for BMD (Figure 14) show a lower value compared to casting since the cooling phase typical of the casting technology is higher than the BMD one. MIM and BMD show comparable hardness values despite the large and well-oriented defects of the latter.

The Results obtained for the BMD_45 and BMD_90 samples, and shown in Figure 14, refer to top-to-bottom YZ plane and to edge-to-core XY plane. The decrease of HRC values for the BMD_90 YZ sample suggests a local anchoring effect of the interface ceramic layer and the raft, which keeps the sample in a fixed position during the furnace sintering. This effect is remarkably influenced by the infill deposition strategy within the first fabricated layers since it is not relevant in the BMD_45 YZ sample nor in the rest of the two YZ samples. A comparison of the obtained values with the literature confirms that this solid-state AM technology is outperforming MIM (ca. 27 HRC, in the as-sintered condition [23]). In particular, comparing experimental values with the literature results concerning other material extrusion AM processes, such as fused filament fabrication (22 HRC [51] to 26 HRC [27]), the performance improvement due to the higher metal-to-binder ratio of the BMD feedstock is remarkable.

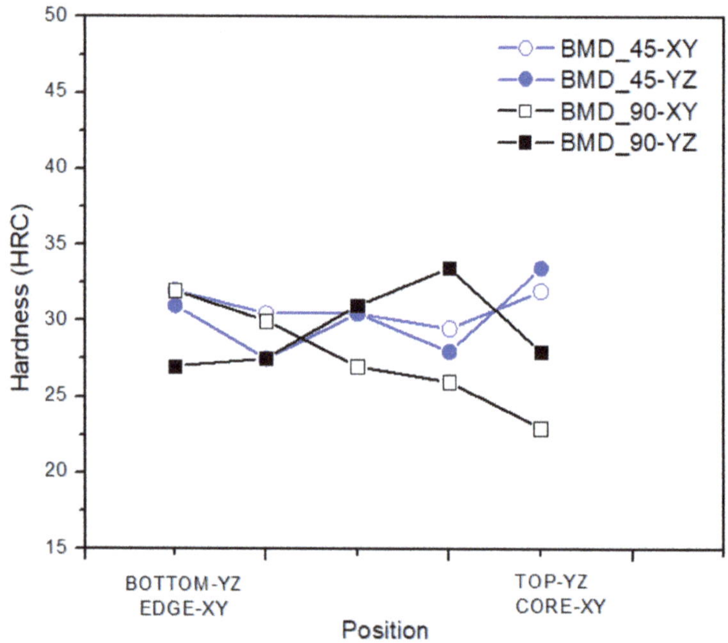

Figure 14. Comparison of HRC hardness values acquired on the growth direction and on the XY plane of the BMD_45 (blue dots) and BMD_90 (black squares) samples.

4. Conclusions

The purpose of the present paper was to characterize the microstructure and the integrity of samples fabricated by the solid-state AM technology called bound metal deposition (BMD) using 17-4PH stainless steel, using two different infill deposition strategies, namely 45° (BMD_45) and 90° (BMD_90). The main conclusions can be listed as follows:

- The composite feedstock rods characterization highlighted the presence of contaminant particles, whose nature shall be further investigated;
- Oxide particles with a high percentage of silicon, oxygen, aluminum, and manganese were found inside the feedstock rod. Their presence was spotted as well on the as-fabricated sample;
- Investigations allowed for the isolation of different defects and to correlate their nature with the AM process and/or the feedstock quality;
- The infill deposition strategy was shown to have a direct influence on the phase composition and the microhardness of the fabricated samples. The 45° infill strategy is shown to be the most reliable option when limited δ-ferrite content and high hardness are required.

Author Contributions: Conceptualization, V.D.P., E.S.; methodology, S.S.; validation, V.D.P., A.S., K.S.; investigation, M.C., A.S., V.D.P. and K.S.; writing—original draft preparation, E.S., V.D.P.; writing—review and editing, K.S., M.C. and S.S. All authors have read and agreed to the published version of the manuscript.

Funding: This research was partially funded by the Grant of Excellence Departments, MIUR-Italy (ARTICOLO 1, COMMI 314–337 LEGGE 232/2016). Financed by the European Union-NextGenerationEU (National Sustainable Mobility Center CN00000023, Italian Ministry of University and Research Decree n. 1033-17/06/2022, Spoke 11- Innovative Materials and Lightweighting), and National Recovery and Resilience Plan (NRRP), Mission 04 Component 2 Investment 1.5-NextGenerationEU, Call for tender n. 3277 dated 30 December 2021. The opinions expressed are those of the authors only and should not be considered representative of the European Union or the European Commission's official position. Neither the European Union nor the European Commission can be held responsible for them.

Data Availability Statement: The data presented in this study are freely available on request from the corresponding author.

Conflicts of Interest: The authors declare no conflict of interest.

References

1. Klocke, F.; Arntz, K.; Teli, M.; Winands, K.; Wegener, M.; Oliari, S. State-of-the-art Laser Additive Manufacturing for Hot-work Tool Steels. *Procedia CIRP* **2017**, *63*, 58–63. [CrossRef]
2. *ASTM F2792-12*; Standard Terminology for Additive Manufacturing Technologies. ASTM International: West Conshohocken, PA, USA, 2012.
3. Khorasani, M.; Loy, J.; Ghasemi, A.H.; Sharabian, E.; Leary, M.; Mirafzal, H.; Cochrane, P.; Rolfe, B.; Gibson, I. A review of Industry 4.0 and additive manufacturing synergy. *Rapid Prototyp. J.* **2022**, *28*, 1462–1475. [CrossRef]
4. Herzog, D.; Seyda, V.; Wycisk, E.; Emmelmann, C. Additive manufacturing of metals. *Acta Mater.* **2016**, *117*, 371–392. [CrossRef]
5. Reza Jandaghi, M.; Pouraliakbar, H.; Shim, S.H.; Fallah, V.; Hong, S.I.; Pavese, M. In-situ alloying of stainless steel 316L by co-inoculation of Ti and Mn using LPBF additive manufacturing: Microstructural evolution and mechanical properties. *Mater. Sci. Eng. A* **2022**, *857*, 144114. [CrossRef]
6. Reza Jandaghi, M.; Pouraliakbar, H.; Iannucci, L.; Fallah, V.; Pavese, M. Comparative assessment of gas and water atomized powders for additive manufacturing of 316 L stainless steel: Microstructure, mechanical properties, and corrosion resistance. *Mater. Charact.* **2023**, *204*, 113204. [CrossRef]
7. Galati, M.; Calignano, F.; Viccica, M.; Iuliano, L. Additive Manufacturing Redesigning of Metallic Parts for High Precision Machines. *Crystals* **2020**, *10*, 161. [CrossRef]
8. Zhai, Y.; Lados, D.A.; LaGoy, J.L. Additive Manufacturing: Making Imagination the Major Limitation. *JOM* **2014**, *66*, 808–816. [CrossRef]
9. DebRoy, T.; Wei, H.L.; Zuback, J.S.; Mukherjee, T.; Elmer, J.W.; Milewski, J.O.; Beese, A.M.; Wilson-Heid, A.; De, A.; Zhang, W. Additive manufacturing of metallic components–Process, structure and properties. *Prog. Mater. Sci.* **2018**, *92*, 112–224. [CrossRef]
10. Bjørheim, F.; La Torraca Lopez, I.M. Tension testing of additively manufactured specimens of 17-4 PH processed by Bound Metal Deposition. *IOP Conf. Ser. Mater. Sci. Eng.* **2021**, *1201*, 012037. [CrossRef]
11. Abe, Y.; Kurose, T.; Santos, M.V.A.; Kanaya, Y.; Ishigami, A.; Tanaka, S.; Ito, H. Effect of Layer Directions on Internal Structures and Tensile Properties of 17-4PH Stainless Steel Parts Fabricated by Fused Deposition of Metals. *Materials* **2021**, *14*, 243. [CrossRef]
12. Tuncer, N.; Bose, A. Solid-State Metal Additive Manufacturing: A Review. *JOM* **2020**, *72*, 3090–3111. [CrossRef]
13. Gabilondo, M.; Cearsolo, X.; Arrue, M.; Castro, F. Influence of Build Orientation, Chamber Temperature and Infill Pattern on Mechanical Properties of 316 L Parts Manufactured by Bound Metal Deposition. *Materials* **2022**, *15*, 1183. [CrossRef]
14. Watson, A.; Belding, J.; Ellis, B.D. Characterization of 17-4 PH Processed via Bound Metal Deposition (BMD). In *TMS 2020, 149th Annual Meeting & Exhibition Supplemental Proceedings*; The Minerals, Metals & Materials Series; Springer: Cham, Switzerland, 2020.
15. Strano, M.; Rane, K.; Briatico Vangosa, F.; Di Landro, L. Extrusion of metal powder-polymer mixtures: Melt rheology and process stability. *J. Mater. Process. Technol.* **2019**, *273*, 116250. [CrossRef]
16. Machaka, R. Metal injection moulding of a 17-4 PH stainless steel: A comparative study of mechanical properties. *IOP Conf. Ser. Mater. Sci. Eng.* **2018**, *430*, 012033. [CrossRef]
17. Sung, H.-J.; Ha, T.K.; Ahn, S.; Chang, Y.W. Powder injection molding of a 17-4 PH stainless steel and the effect of sintering temperature on its microstructure and mechanical properties. *J. Mater. Process. Technol.* **2002**, *130–131*, 321–327. [CrossRef]
18. German, R.M. MIM 17–4 PH stainless steel: Processing, properties and best practice. *Powder Inject. Mould. Int.* **2018**, *12*, 49–76.
19. Di Pompeo, V.; Santoni, A.; Santecchia, E.; Spigarelli, S. On the Short-Term Creep Response at 482 °C (900 °F) of the 17-4PH Steel Produced by Bound Metal Deposition. *Metals* **2022**, *12*, 477. [CrossRef]
20. Specialty Steel Supply. 17-4PH Stainless Steel. Available online: http://www.specialtysteelsupply.com/17-4ph-stainless-steel (accessed on 16 May 2022).
21. Markforged. 17-4PH Stainless Steel Datasheet. Available online: https://markforged.com/materials/metals/17-4-ph-stainless-steel (accessed on 8 May 2022).
22. Zai, L.; Zhang, C.; Wang, Y.; Guo, W.; Wellmann, D.; Tong, X.; Tian, Y. Laser Powder Bed Fusion of Precipitation-Hardened Martensitic Stainless Steels: A Review. *Metals* **2020**, *10*, 255. [CrossRef]

23. Pellegrini, A.; Lavecchia, F.; Guerra, M.G.; Galantucci, M.L. Influence of aging treatments on 17–4 PH stainless steel parts realized using material extrusion additive manufacturing technologies. *Int. J. Adv. Manuf. Technol.* **2023**, *126*, 163–178. [CrossRef]
24. Gonzalez-Gutierrez, J.; Arbeiter, F.; Schlauf, T.; Kukla, C.; Holzer, C. Tensile properties of sintered 17–4PH stainless steel fabricated by material extrusion additive manufacturing. *Mater. Lett.* **2019**, *248*, 165–168. [CrossRef]
25. Galati, M.; Minetola, P. Analysis of Density, Roughness, and Accuracy of the Atomic Diffusion Additive Manufacturing (ADAM) Process for Metal Parts. *Materials* **2019**, *12*, 4122. [CrossRef] [PubMed]
26. Akessa, A.D.; Tucho, W.M.; Lemu, H.G.; Grønsund, J. Investigations of the Microstructure and Mechanical Properties of 17-4 PH ss Printed Using a MarkForged Metal X. *Materials* **2022**, *15*, 6898. [CrossRef]
27. Kedziora, S.; Decker, T.; Museyibov, E.; Morbach, J.; Hohmann, S.; Huwer, A.; Wahl, M. Strength Properties of 316L and 17-4 PH Stainless Steel Produced with Additive Manufacturing. *Materials* **2022**, *15*, 6278. [CrossRef] [PubMed]
28. Desktop Metal. 17-4PH Datasheet. Available online: https://www.desktopmetal.com/uploads/BMD-SPC-MDS-17-4ph-211112_c.pdf (accessed on 10 May 2022).
29. Rasband, W.S. *ImageJ*; U.S. National Institutes of Health: Bethesda, MD, USA, 2016.
30. *ASTM E562-19e1*; Standard Test Method for Determining Volume Fraction by Systematic Manual Point Count. ASTM International: West Conshohocken, PA, USA, 2022.
31. Döbelin, N.; Kleeberg, R. Profex: A graphical user interface for the Rietveld refinement program BGMN. *J. Appl. Crystallogr.* **2015**, *48*, 1573–1580. [CrossRef]
32. Hausnerova, B.; Mukund, B.N.; Sanetrnik, D. Rheological properties of gas and water atomized 17-4PH stainless steel MIM feedstocks: Effect of powder shape and size. *Powder Technol.* **2017**, *312*, 152–158. [CrossRef]
33. Contreras, J.M.; Jiménez-Morales, A.; Torralba, J.M. Influence of the Morphology and Particle Size on the Processing of Bronze 90/10 Powders by Metal Injection Moulding (MIM). *Mater. Sci. Forum* **2007**, *534–536*, 365–368. [CrossRef]
34. Seerane, M.; Ndlangamandla, P.; Machaka, R. The influence of particle size distribution on the properties of metal injection-moulded 17-4 PH stainless steel. *J. S. Afr. Inst. Min. Metall.* **2016**, *116*, 935–940. [CrossRef]
35. German, R.M.; Bose, A. *Injection Molding of Metals and Ceramics*; Metal Powder Industries Federation: Princeton, NJ, USA, 1997.
36. Forcellese, P.; Mancia, T.; Simoncini, M.; Bellezze, T. Investigation on Corrosion Resistance Properties of 17-4 PH Bound Metal Deposition As-Sintered Specimens with Different Build-Up Orientations. *Metals* **2022**, *12*, 588. [CrossRef]
37. Nyborg, I.; Tunberg, T.; Wang, P.X. Surface product formation during water atomization and sintering of austenitic stainless steel powder. *Met. Powder Rep.* **1990**, *45*, 750–753. [CrossRef]
38. Wu, M.-W.; Huang, Z.-K.; Tseng, C.-F.; Hwang, K.-S. Microstructures, Mechanical Properties, and Fracture Behaviors of Metal-Injection Molded 17-4PH Stainless Steel. *Met. Mater. Int.* **2015**, *21*, 531–537. [CrossRef]
39. Suwanpreecha, C.; Manonukul, A. On the build orientation effect in as-printed and as-sintered bending properties of 17-4PH alloy fabricated by metal fused filament fabrication. *Rapid Prototyp. J.* **2022**, *28*, 1076–1085. [CrossRef]
40. Li, K.; Sridar, S.; Tan, S.; Xiong, W. Effect of homogenization on precipitation behavior and strengthening of 17-4PH stainless steel fabricated using laser powder bed fusion. *arXiv* **2021**, arXiv:2112.06289.
41. Stoudt, M.R.; Ricker, R.E.; Lass, E.A.; Levine, L.E. The Influence of Post-Build Microstructure on the Electrochemical Behavior of Additively Manufactured 17-4 PH Stainless Steel. *JOM* **2017**, *69*, 506–515. [CrossRef]
42. Wu, Y.; German, R.M.; Blaine, D.; Marx, B.; Schlaefer, C. Effects of residual carbon content on sintering shrinkage, microstructure and mechanical properties of injection molded 17-4 PH stainless steel. *J. Mater. Sci.* **2022**, *37*, 3573–3583. [CrossRef]
43. Huber, D.; Vogel, L.; Fischer, A. The effects of sintering temperature and hold time on densification, mechanical properties and microstructural characteristics of binder jet 3D printed 17-4 PH stainless steel. *Addit. Manuf.* **2011**, *46*, 102114. [CrossRef]
44. Kazior, J. Influence of Sintering Atmosphere, Temperature and the Solution-Annealing Treatment on the Properties of Precipitation-Hardening Sintered 17-4 PH Stainless Steel. *Materials* **2023**, *16*, 760. [CrossRef]
45. Abbaschian, R.; Abbaschian, L.; Reed-Hill, R.E. *Physical Metallurgy Principles*, 4th ed.; Cengage Learning: Stamford, CA, USA, 2009.
46. Hutchinson, B.; Hagström, J.; Karlsson, O.; Lindell, D.; Tornberg, M.; Lindberg, F.; Thuvander, M. Microstructures and hardness of as-quenched martensites (0.1–0.5% C). *Acta Mater.* **2011**, *59*, 5845–5858. [CrossRef]
47. Gyhlesten Back, J.; Babu Surreddi, K. Microstructure analysis of martensitic low alloy carbon steel samples subjected to deformation dilatometry. *Mater. Charact.* **2019**, *157*, 109926. [CrossRef]
48. Lou, J.; He, H.; Li, Y.; Zhang, H.; Fang, Z.; Wei, X. Effects of Trace Carbon Contents on Lattice Distortion and Nano-Copper Phase Precipitation in Metal Injection-Molded 17-4PH Stainless Steel. *JOM* **2019**, *71*, 1073–1081. [CrossRef]
49. Schroeder, R.; Hammes, G.; Binder, C.; Klein, A.N. Plasma Debinding and Sintering of Metal Injection Moulded 17-4PH Stainless Steel. *Mater. Res.* **2011**, *14*, 564–568. [CrossRef]
50. Saboori, A.; Piscopo, G.; Lai, M.; Salmi, A.; Biamino, S. An investigation on the effect of deposition pattern on the microstructure, mechanical properties and residual stress of 316L produced by Directed Energy Deposition. *Mater. Sci. Eng. A* **2020**, *780*, 139179. [CrossRef]
51. Zhang, Y.; Roch, A. Fused filament fabrication and sintering of 17-4PH stainless steel. *Manuf. Lett.* **2022**, *33*, 29–32. [CrossRef]

crystals

MDPI

Article

Formation and Evolution of Interfacial Structure in Al–Si–Mg/Stainless Steel Bimetals during Hot-Dipping Process

Byung-Joo Kim [1], Ha-Yoon Lim [1], Saif Haider Kayani [1], Yun-Soo Lee [1,*], Su-Hyeon Kim [1] and Joon-Hyeon Cha [2]

[1] Metallic Materials Division, Korean Institute of Material Science, Changwon 51508, Republic of Korea; kbj@kims.re.kr (B.-J.K.); hylim817@kims.re.kr (H.-Y.L.); saifkayani@kims.re.kr (S.H.K.); shawnkim@kims.re.kr (S.-H.K.)

[2] Steel Supporting Center Task Force Team, Pohang Institute of Metal Industry Advancement, Pohang-si 37666, Republic of Korea; chajh@pomia.or.kr

* Correspondence: yslee@kims.re.kr; Tel.: +82-055-280-3293

Abstract: Understanding trends in the formation of the intermetallic compound (IMC) layer in Al/Fe bimetallic composites can aid in significantly improving their mechanical properties. However, it is currently challenging to predict IMC layer formation during hot-dip aluminizing. Furthermore, the results from previous studies are difficult to compare owing to the variation in the process parameters used. Therefore, to understand how temperatures and holding times affect the thickness and hardness properties of IMC layers, we investigated the interfacial properties of aluminized stainless steel in molten Al-Si-Mg. AISI 420 stainless steel was hot-dip aluminized in an Al–Si–Mg alloy melt for 10–120 min at four different temperatures: 700, 750, 800, and 850 °C. Morphology, type, and element distribution of the phases formed in the reaction layer and the reduction rate of the aluminizing process were studied. Notably, while the reaction layer thickness increased with increasing aluminizing temperature when the holding time was low, long-term reaction caused the reaction layer to become thicker at lower temperatures. The mechanism of this morphological transformation is discussed. The results demonstrated effective trends in controlling the morphology of the intermetallic compound layer with respect to various hot-dip Al plating process parameters.

Keywords: Al–Si–Mg/stainless steel bimetal; hot-dip aluminizing process; phase transformation; hardness properties

1. Introduction

Al/Fe bimetallic composites combine the high strength and wear resistance of Fe-based materials such as cast iron and stainless steel (STS) with the heat conductivity, corrosion resistance, and light weight of Al alloys [1], making them highly promising for industrial applications. They have thus attracted significant attention for use in automobile parts such as engine blocks and pistons [2–4].

The liquid–solid composite method is often used to prepare Al/Fe bimetallic composites because of its straightforward process and versatile material choices [5]. Though it is challenging to achieve robust metallurgical bonding owing to the physicochemical differences between Al and Fe [6,7], the removal of the Fe substrate from molten Al results in a pure Al or Al alloy layer adhering to the surface of the intermetallic compound (IMC) layer that forms during the aluminizing process [8]. The adhesion between Al and Fe is influenced by the continuous distribution of this Fe–Al IMC layer [9].

The interfacial bond strength of Al/Fe bimetal is closely related to the thickness of the interlayer. The thickness of the IMC layer is closely related to the interfacial bonding strength of the Al/Fe bimetal. Because of the brittle nature of IMCs, thin reaction layers provide better bonding strength than thicker reaction layers [2,10,11]. Several researchers have therefore attempted to minimize the thicknesses of IMCs in Al/Fe (steel) bimetals by adding alloying elements to the Al melts. Si atoms are particularly pivotal elements. As the Si content in the Al alloy melt increases, the solid solution of Si in the Fe_2Al_5 phase increases correspondingly [12]. Si atoms then fill the vacancies on the C-axis of the Fe_2Al_5 phase, leading to the formation of ternary Al–Fe–Si IMCs [5]. When Si atoms occupy the vacancies along the C-axis of the Fe_2Al_5 phase, they effectively block the preferred diffusion pathways for Al atoms. This occupation disrupts the facile diffusion of aluminum along the C-axis. These ternary Al–Fe–Si IMCs restrict the diffusion of Al atoms into Fe_2Al_5, causing a reduction in the Fe_2Al_5 layer thickness. Elements such as Cr and Ni also affect the formation of the reaction layer in STS [5]. However, predicting the formation of the reaction layer in Fe–Cr/Al–Si bimetals remains challenging.

Although numerous studies have been conducted on the development of the aluminizing process, controversy remains regarding phase formation during interfacial reactions in hot-dip aluminizing. The inhibition of the STS/Al reaction by Si is a complex process, and the mechanism underlying the formation of the reaction layer is not fully understood. Furthermore, it is difficult to compare various parameters—such as the composition, reaction temperature, and holding time of the Al alloy and Fe substrate—across studies. Recent studies have often used relatively short hot-dipping times, ranging from 5 to 1800 s [5,12,13]. Considering the potential effects of exposing STS to Al–Si alloys within the aluminizing temperature range of 700–850 °C, it is important to understand how the temperatures and holding time affect the layer thickness and reduction rates. In this study, we investigated the interfacial characteristics of hot-dip aluminized STS in an Al-Si-Mg alloy melt, with a focus on the effects of reaction temperature and holding time on the reaction layer.

2. Materials and Methods

The hot-dip aluminizing process was carried out by immersing the as-received AISI 420 stainless steel rods (hereafter referred to as STS420), which had diameters of 10 mm and lengths of 300 mm (Figure 1a), into a commercial Al–7Si–0.4Mg alloy melt (hereinafter referred to as AC4C). The AC4C alloy, the weight of which was approximately 20 kg (Figure 1a), was melted in an electrical resistance furnace using a crucible with a 30 kg capacity. Table 1 summarizes the chemical compositions of the AC4C and STS420 rods, as determined by optical emission spectrometry (OES; Foundry-Master UV, Oxford Instruments, Abingdon, England). Although the aluminizing process for Al/Fe bimetals is typically performed in the temperature range 700–1100 °C [13,14], excessively high aluminizing temperatures can partially melt the solid material [15]. Hence, the temperature of the molten AC4C alloy was sequentially adjusted to 700, 750, 800, and 850 $\pm$ 5 °C for each reaction. Each STS420 rod interacted with the AC4C alloy melt for 10 to 120 min and was then extracted, as illustrated in Figure 1b. Figure 1c exhibits the morphologies of the AC4C/STS420 bimetal samples aluminized at various holding times and reaction temperatures. Under all the aluminizing conditions, the AC4C alloy adequately coated the STS420 rods. An X-ray diffraction (XRD) measurement was used to identify the phases of the STS420 rod and Al-Si-Mg/STS420 bimetals. Data were collected using an XRD instrument (Rigaku International Corp., D/Max 2500VL/PC, Tokyo, Japan) with Cu Kα radiation from 20 to 80 (2θ) at a scan rate of 2°/min. Figure 1a shows the XRD result and BSE image of the STS420 rod. In the XRD pattern of the STS 420 rod, peaks corresponding to the (111), (200), and (211) planes are clearly observed, and no oxide layer is observed at the interface region (BSE image in Figure 1a).

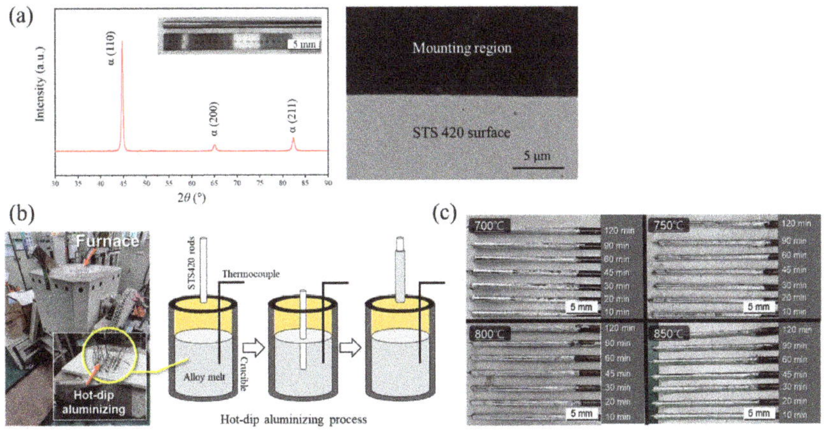

Figure 1. (**a**) XRD pattern and BSE images of STS 420 rod; (**b**) hot-dip aluminizing equipment and process schematic; (**c**) photographs of AC4C/STS420 bimetal specimens aluminized under various process conditions.

Table 1. Chemical compositions of AC4C and STS 420 materials.

Alloy	Composition (wt. %)										
	Al	Fe	Cr	Si	Mg	Mn	Zn	P	Ti	Pb	C
AC4C	Bal.	0.151	-	7.630	0.410	<0.003	0.019	-	0.046	0.013	<0.003
STS420	<0.003	Bal.	13.53	0.451	-	0.543	<0.003	0.024	<0.003	<0.003	0.341

The metallographic specimens were taken from a cross-section 30 mm from the bottom of AC4C/STS420 bimetal and mechanically polished using standard procedures up to colloidal silica for 20 min. The interfacial structure in the AC4C/STS420 bimetal was characterized by an optical microscope (OM; Nikon Eclipse MA200, Tokyo, Japan), field-emission backscattered electron/energy-dispersive spectroscopy (FE-SEM/EDS; JSM-7001F, JEOL Ltd., Tokyo, Japan), and atomic distribution and line scans of the interface region were measured using electron probe microanalysis (EPMA; JXA-8530F, JEOL Ltd., Tokyo, Japan). The thicknesses of the interfacial layer were performed using IMT Isolution DT (ver. 26.5, British Columbia, Canada). Micro-Vickers hardness testing was performed on the interface region of AC4C/STS420 bimetal using a Mitutoyo HM-102 hardness tester (Mitutoyo, Kanagawa, Japan), with a load of 10 g (HV0.1) and a dwell time of 15 s.

3. Results and Discussion

Figure 2 shows the XRD pattern of the AC4C/STS420 bimetals reacted at 700–850 °C for 30 min. For all four bimetals, peaks corresponding to the Fe, Al, Fe_2Al_5 and Al_8Fe_2Si, and $FeAl_3$ phases were identified. Phase identification at the interface of AC4C/STS420 bimetals was further investigated using EPMA. Figure 3 shows the Fe, Cr, Al, and Si distributions on the interface regions of AC4C/STS420 bimetal aluminized at 700 °C and 850 °C for 60 min, as determined by EPMA. Figure 3c,d exhibit the EPMA line scanning analyses corresponding to the IMC layer identified in Figure 3a and b, respectively. The results confirmed that the Fe and Cr contents in the IMCs increased with increasing proximity to the STS420 surface. The IMC layers stacked on the STS420 surface comprised Fe_2Al_5 and Al_8Fe_2Si phases with about 5 wt.% Cr. The Fe_2Al_5 layer formed parallel to the STS420 surface and exhibited uniform layer thickness. The Al_8Fe_2Si layer, which formed after the Fe_2Al_5 layer, exhibited different morphologies depending on the aluminized temperature. The cross-sectional BSE micrographs of the reaction layers are given in Figure 4. At a reaction temperature of 700 and 750 °C, a flat and linear Al_8Fe_2Si layer was observed between Fe_2Al_5 and AC4C (Figure 4a,b). When increasing the reaction temperature to 800 and 850 °C, block-shaped Al_8Fe_2Si structures formed on the Fe_2Al_5 layer (Figure 4c,d). This formation of block-shaped Al_8Fe_2Si phases has also been reported in previous aluminizing studies [12,16]. These microstructures suggest that the Al_8Fe_2Si phase grew from the Fe_2Al_5 phase during solidification. The phase compositions, microstructures, and growth sequences of the IMCs formed at the interface were similar to the results obtained in previous studies of Al-Si/(Fe or steel) bimetals [4,7]. Wang et al. also confirmed that the IMC layer of Al-10Si/Fe bimetals aluminized at 700 °C for up to 30 min contains Fe_2Al_5, Al_8Fe_2Si, and $Al_8Fe_2Si_2$ [12]. During the aluminizing process in Al-Si melt, the $Al_{13}Fe_4$ phase that forms after the Fe_2Al_5 phase transitions to Al_8Fe_2Si with increasing Si content. The Si content of AC4C (in this study) is approximately 7.6 wt.%, which may explain why other types of Al-Fe compounds are not observed. Additionally, the plate-like $Al_9Fe_2Si_2$ phase commonly reported in Al-Si/Fe bimetals [5,12] was not observed in this study, which may be due to Cr diffusing from STS 420 into the AC4C melt. Elements such as Cr [17], Mn [18], and Co [19] have been reported to induce the transformation of the plate-like Al-Fe-Si phase into the Al(Fe,Cr)Si phase, resulting in the absence of the $Al_9Fe_2Si_2$ phase in several cast Al-Si-Fe-Cr alloys [20]. The Al_8Fe_2Si phase reacts with the liquid to form $Al_9Fe_2Si_2$ by a peritectic reaction at 640 °C: Al_8Fe_2Si + liquid → $Al_9Fe_2Si_2$ + Al [12]. The diffused Cr likely prevented the formation of $Al_9Fe_2Si_2$, which could solidify during the cooling process after hot-dip aluminizing.

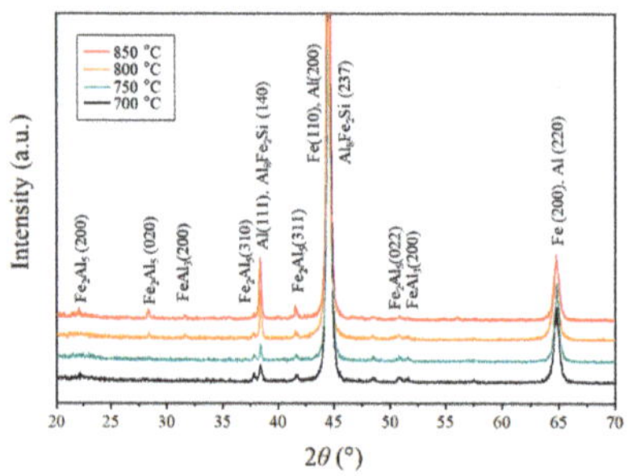

Figure 2. XRD pattern of AC4C/STS 420 bimetals aluminized at different temperatures for 30 min.

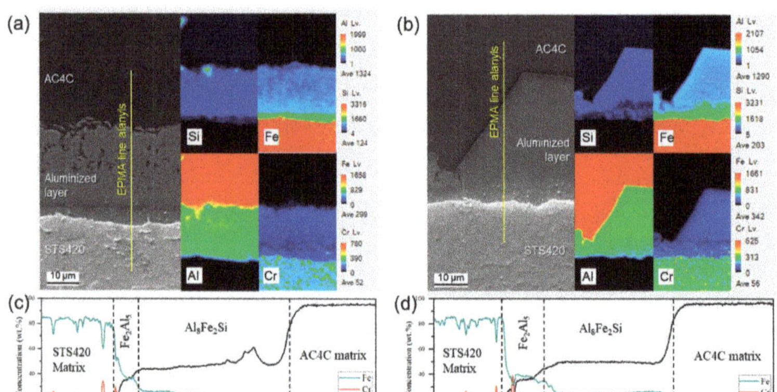

Figure 3. BSE/EPMA mapping and line scanning analysis of the interface region in AC4C/STS420 bimetals aluminized at 700 °C (**a**,**c**) and 850 °C (**b**,**d**) for 60 min.

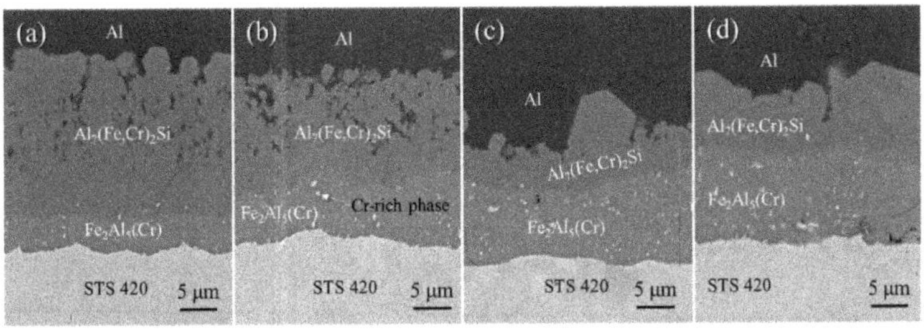

Figure 4. BSE images of the interfacial region of AC4C/STS 420 bimetals reacted at different temperatures for 30 min: 700 °C (**a**), 750 °C (**b**), 800 °C (**c**), and 850 °C (**d**).

Figure 5 shows high-magnification images of the interfacial region in AC4C/STS420 after reacting at 800 °C for 120 min. The results of the EDS analysis of points 1–7 in Figure 5 are listed in Table 2. This helped further identify each phase in the IMC layer, which was layered with the Fe_2Al_5 and Al_8Fe_2Si phases. The aluminized layer next to the STS420 substrate was broad and primarily comprised the Fe_2Al_5 phase, within which fine Cr-rich precipitates (point 3, indicated in Figure 5a) were uniformly distributed. Cr can dissolve up to 6 wt.% in the Fe_2Al_5 phase, and excess Cr can precipitate out of this phase. Since the Cr content of STS 420 is about 13.4 wt.%, the white particles present in the Fe_2Al_5 phase in Figure 4 are assumed to be precipitated by excess Cr. The Fe_2Al_5 phase containing similar Cr particles was also previously reported in aluminized STS 430 [16]. In the Al-Cr-Fe ternary system [21], Al_8Cr_5 may coexist with the Fe_2Al_5 phase among the Cr-rich precipitates.

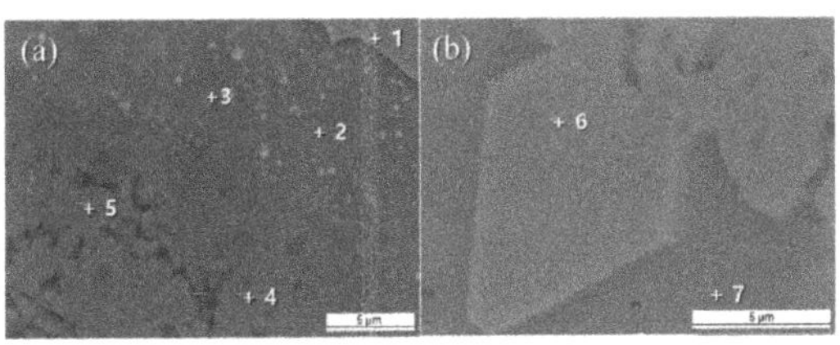

Figure 5. BSE images of different locations in the interface region in AC4C/STS420 bimetal fabricated at 800 °C for 120 min; (**a**) adjacent to the STS420 substrate and (**b**) adjacent to the Al top coat.

Table 2. EDS analysis results of the IMC layers in AC4C/STS420 bimetals (points 1–7 in Figure 5).

Point	Composition					Corresponding Phase
	Al (at.%)	Si (at.%)	Cr (at.%)	Fe (at.%)	Mn (at.%)	
1	-	0.73	11.90	87.37	0.82	Fe
2	63.64	5.96	4.30	26.10	0.20	Fe_2Al_5(Cr)
3	48.57	3.51	23.96	23.96	0.48	Cr rich phase
4	69.21	11.60	3.59	15.60	-	$Al_8(Fe,Cr)_2Si$
5	70.98	10.55	3.84	14.63	-	$Al_8(Fe,Cr)_2Si$
6	72.05	9.14	3.41	15.40	-	$Al_8(Fe,Cr)_2Si$
7	98.96	1.04	-	-	-	Al

Figure 6a shows the OM images of the interface regions of AC4C/STS420 bimetals prepared at 700, 750, 800, and 850 °C under various aluminizing times. Increasing the temperature and aluminizing time resulted in a clear systematic change in the surface morphology. At reaction temperatures of 700 and 750 °C, distinct Fe_2Al_5 and Al_8Fe_2Si

layers with comparable thicknesses were observed to be aligned with the STS420 surface. When hot-dipping aluminized at 700 °C for 10 min, the average thicknesses of the Fe_2Al_5 and Al_8Fe_2Si layers were 3.1 and 11.8 μm, respectively. These thicknesses increased with holding time; extending the duration to 120 min caused the average thicknesses of the Fe_2Al_5 and Al_8Fe_2Si layers to increase to 19.5 and 22.2 μm, respectively. These results agree with those in a study by Wang et al., who reported that in Al–Si/Fe bimetals aluminized at 700 °C, the IMCs Fe_2Al_5 and Al_8Fe_2Si form sequentially, and that the thickness of the reaction layer increases with increasing reaction time [12]. Meanwhile, when the reaction temperature was increased from 700 to 750 °C and maintained for 120 min, the trend changed: the thickness of the Fe_2Al_5 layer increased from 19.5 to 23.0 μm, but that of the Al_8Fe_2Si layer reduced from 22.2 to 13.2 μm. When the reaction temperature was further increased to 800 and 850 °C, the Fe_2Al_5 layer changed into a wavy structure and its thickness remained relatively similar (approximately 7–10 μm), even when the reaction time was increased. Though a small amount of block-shaped Al_8Fe_2Si phase was observed even at a reaction temperature of 750 °C when a flat reaction layer formed, the block-shaped Al_8Fe_2Si phase became relatively coarser and more irregular when formed at reaction temperatures of 800 and 850 °C, making it challenging to measure the average thickness of the layer. While the formation of block-shaped Al_8Fe_2Si phases at 650 [16] and 700 °C [12] has been reported in previous aluminizing studies, in the present study, the block-shaped IMC began to form at 750 °C, with major growth occurring at temperatures exceeding 800 °C.

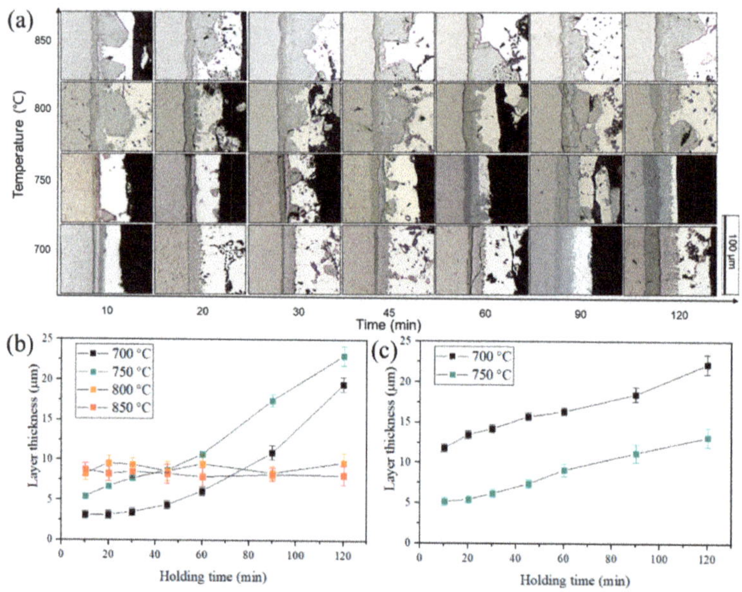

Figure 6. (**a**) OM images of the interfaces of AC4C/STS420 bimetal specimens fabricated at various aluminizing conditions and change in thickness of the IMC layers as influenced by reaction temperature and holding time: (**b**) Fe_2Al_5 and (**c**) Al_8Fe_2Si.

Figure 7 shows a schematic of the interfacial reaction of the AC4C/STS420 bimetal. During hot-dip aluminizing, a solute gradient region is created in the Al alloy melt by Fe and Cr atoms diffusing from the STS surface (Figure 7a). At a reaction temperature of 700 °C, the Fe_2Al_5 reaction layer grew uniformly as Al and Si atoms diffused into the STS substrate (Figure 3a). Previous studies have reported that the addition of Si to a molten Al alloy hinders the growth of Fe_2Al_5, resulting in a flat reaction layer [10,12]; the uniformity of the flat Fe_2Al_5 layer is influenced by the Si concentration in the molten Al alloy [10].

Chemical composition analysis showed that the STS and Fe_2Al_5 substrates contained Si concentrations of approximately 0.73 and 3.51 wt.%, respectively (Figure 5 and Table 2), indicating that the Si atoms in the AC4C melt diffused easily into the STS substrate. As the aluminizing temperature increases, the IMC layer is expected to transform more rapidly due to the higher diffusivity of atoms at elevated temperatures [11]. When the reaction temperature was 850 °C, the Fe_2Al_5 layer grew in an unstable wavy shape owing to the relatively high diffusivity of Fe and Al atoms (Figure 7c). During the cooling process after aluminizing, the diffusion of alloying elements in the remaining molten metal provides the necessary driving force for the formation of intermetallic phases [12].

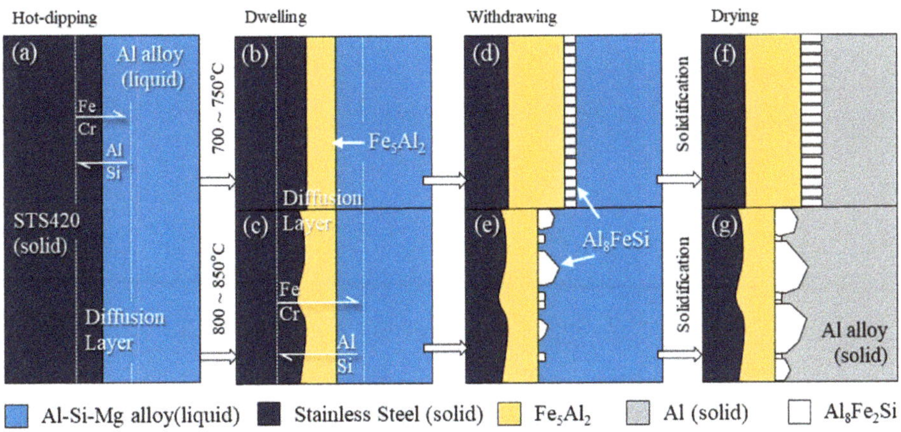

Figure 7. Schematics of phase formation sequence in AC4C/STS420 interface during aluminizing process. (**a**) Diffusion dissolution layer forms during the initial stages of the aluminizing process. The Fe_2Al_5 layer forms during the reaction step: reaction temperatures of (**b**) 700 °C and (**c**) 850 °C. The Al_8Fe_2Si phase nucleates and grows on the Fe_2Al_5 layer until solidification is complete: reaction temperatures of (**d**,**f**) 700 °C and (**e**,**g**) 850 °C.

The morphology of the Al_8Fe_2Si phase changed from flat (Figure 7d) to block-shaped (Figure 7e) as the melting temperature increased. The Al_8Fe_2Si phase formed after the Fe_2Al_5 layer (Figures 3 and 4) and appeared to grow in the Fe-rich region of the Al melt during solidification. The Al_8Fe_2Si phase was heterogeneously nucleated on the Fe_2Al_5 phase and grew to an Fe diffusion layer in the Al melt (Figure 7d,e). As the aluminizing temperature increased to 800 and 850 °C, the block-shaped Al_8Fe_2Si phase was preferentially formed. As the temperature increased, the diffusion coefficient also increased, leading to a more active dissolution of the Fe_2Al_5 layer in the Al melt. The diffusion of Al atoms toward the STS420 substrate was dominant at 700 and 750 °C, and the thickness of the reaction layer increased with increasing reaction time. However, at 800 °C and 850 °C, the thickness of Fe_2Al_5 remained relatively constant, depending only on the holding time. During the interaction of STS420 with the molten AC4C melt, an Fe_2Al_5 layer formed and simultaneously dissolved into the molten AC4C. The simultaneous dissolution into the Al melt during the growth of the Fe_2Al_5 phase was confirmed using synchrotron radiation techniques by Zhang et al. [22]. Therefore, the thickness of the Fe_2Al_5 layer is determined by the competition between growth and dissolution. At reaction temperatures of 700 and 750 °C, the formation of the Fe_2Al_5 phase driven by atomic diffusion may be more dominant, and its thickness is believed to increase with reaction time. However, at temperatures above 800 °C, the crystal structure of STS420 (Fe–13Cr) can transform from a body-centered cubic (BCC) to a face-centered cubic (FCC) structure, in which atomic diffusion is relatively challenging [23]. The thickness of the reaction layer developed during hot-dipping or annealing depends on the atomic diffusion rate. Al diffusion is slower in FCC Fe than in BCC Fe by two orders of magnitude [13]. This change in the crystal structure resulted in a balance between the formation of the Fe_2Al_5 phase and its dissolution in the Al melt, thus maintaining the thickness of the Fe_2Al_5 layer despite the elevated aluminizing temperature.

Figure 8a shows the variations in the cross-section of the STS420 rods in AC4C/STS420 bimetals under various hot-dip aluminizing conditions. The reduction rate R for the STS420 rods was determined using the equation provided [11]:

$$R = A_0 B / t \tag{1}$$

where A_0 is the initial area of the STS420 rod cross-section (78.5 mm^2), R is the reduction rate (Figure 8b), and t is the hot-dip aluminizing time. As the reaction temperature increased from 700 °C to 850 °C, the area reduction rate increased from 0.09 to 0.33 mm^2 min^{-1}.

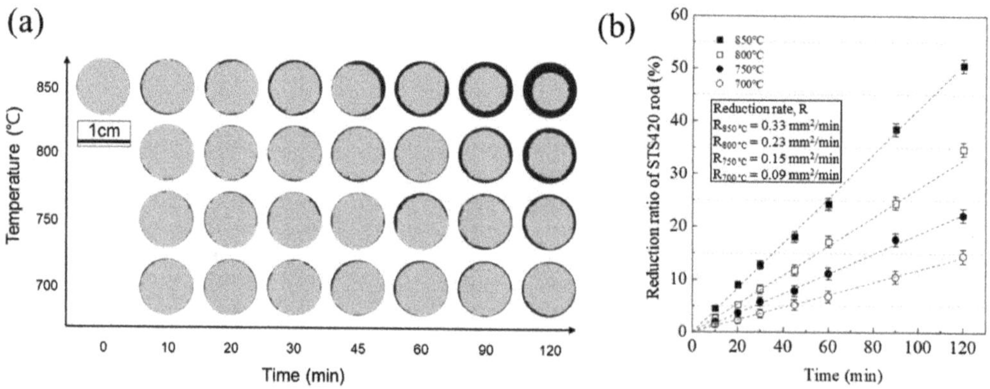

Figure 8. (**a**) Change in cross-section of STS rods in AC4C/STS420 bimetals with different aluminizing conditions; (**b**) area reduction ratio of STS420 rod under varying reaction temperatures and holding times.

The cross-sectional area of the STS420 rod decreased proportionately with the increases in reaction temperature and holding time. While approximately 14% of the STS420 rods dissolved during reaction at 700 °C for 120 min, approximately 50% of the STS420 rods dissolved when the reaction temperature was increased to 850 °C. The above results indicate that the high reaction temperature enhances the diffusion of Cr and Fe elements from the STS 420 rod into the molten aluminum alloy. At relatively low reaction temperatures of 700 and 750 °C, both the aluminumizing time and thickness of the reaction layer (Fe_2Al_5 and Al_8Fe_2Si) increased simultaneously. This is consistent with previous studies that reported that the IMC thickness increases proportionally with reaction temperature and time [24,25]. However, at relatively high reaction temperatures of 800 and 850 °C, the Fe_2Al_5 layer formed during the hot-dipping process of the Al–Si/STS bimetal maintained a relatively constant thickness, even upon increasing the holding time (Figure 6b). For short reaction times (10–20 min), the thickness of the Al–Si/STS reaction layer increased with increasing reaction temperature because of the increasing diffusion coefficient [26,27], which in turn was attributed to an increase in the diffusion layer with increasing temperature. Therefore, sufficient heating and holding time are required to ensure sufficient atomic diffusion between the interacting liquid and solid components [21]. However, during the formation of Fe_2Al_5 layers in the course of the hot-dipping process, these IMCs lost their atomic components to the Al melt and eventually dissolved. This decomposition of the Fe_2Al_5 phases occurred more vigorously at the relatively higher temperatures of 800 and 850 °C, which can be attributed to the greater thickness reduction in the STS rods at higher temperatures owing to greater diffusion. In addition, at the same time, the solubility of the Fe_2Al_5 layer in the molten Al alloy increased, creating a high-Fe diffusion region near the STS420 substrate during the aluminizing process. As a result, the thickness of the IMC layer was maintained by balancing its formation in the high-Fe diffusion region (increasing thickness) near the STS420 rod and dissolution into the Al melt (decreasing thickness). Therefore, even when the reaction time was prolonged, the thickness of the Fe_2Al_5 phase formed during the immersion process did not change significantly, whereas the Al_8Fe_2Si phase formed during the solidification process showed coarser growth owing to an increase in the Fe content diffused near the STS420 substrate. Previous studies [10,28] have also reported that a coarse blocky Al–Fe–Si phase with a high concentration of atomic diffusion regions forms in Al–Si alloy melts during the hot-dip aluminizing process.

Figure 9 displays the hardness values of the AC4C/STS420 bimetals interface formed under different aluminizing conditions (700 and 850 °C for 10 and 120 min). The hardness of the IMCs exhibits much higher values than those of the STS420 and AC4C matrix. The high hardness of the aluminized layer indicates the formation of hard intermetallic phases. The STS, Fe_2Al_5, Al_8Fe_2Si, and Al phases exhibited hardness values of approximately 300, 1260, 1130, and 75 HV, respectively. Therefore, it is apparent that the significantly higher hardness of the AC4C/STS420 interface compared with the metallic materials STS420 and AC4C was due to the formation of Fe_2Al_5 and Al_8Fe_2Si phases. Slightly differing from these results, the Fe_2Al_5 phase of Al/Fe bimetals typically exhibits a microhardness value of approximately 1000 HV [7]. Troysi et al. [29] have reported that the Fe_2Al_5 phase of aluminized 1020 steel (with a Cr composition of approximately 0.03 wt.%) exhibits a hardness value of 1070 HV. However, in this study, the Fe_2Al_5(Cr) layer in Al-Si-Mg/STS420 bimetals exhibits a hardness of approximately 1260 HV, which is about 20% higher than that of the Fe_2Al_5 layer in Al/Fe bimetals. This hardness increase is probably due to the strengthening effect of Cr in Fe_2Al_5 (Cr) and Cr-rich precipitates. Similarly, the hardness values of the Fe_2Al_5(Cr) phase of Al/STS bimetals containing Cr-rich precipitates were reported to range approximately from 1143 to 1250 HV [30]. A rapid increase in the hardness at the interface between the metallic materials and the IMCs leads to stress singularities [31,32]. At the interface between the STS420 matrix and top Al coat, a gradual change in hardness was observed; as the thickness of the reaction layer increased, the hardness gradient also increased (Figure 9). Therefore, the hardness gradient corresponds to the change in the thickness of the reaction layer depending on the aluminizing temperature and holding time, as shown in Figure 6. Previous studies have also reported that the thickness and hardness gradient of the Al/Fe reaction layer increase with increasing temperature [33,34]. The reaction layer was thicker when reacted at 700 °C for 90 min and 750 °C for 60 min than when reacted at higher temperatures (800–850 °C). In addition, at 800–850 °C, block-shaped IMCs formed, which are prone to stress concentration [35], and the shape of the reaction layer changed from linear to wavy. In our study, the hardness profile of the aluminized Fe_2Al_5 reaction layer gradually increased with increasing aluminizing time at 700 °C (Figure 9a) but remained relatively uniform at 850 °C (Figure 9b). Thinner reaction layers deform more easily than thicker reaction layers, and flatter reaction layers impart Al/Fe bimetals with better hardness properties by relieving stress concentrations during the aluminizing process [2,10,36]. Therefore, AC4C/STS420 bimetals aluminized for short periods at low temperatures (specifically, 700 °C for 10~20 min in this study) are expected to exhibit superior hardness properties to those aluminized at higher temperatures.

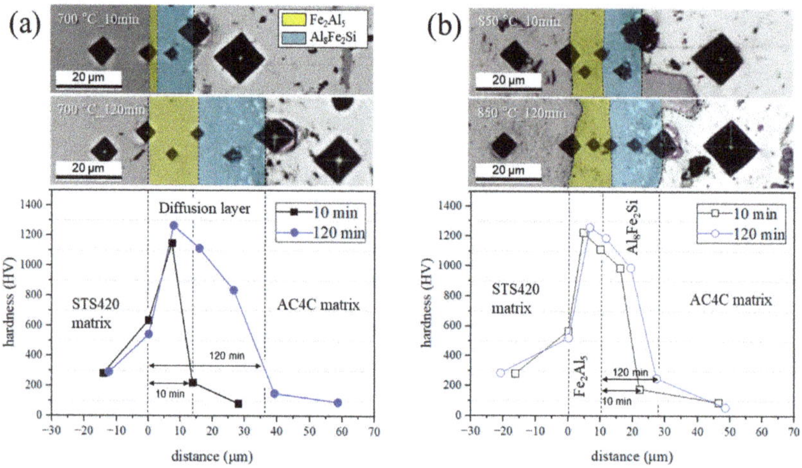

Figure 9. Vickers hardness values of AC4C/STS420 bimetals fabricated at different aluminizing temperatures for 10 and 120 min: (**a**) 700 °C and (**b**) 850 °C.

4. Conclusions

In this study, we investigated the interfacial region of aluminized STS420 rods in Al-Si-Mg alloy melt, including the formation mechanism of IMC phase and the effects of the aluminizing parameters on the IMC morphologies and hardness. The key findings of our investigation are as follows.

The proposed formation mechanism of AC4C/STS420 IMC during the aluminizing process is as follows: Fe_2Al_5 initially appears during the aluminizing process in the AC4C melt, and Al_8Fe_2Si is grown after the aluminized STS rods are lifted out of the AC4C melt.

As the reaction temperature increased, the uniform flat Fe_2Al_5 and Al_8Fe_2Si layers changed to exhibit wavy and blocked morphologies, respectively. At low reaction temperatures (700 and 750 °C), the thickness of the reaction layer gradually increased with increasing reaction time, whereas it remained relatively constant at high temperatures (800 and 850 °C).

The thickness and morphology of the IMC layer influence the hardness profile of the AC4C/STS420 bimetals interface. The hardness value of the IMCs layer was much higher than that of the metal matrix (STS420 and AC4C), and as the thickness of the reaction layer increased, the hardness gradient of the layer also increased. Therefore, this study suggested that the aluminizing conditions to form a fine and uniform reaction layer should be 10 to 20 min at 700 °C.

Author Contributions: Conceptualization, Y.-S.L.; methodology, H.-Y.L.; investigation, H.-Y.L. and S.H.K.; resources, S.-H.K.; data curation, B.-J.K.; writing—original draft preparation, B.-J.K.; writing—review and editing, B.-J.K. and Y.-S.L.; visualization, B.-J.K.; supervision, S.-H.K.; project administration, Y.-S.L.; funding acquisition, Y.-S.L., J.-H.C. and S.-H.K. All authors have read and agreed to the published version of the manuscript.

Funding: This research was funded by the Fundamental Research Program of the Korea Institute of Materials Science, grant number PNK9840, and National Research Foundation of Korea (NRF) grants funded by the Korean government (MSIP), grant number NRF-2021M3H4A1A04091999.

Data Availability Statement: The raw data supporting the conclusions of this article will be made available by the authors on request.

Conflicts of Interest: The authors declare no conflicts of interest.

References

1. Wang, C.; Wang, Z.; Xu, H.; Zhang, G. Microstructure and Age Hardening Behavior of Al/Fe Bimetal Prepared by One-Step Compound Casting. *J. Alloys Compd.* **2022**, *905*, 164170. [CrossRef]
2. Chen, Q.; Chang, X.; Qi, L.; Zheng, J.; Xie, C.; Chen, G. Interface Heterogeneity of Aluminum-Steel Bimetal Parts Manufactured via Thixotropic-Core Compound Forging. *J. Mater. Process. Technol.* **2022**, *306*, 117648. [CrossRef]
3. Chang, X.; Chen, G.; Wang, B.; Chen, Q.; Zhang, H. Thixotropic-Core Compound Forging for Aluminum-Steel Bimetal Gears. *J. Mater. Process. Technol.* **2022**, *299*, 117371. [CrossRef]
4. Jiang, W. Effects of Hot-Dip Galvanizing and Aluminizing on Interfacial Microstructures and Mechanical Properties of Aluminum/Iron Bimetallic Composites. *J. Alloys Compd.* **2016**, *688*, 742–751. [CrossRef]
5. Mao, F.; Zhang, P.; Wei, S.; Chen, C.; Zhang, G.; Xiong, M.; Wang, T.; Guo, J.; Wang, C. Interface Microstructure and Mechanical Properties of Al/Steel Bimetallic Composites Fabricated by Liquid-Solid Casting with Rare Earth Eu Additions. *Materials* **2022**, *15*, 6507. [CrossRef] [PubMed]
6. Kepa, T.; Pedraza, F.; Rouillard, F. Intermetallic Formation of Al-Fe and Al-Ni Phases by Ultrafast Slurry Aluminization (Flash Aluminizing). *Surf. Coat. Technol.* **2020**, *397*, 126011. [CrossRef]
7. Dey, P.P.; Sahu, S.; Banerjee, P.S.; Ghosh, M. A Review on Metallurgical Features of Hot-Dip Aluminized Steel. *Eng. Res. Express* **2023**, *5*, 012002. [CrossRef]
8. Liberski, P.; Gierek, A.; Kania, H.; Podolski, P.; Tatarek, A. Formation of Coatings from a Liquid Phase on the Surface of Iron-Base Alloys. *Arch. Foundry Eng.* **2008**, *8*, 93.
9. Khalid, M.Z.; Friis, J.; Ninive, P.H.; Marthinsen, K.; Ringdalen, I.G.; Strandlie, A. Modified Embedded Atom Method Potential for Fe-Al Intermetallics Mechanical Strength: A Comparative Analysis of Atomistic Simulations. *Phys. B Condens. Matter* **2021**, *618*, 413157. [CrossRef]
10. Cheng, W.-J.; Wang, C.-J. Effect of Silicon on the Formation of Intermetallic Phases in Aluminide Coating on Mild Steel. *Intermetallics* **2011**, *19*, 1455–1460. [CrossRef]
11. Kim, B.-J.; Cheon, H.-S.; Lee, Y.-H.; Kim, W.-K.; Lee, Y.-S.; Kim, S.-H. Effect of Reaction Temperature and Holding Time on the Interfacial Microstructure of Al–Si–Mg/Cu Bimetals Manufactured by Hot-Dip Aluminizing. *Mater. Chem. Phys.* **2024**, *313*, 128758. [CrossRef]
12. Wang, H.; Sun, S.; Li, X.; Wang, J.; Su, X. Effect of Silicon on Interfacial Reaction and Morphology of Hot-Dip Aluminizing. *J. Mater. Res. Technol.* **2022**, *20*, 3723–3734. [CrossRef]
13. Prasanthi, T.N.; Sudha, C.; Reddy, P.S. Hot Dip Aluminization of 304L SS and P91 Ferritic-Martensitic Steel—Comparison of Interface Morphology and Growth Kinetics of Reaction Zones. *Surf. Coat. Technol.* **2022**, *440*, 128465. [CrossRef]
14. Chang, Y.-Y.; Tsaur, C.-C.; Rock, J.C. Microstructure Studies of an Aluminide Coating on 9Cr-1Mo Steel during High Temperature Oxidation. *Surf. Coat. Technol.* **2006**, *200*, 6588–6593. [CrossRef]
15. Divandari, M.; Vahid Golpayegani, A.R. Study of Al/Cu Rich Phases Formed in A356 Alloy by Inserting Cu Wire in Pattern in LFC Process. *Mater. Des.* **2009**, *30*, 3279–3285. [CrossRef]
16. Liu, B.; Yang, Q.; Wang, Y. Intereaction and Intermetallic Phase Formation between Aluminum and Stainless Steel. *Results Phys.* **2019**, *12*, 514–524. [CrossRef]
17. Yang, Y.; Zhong, S.-Y.; Chen, Z.; Wang, M.; Ma, N.; Wang, H. Effect of Cr Content and Heat-Treatment on the High Temperature Strength of Eutectic Al–Si Alloys. *J. Alloys Compd.* **2015**, *647*, 63–69. [CrossRef]
18. Shabestari, S. The Effect of Iron and Manganese on the Formation of Intermetallic Compounds in Aluminum–Silicon Alloys. *Mater. Sci. Eng. A* **2004**, *383*, 289–298. [CrossRef]
19. Sha, M.; Wu, S.; Wang, X.; Wan, L. Effects of Co Addition on Fe-Bearing Intermetallic Compounds and Mechanical Properties of AlSi20Cu2Ni1Fe0.7–1 Alloys. *J. Alloys Compd.* **2013**, *551*, 468–474. [CrossRef]
20. Hosseinifar, M.; Malakhov, D.V. The Sequence of Intermetallics Formation during the Solidification of an Al-Mg-Si Alloy Containing La. *Met. Mater Trans A* **2011**, *42*, 825–833. [CrossRef]
21. Palm, M. The Al–Cr–Fe System–Phases and Phase Equilibria in the Al-Rich Corner. *J. Alloys Compd.* **1997**, *252*, 192–200. [CrossRef]
22. Zhang, N.; Hu, Q.; Yang, F.; Lu, W.; Ding, Z.; Cao, S.; Yu, L.; Ge, X.; Li, J. Effect of Si on the Growth Behavior of the Fe2Al5 Phase at Al-xSi(Liquid)/Fe(Solid) Interface During Holding by In-Situ Synchrotron Radiography. *Met. Mater Trans A* **2020**, *51*, 2711–2718. [CrossRef]
23. Xiong, W.; Selleby, M.; Chen, Q.; Odqvist, J.; Du, Y. Phase Equilibria and Thermodynamic Properties in the Fe-Cr System. *Crit. Rev. Solid State Mater. Sci.* **2010**, *35*, 125–152. [CrossRef]
24. Bakke, A.O.; Arnberg, L.; Løland, J.-O.; Jørgensen, S.; Kvinge, J.; Li, Y. Formation and Evolution of the Interfacial Structure in al/Steel Compound Castings during Solidification and Heat Treatment. *J. Alloys Compd.* **2020**, *849*, 156685. [CrossRef]
25. Dey, P.P.; Modak, P.; Ghosh, A.; Chakrabarti, D.; Banerjee, P.S.; Ghosh, M. Investigation of Phase Evolution of Al–Si–Mg Coating on Hot Dipped Interstitial-Free Steel. *Results Mater.* **2020**, *6*, 100078. [CrossRef]
26. Liu, T.; Wang, Q.; Sui, Y.; Wang, Q.; Ding, W. An Investigation into Interface Formation and Mechanical Properties of Aluminum–Copper Bimetal by Squeeze Casting. *Mater. Des.* **2016**, *89*, 1137–1146. [CrossRef]
27. Mao, A.; Zhang, J.; Yao, S.; Wang, A.; Wang, W.; Li, Y.; Qiao, C.; Xie, J.; Jia, Y. The Diffusion Behaviors at the Cu-Al Solid-Liquid Interface: A Molecular Dynamics Study. *Results Phys.* **2020**, *16*, 102998. [CrossRef]

28. Zarei, F.; Nuranian, H.; Shirvani, K. Effect of Si Addition on the Microstructure and Oxidation Behaviour of Formed Aluminide Coating on HH309 Steel by Cast-Aluminizing. *Surf. Coat. Technol.* **2020**, *394*, 125901. [CrossRef]
29. Troysi, F.D.; Brito, P.P. Development and Characterization of an Iron Aluminide Coating on Mild Steel Substrate Obtained by Friction Surfacing and Heat Treatment. *Int. J. Adv. Manuf. Technol.* **2020**, *111*, 2569–2576. [CrossRef]
30. Kishore, K.; Chhangani, S.; Prasad, M.J.N.V.; Bhanumurthy, K. Microstructure Evolution and Hardness of Hot Dip Aluminized Coating on Pure Iron and EUROFER 97 Steel: Effect of Substrate Chemistry and Heat Treatment. *Surf. Coat. Technol.* **2021**, *409*, 126783. [CrossRef]
31. Chi, S.-H.; Chung, Y.-L. Mechanical Behavior of Functionally Graded Material Plates under Transverse Load—Part I: Analysis. *Int. J. Solids Struct.* **2006**, *43*, 3657–3674. [CrossRef]
32. Bhavar, V.; Kattire, P.; Thakare, S.; patil, S.; Singh, R. A Review on Functionally Gradient Materials (FGMs) and Their Applications. *IOP Conf. Ser.: Mater. Sci. Eng.* **2017**, *229*, 012021. [CrossRef]
33. Zhang, D.; Zhang, G.; Yu, H.; Lv, W.; Wen, K.; Xu, H. Controlling Interfacial Composition and Improvement in Bonding Strength of Compound Casted Al/Steel Bimetal via Cr Interlayer. *J. Mater. Res. Technol.* **2023**, *23*, 4385–4395. [CrossRef]
34. Lou, B.-S.; Chen, Y.-Y.; Wu, Z.-Y.; Kuo, Y.-C.; Duh, J.-G.; Lee, J.-W. $(Fe,Mn)_3AlC_x$ κ-Carbide Formation and Characterization in Pack Aluminization of Fe–29Mn–9Al–0.9C Lightweight Steel. *J. Mater. Res. Technol.* **2022**, *20*, 1524–1532. [CrossRef]
35. Kim, B.J.; Dahle, A.K.; Park, Y.H.; Lee, Y.C. The Effect of Sr Additions on Al–Cu–Si Ternary Eutectic Alloys for a High-Ductility Bimodal Microstructure. *Mater. Sci. Eng. A* **2022**, *833*, 142547. [CrossRef]
36. Guo, Z.; Liu, M.; Bian, X.; Liu, M.; Li, J. An Al–7Si Alloy/Cast Iron Bimetallic Composite with Super-High Shear Strength. *J. Mater. Res. Technol.* **2019**, *8*, 3126–3136. [CrossRef]

crystals

MDPI

Article

Electrochemical Study and Mechanical Properties of Ti-Zr Alloy for Biomedical Applications

Iosif Hulka [1], Ion-Dragoș Uțu [2], Santiago Brito-Garcia [3], Amparo Verdu-Vazquez [4] and Julia C. Mirza-Rosca [3,5,*]

1 Research Institute for Renewable Energie, Politehnica University of Timişoara, G. Muzicescu 138, 300501 Timişoara, Romania; iosif.hulka@upt.ro
2 Department of Materials and Manufacturing Engineering, Faculty of Mechanical Engineering, Politehnica University of Timișoara, Blvd. Mihai Viteazu, 300222 Timișoara, Romania; dragos.utu@upt.ro
3 Department of Mechanical Engineering, Las Palmas de Gran Canaria University, 35017 Tafira, Spain; santiago.brito@ulpgc.es
4 Building Technology Department, Madrid Polytechnic University, 28040 Madrid, Spain; amparo.verdu@upm.es
5 Materials Engineering and Welding Department, Transilvania University of Brasov, 500036 Brasov, Romania
* Correspondence: julia.mirza@ulpgc.es

Abstract: In response to concerns of potential cytotoxicity and adverse tissue reactions caused by vanadium and aluminum in the currently used biomaterial Ti-6Al-4V, the Ti–20Zr alloy was evaluated in this study because it has been suggested as a candidate for human body implant material. The Ti-20Zr alloy was obtained by vacuum-melting, followed by heat treatment at 1000 °C for 1 h, and then air-cooled. Optical and scanning electron microscopy revealed that the sample had an α and β lamellar microstructure. Analysis showed that the mechanical properties, in terms of hardness measurements performed at low loads, were significantly different between the two phases. Thus, it was found out that the α phase is softer by about 30% compared to the β phase. The Electrochemical Impedance Spectroscopy technique (EIS) was employed to study the electrochemical behavior in simulated body fluid (SBF). The electrochemical behavior demonstrated that Ti-20Zr alloy exhibits excellent corrosion resistance due to the stable oxide layer formed on its surface. SEM and EDS investigations showed that the surface topography, after electrochemical studies, is characterized by a porous film with increased oxygen content, which might be suitable for the osteoinductive growth of bone.

Keywords: implant; biomaterial; Ti-Zr; microstructure; EIS; microhardness

1. Introduction

Dental implant treatments have gained popularity because they preserve the adjacent tooth structure and bone compared to other treatments; thus, this method has become an important option for replacing missing teeth [1]. To manufacture implants, many metallic alloys have been studied and employed. Among them, the most common are Co-Cr alloys, Co-Cr-Mo alloys, stainless steel, commercially pure Ti (cpTi), and Ti-based alloys. Titanium is used for the manufacturing of implants because it is biologically inert, has increased biocompatibility, and has good bonding with osteoblasts. Moreover, when exposed to the environment it can spontaneously develop a TiO_2 oxide film on its surface [2,3] in the order of nanometers, which can absorb phosphate- and calcium-inducing proteins to form apatite, thus promoting osseointegration [4]. Nevertheless, the oxide film developed on its surface is very thin [5,6] and is prone to damage, which leads to the release of Ti ions [7]. Also, cpTi has low wear resistance, which affects the mechanical properties of the implant. In order to overcome some of its drawbacks, Ti is alloyed with a variety of elements to enhance its corrosion resistance, to improve its mechanical properties, and to lower its modulus of elasticity. Ti-based alloys are widely used as biomaterials in dentistry for manufacturing abutments, orthodontic wires, and prostheses [4]. Among them, probably one of the most popular materials used for implants is Ti-6Al-4V due to its advantages like good corrosion resistance and mechanical strength combined with a lower modulus of elasticity compared to cpTi. However, it has been reported that aluminum and vanadium ions can lead to neurological problems and adverse reactions in the human body over an extended period of time [8–10]. To further improve the mechanical properties and biocompatibility of Ti, Ti-xZr alloys have been developed [11]. These alloys are of interest since they have a lower Young's modulus, increased hardness, and in some cases better biocompatibility compared to Ti-6Al-4V and cpTi. Like titanium, zirconium belongs to the IV group of the periodic table and is recognized as having similar chemical properties. Alloys with different concentrations of zirconium, such as Ti-10Zr, Ti-20, Ti-30Zr, and Ti-50Zr, have been studied in specific conditions for dental implant applications [12–14]. Researchers have conducted in vitro, animal, and clinical studies on commercial Roxolid alloys, which are Ti-based alloys with 13–17 wt.% Zr [15]. It was found out that the alloys possess increased biocompatibility compared to cpTi, proving that they are a great candidate for small-diameter implants. Alternate outcomes were revealed during the examination of the electrochemical behavior of the alloys with titanium and zirconium [16,17]. These studies showed that the alloy passivated more easily in Ringer solution compared to cpTi, having a more stable passive film than Ti [18,19]. The influence of Zr concentration on Ti-xZr alloys (x = 5%, 10%, 15%) was studied and it was concluded that the alloy presents no cytotoxic effect on osteoblastic cells [20], confirming its safe use for medical devices [21,22]. The authors of the present work also investigated the electrochemical behavior of the Ti-Zr alloy in Ringer's solution, and it was found that the alloy has a superior corrosion resistance compared to cpTp [23]. Furthermore, Zr addition results in microstructural refinement without a significant reduction in ductility [24]. Considering these aspects, from the literature one can notice that Ti-Zr is of particular interest in dentistry; however, its performance can still be improved, and this can be achieved by changing the alloy's Zr content to produce an alloy with a defect-free microstructure, improving osseointegration by passivation in different solutions, heat treatments, etc.

Thus, based on the literature and on the author's background, the present paper aims to further study the Ti-20Zr alloy, manufactured by vacuum-melting followed by heat treatment and air-cooling. The microstructure and mechanical properties in terms of hardness, as well as in vitro studies performed in artificial extracellular fluids, were investigated. Studies were performed to find an adequate range for the stability of the heat-treated Ti-20Zr alloy and to quantify the resistance of the passive film developed on it. The present work is a basic research study to investigate the influence of heat treatment on this particular alloy and to gather pertinent information for the design of novel bio-medical material.

2. Materials and Methods

2.1. Ti-20Zr Alloy Preparation

The Ti-20Zr alloy was synthesized using a double-electron-beam melting furnace, namely, the EMO 80 model manufactured by ZIROM S.A. in Giurgiu, Romania. The sample was made from Ti and Zr, both of which had a bulk purity greater than 99.5%. The ingots were rolled over and remelted at least six times to prevent macroscopic defects caused by insufficient mixing. The sample was made as an ingot with a 20 mm diameter and 30 mm length. To prevent segregation, homogenization was utilized by subjecting the material to heat treatment in a tubular furnace at a temperature of 1000 °C with a heating rate of 10 °C per minute, followed by cooling in ambient air until the samples reached room temperature. From our experience, we have observed that if we use 1000 °C for heat treatment, we achieve homogenization within the material without changing the α-β phases. If we go much higher than the beta-transus temperature (980 °C), where homogenization in the β phase takes place, we notice the formation of an increased volume of non-equilibrium martensitic phases formed during cooling, such as α'', which make the material more fragile. Therefore, we wanted to maintain a balance between mechanical properties and corrosion resistance. Specimens were cut and were specifically prepared for each of the following tests: scanning electron microscopy (SEM), microhardness and electrochemical measurements.

2.2. Microscopic Observations

The metallographic preparation used to study the alloy consisted of (i) cutting the specimens, (ii) hot-mounting them in carbon-based resin, (iii) grinding the samples up to 2400-grit SiC paper, and subsequently (iv) polishing the surfaces using alpha alumina with a particle size of 0.1 μm until a highly reflective mirror-like appearance was achieved. The samples were previously ultrasonically cleaned with deionized water and then etched for 15 s in Kroll's reagent, which contains 10 mL of hydrofluoric acid, 5 mL of nitric acid, and 85 mL of water. When etching Ti or Ti alloys with Kroll's reagent, the colors of the β phase become dark brown. For microscopic observations, an Olympus PME 3-ADL metallographic microscope (Olympus, Tokyo, Japan) was employed. To investigate the microstructure before and after the electrochemical tests, elemental analysis was conducted using a Carl Zeiss Sigma 300 VP scanning electron microscope (FE-SEM) from Zeiss, Jena, Germany, which was equipped with an energy-dispersive X-ray spectrometer (EDS), from Zeiss, Jena, Germany. The SEM micrographs were acquired in high-vacuum mode with a cathode voltage of 15 kV and a working distance of approximately 10 mm. The sample was analyzed using a backscattered electron detector (BSD) (Zeiss, Jena, Germany) to identify the various phases present based on the Z contrast, and the secondary electron (SE) (Zeiss, Jena, Germany) was used to investigate the topography of the surface after the electrochemical tests. EDS was performed for chemical characterization by positioning the sample at eucentric height, which was at 10 mm working distance from the surface of the sample.

2.3. Microhardness and Indentation Depth

Vickers microhardness measurements were performed on the polished surface using a Remet HX-1000 hardness tester. The microhardness measurements were carried out tangentially to the surface and the indents were made at intervals of 0.5 mm along the diameter of the sample. A variety of weights, including 0.5, 1, 2, 3, 4, 5, 10, 20, 50, 100, and 200 g, were utilized, along with a dwell time of 15 s to study the hardness variation. Ten indents were taken for each load and the average value for each sample expressed as Vickers hardness (HV) was calculated. Moreover, based on the measurements, the corresponding indentation depth was calculated. Given the geometry of the Vickers indenter, the depth of the indentation can be estimated by employing the subsequent equation derived from the overarching Vickers equation:

$$\delta = \sqrt{\frac{1.854 \times F}{49 \times HV}} \quad (1)$$

where δ is indentation depth [μm], F is applied load, and HV is hardness value.

2.4. Electrochemical Behavior

Under simulated physiological conditions, Ti-20Zr alloy's electrochemical behavior was examined by open-circuit potential (OCP), cyclic voltammetry (CV), linear polarization (LP) and Electrochemical Impedance Spectroscopy (EIS). These techniques were performed using a BioLogic Essential SP-150 Potentiostat from Seyssinet-Pariset in France. The experimental solution utilized was a simulated body fluid (SBF) consisting of the following components (in grams per liter): NaCl-6.8; KCl-0.4; $CaCl_2$-0.2; $MgSO_4.7H_2O$-0.2; $NaH_2PO_4.H_2O$-0.15; $NaHCO_3$-1.1; glucose-1.2. The electrochemical tests were conducted at a temperature of around 22 °C using a standard glass cell that held 80 cc of electrolyte. The working electrode's potential was measured relative to a saturated calomel electrode (SSCE) immersed in a solution of NaCl. The measured potentials were recorded relative to the SSCE electrode, with a Pt gauze serving as the counter electrode. Before conducting the studies, each electrode was subjected to a voltage of −1 V (SSCE) for a duration of 10 min in order to reduce any oxides that may have formed on the surface of the sample. The cyclic voltammetry measurements were performed using a Biologic SP-150 (BioLogic Science Instruments, Seyssinet-Pariset, France) potentiostat/galvanostat. The potential range was from −1.5 V to 2 V with scan rates of 10, 50, 100, and 200 mV/s. Furthermore, the impedance data were collected at different potentials using the Biologic SP-150 module. Using an alternative voltage amplitude of 10 mV in the frequency range between 10^{-1} and 10^5 Hz, single sine wave records were made. The collected EIS data were analyzed, employing ZView 2 software by Scribner Associates, Inc., Southern Pines, NC, USA. The collected data were modeled with an advanced non-linear least squares Levenberg–Marquardt method, employing an analogous electrical circuit. Furthermore, by continually polarizing the electrodes and allowing the system to equilibrate for 600 s at each potential, impedance spectra in the range of −0.5 V to 1.5 V vs. SCE with a step of 0.2 V were recorded in order to characterize the oxide layer. For the linear polarization measurements (LP), the samples were polarized within a potential range of −1.5 V to 3.0 V with a scan rate of 1 mV/s.

3. Results and Discussions

3.1. Microstructure

The microstructure of the analyzed alloy was significantly impacted by the manufacturing process and the heat treatment. In this study, after manufacturing, the alloy was held for 60 min at 1000 °C, above the β transus temperature (the temperature at which a change occurs from the α phase to the β phase), and afterward air-cooled till the sample reached room temperature. A standard procedure to prepare Ti alloys for metallographic investigations is to etch the samples after polishing with Kroll's reagent [25]. The surface of the sample etched for 15 s was observed using a metallographic microscope, and the microstructure at two different magnifications is presented in Figure 1. The sample, air-cooled from the annealing temperature, produced fine-structured lamellar α and β phases due to the faster cooling, which was revealed by the reagent. This is because the β phase is colored preferentially compared to the α phase. One can observe that the microstructure reveals very clearly the α lamellar microstructure which forms due to solidification after homogenization in the β phase. Furthermore, the α phase was noticed at the grain boundaries, as presented in Figure 1b. Grain boundary α is an allotriomorph crystal structure generally located at the β grain boundaries [26] which form due to late crystallization. A similar microstructure was observed by Jose Moreno et al. when studying the microstructure and corrosion behavior of Ti-20Zr alloys in NaF-doped artificial saliva [27]. Thus, it is believed that the α′ phase is present as well in the microstructure since the acicular lamellas in the nearby grain boundaries are much finer compared to the ones observed in Figure 1a. This might be attributed to the heat treatment to which the sample was subjected, followed by air-cooling. The α′ phase is like the α phase, and it is difficult to identify it. However, Chávez et al. identified the α′ phase in Ti-30Zr alloys; it was obtained by arc-melting and formed due to the distortion of the α phase caused by a complete solid solution of Zr combined with fast cooling [28]. The needle microstructure is a typical α′ martensitic phase of Ti, which in our case is attributed to the substitutional Zr added to the alloy's composition [20].

From the SEM images presented in Figure 2, it is evident that the Ti-20Zr alloy is a solid solution with small segregations. This is attributed to a homogeneous composition, to the similar physical properties of Ti and Zr, and to a high cooling rate, which inhibits the formation of segregations [12]. Moreover, the examined region exhibits areas of dark gray and lighter shades. SEM tests using a backscattered electron (BSD) detector reveal elements with higher atomic numbers, as they appear brighter in micrographs. The bright areas correlate to the regions with a high concentration of Zr. Zirconium (Zr) atoms possess a greater atomic number, resulting in a stronger scattering of electrons towards the detector compared to titanium (Ti) atoms. As a result, Zr atoms seem more luminous in the SEM micrographs. This is further corroborated by the EDX spectra, as the Zr peak has a greater intensity when obtained from the illuminated region (Figure 2b, labeled Area 1). These data are supported by the quantification as well.

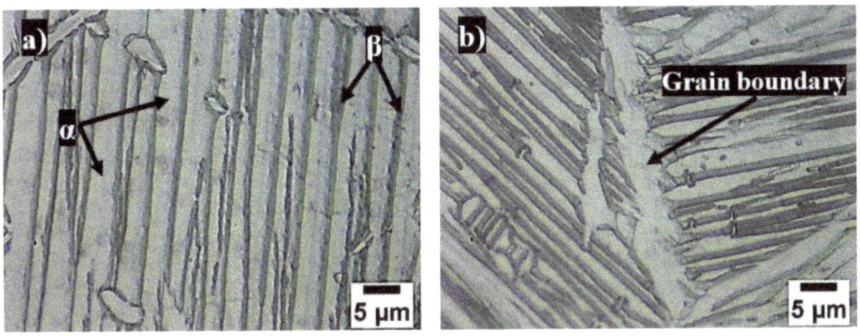

Figure 1. Optical micrographs of Ti-20Zr alloy at 100× magnification: (**a**) lamellar structure and (**b**) grain boundary.

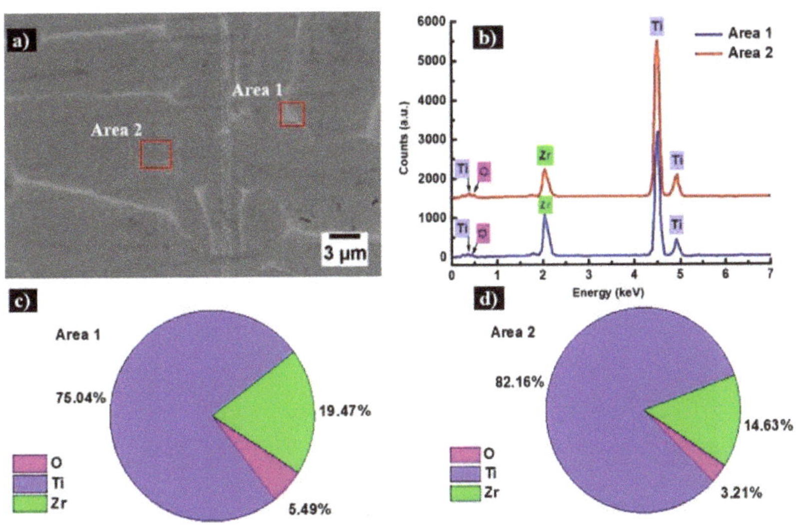

Figure 2. Microstructure of Ti-20Zr sample: (**a**) high magnification micrograph collected using BSD detector at 3.000x; (**b**) EDX spectra with associated quantifications in wt% (**c**,**d**).

3.2. Microhardness and Indentation Depth

The microhardness values recorded as measurements from the surface of the samples under different loads are summarized in Table 1. Considering our observations, we decided to divide the results into the functions of the effects of phase and load on the hardness values. It was noticed that the microhardness values varied when using loads up to 20 gf. It is assumed that the variation is caused by the fact that the indents were collected in soft and hard phases individually. Thus, these phases are ascribed to the α and β phases identified within the microstructure. For loads in the range of 0.5–20 gf, five measurements were performed in the soft phase and five measurements in the hard phase, and the calculated average values for each phase are listed in Table 1. According to the observations, the α phase is somewhat softer (about 30% lower) than the β phase. Comparable results were documented by Min et al. when investigating the mechanical characteristics of Ti-Mo-Fe and Ti-Mo alloys after heat treatment [29]. Starting with a 50 gf load, it was not possible to select the phases and the average microhardness value was obtained as the mean of ten indentations randomly performed on the surface of the sample. It was observed that with the starting load of 50 gf, there were no significant differences in the hardness values when using the same load. Thus, this is attributed to the fact that the load starts to press the indent in a way that makes it no longer possible to measure only one phase, but both simultaneously. Regarding the effect of load on the microhardness, in our study, the HV values generally increased with the applied load for indentation loads up to 20 gf. This is a distinctive feature for a low-load range of loads, and it was encountered in our previous study as well [30]. Generally, when a very low load is employed for measurements, the hardness decreases with the increase in the testing load, which is attributed to the indentation size effect (ISE). In contrast to the ISE, it might happen that the microhardness increases with the increasing load, a phenomenon known as reverse ISE [31]. Analyzing the data presented in Table 1, in our study a reserve ISE dominates the measurements when loads up to 20 gf are employed. On the other hand, by applying loads over 50 gf, one can notice that the hardness values approach a steady state at around 233 HV; thus, the hardness does not vary that much compared to the HV values obtained when using lower loads. However, there were some small differences depending on the hard or soft region where the tip of the intender was fixed and the indentation was performed. From the measurements, the applied load had a detrimental effect on the hardness values, especially when lower loads were employed in the study. According to the measurements, the Ti-20Zr alloy exhibits a hardness that is 1.2 times greater than that of commercially pure Ti, which confirms the superior mechanical strength of the alloy compared to cp-Ti [32–34]. The measured microhardness values were used to calculate the indentation depth (in μm) using the formula presented in the experimental part of this work. Analyzing the results presented in Table 1, one can notice that the indentation depth increases with the applied load.

Table 1. Ti-20Zr microhardness values obtained with different loads and the corresponding indentation depths.

	Soft and Hard Phases in Ti-20Zr		
Load (gf)	**Phase**	**Hardness (HV)**	**Indentation Depth (μm)**
0.5	Soft Hard	37.2 49.9	0.7 0.6
5	Soft Hard	163.8 288.5	1.1 0.8
10	Soft Hard	193.9 241.7	1.4 1.2
20	Soft Hard	202.6 299.1	1.9 1.6
50	Average	234.1	2.8
100	Average	228.7	4.1
200	Average	239.6	5.6

3.3. *Electrochemical Characterization*

3.3.1. DC Electrochemical Test

Different tests were applied: open-circuit potential (OCP), cyclic voltammetry (CV), and linear polarization. The OCP goal (see Figure 3) is to measure the potential of the sample without affecting, in any way, the electrochemical reactions which take place on the surface of the alloy. It can be observed that after around 4 h, the potential tends towards a cvasi-steady-state value. To determine the potential ranges and kinetics of the various electrochemical reactions of metallic samples in SBF and to choose the most suitable conditions for the potentiostatic study of corrosion reactions, cyclic voltammograms (CVs) were performed on a Ti-20Zr sample at various potential sweep rates. Figure 4 depicts the

CVs of Ti-20Zr in simulated body fluid recorded by the potential cycling of the working electrode at 10, 50, 100, and 200 mV/s between −1.50 and +1.50 V vs. SCE.

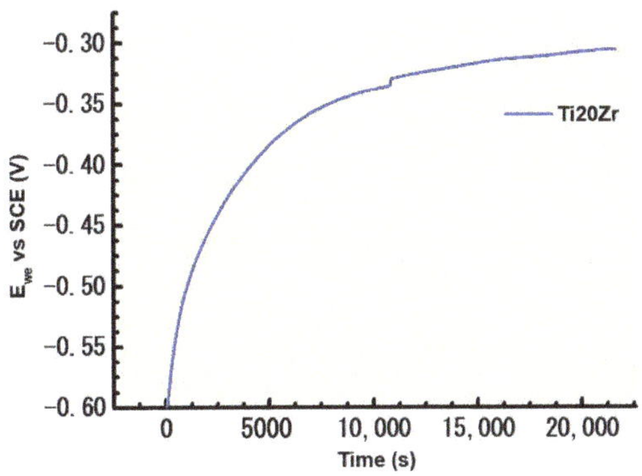

Figure 3. Open-circuit potential of Ti20Zr during 6 h of immersion in SBF.

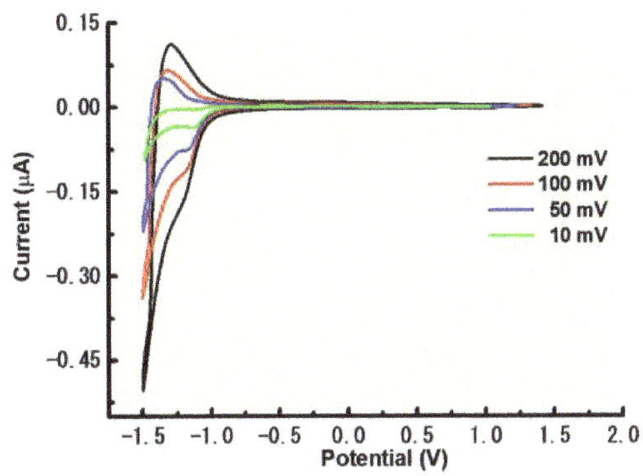

Figure 4. Cyclic voltammetry spectra for Ti-20Zr tested in simulated body fluid.

The CVs show the onset of the hydrogen evolution reaction at potentials below −1.5 V vs. SCE. An anodic current peak at about −1.25 V vs. SCE is observed during the positive-going scan due to the formation of a TiO2 film on the surface. On the other hand, when the potential excursion in the positive direction exceeds −0.5 V vs. SCE, the current is almost zero μA during the rest of the domain of potential, indicating that the passive layer growth on the surface of the sample is very stable and resistant to corrosion. A broad cathodic peak can be observed in the negative-going potential scan, involving the electroreduction of the passive film formed on the metallic surface.

The polarization curve for potentials from −1.5 V to 3 V vs. SCE is presented in Figure 5. According to earlier studies, the passivation behavior in the anodic branch of the polarization curve is seen, and is related to the protectiveness of the corrosion products. According to Zhang et al. [35], the formation of TiO_2 and ZrO_2 when the alloy contains the right amount of Zr can increase passivation capability, which in turn improves the alloy's corrosion resistance. Even still, the alloy continues to dissolve as the anodic current density rises, with anodic potential reaching beyond 2 V vs. SCE. The kinetic of the cathodic reaction—oxygen reduction—is suggested by the slope of the polarization curve's cathodic portion. Current density stability was observed in the anodic region of the polarization curve for potential values greater than approximately −0.75 V, which indicates complete passivation. The anodic portion of the polarization curve shows a significant plateau of passive current.

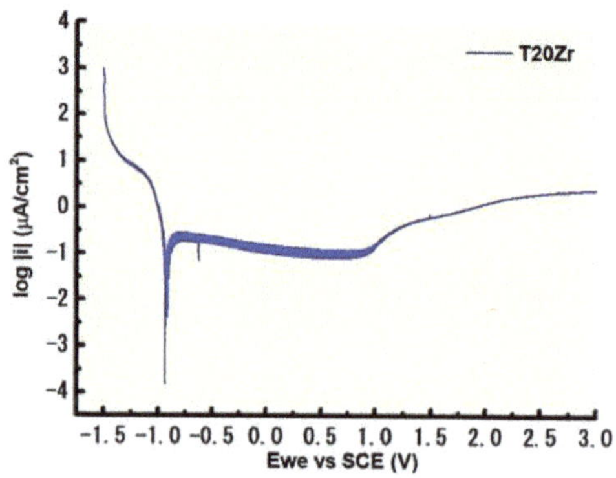

Figure 5. Polarization curve of Ti-20Zr tested in simulated body fluid.

3.3.2. AC Electrochemical Test

Electrochemical Impedance Spectroscopy (EIS) is a potent analytical method employed to analyze the electrochemical processes that occur at the interface of a material immersed in an electrolyte solution. In the context of corrosion analysis, EIS is employed to understand the behavior of metals and alloys as they degrade due to electrochemical reactions. The Nyquist plot (Figure 6a) is a key graphical representation in Electrochemical Impedance Spectroscopy (EIS) analysis. It provides valuable insights into the electrochemical behavior of a system by displaying the impedance data in the complex plane. It can be observed that from −0.5 V to +0.3 V, the radius (and implicitly, the diameter) of the semi-circle is increasing. The diameter of the semi-circle represents the charge transfer resistance (Rct) of the system. A larger diameter indicates a higher charge transfer resistance, which is due to the formation of a protective film on the surface. The center of the semicircle is usually close to the real axis, indicating a negligible solution resistance (Rs). In the positive direction, starting at 0.3 V, the diameter of the semi-circle decreases, and at 1.5 V and lower frequencies, a sloping line or tail might be observed that deviates from the semi-circle. This is called the Warburg tail and is associated with mass transport limitations at the electrode surface, especially in systems involving diffusion-controlled reactions. The slope of the Warburg tail can provide information about the diffusion coefficient of the species involved. The growth of the passive film on the alloy's surface increases the corrosion resistance, as seen in the Bode IZI diagrams (Figure 6b), where the impedance module shifts toward higher values as potential increases. An inherent characteristic of the initial formation of a passive film on the metal's surface is shown by the corrosion potential of the Bode-phase diagrams (Figure 6c). The produced film thickens with increasing potential and exhibits a capacitive response, as evidenced by a phase angle near to 90 degrees throughout a broad frequency range. An increase in the capacity (C), which is associated with a rise in the total surface area, is linked to this occurrence.

The resulting experimental data were adapted to an analogous electrical circuit after the profiles of the impedance spectra were analyzed. A set of passive components (resistances, capacitances, inductors, and other types of distributed impedance) that exhibit corrosion-like behavior in the frequency range being studied is known as an equivalent circuit. Better results are obtained using the Constant Phase Element (CPE) introduced by Boukamp [36]; this is a versatile element in electrical equivalent circuits that can capture a wide range of electrochemical behaviors in impedance spectroscopy, especially those involving non-ideal capacitive, inductive, and resistive characteristics. Its parameter "n" allows for the continuous adjustment of the phase angle, enabling a more accurate representation of complex electrochemical systems. Because the value of n is closer to the unit and the surface is more uniform, the response of the real system is closer to the ideal. Therefore, when n equals zero, the CPE element behaves as a basic resistance and a capacitor with capacitance Y^0 for n = 1. To fit the EIS experimental data, we used the circuit presented in Figure 7, similar to that used by Wang et al. [34], with the following elements:

- Rs: Solution resistance. The resistance due to the electrolyte solution through which the current passes.
- Qp: CPE corresponding to the porous external passive layer, characterized by $Y_p{}^0$ and n_p.
- Rp: Resistance attributed to the external porous layer.
- Qc: CPE corresponding to the inner passive layer. characterized by Yc 0 and nc.
- Rc: Polarization resistance of the alloy.

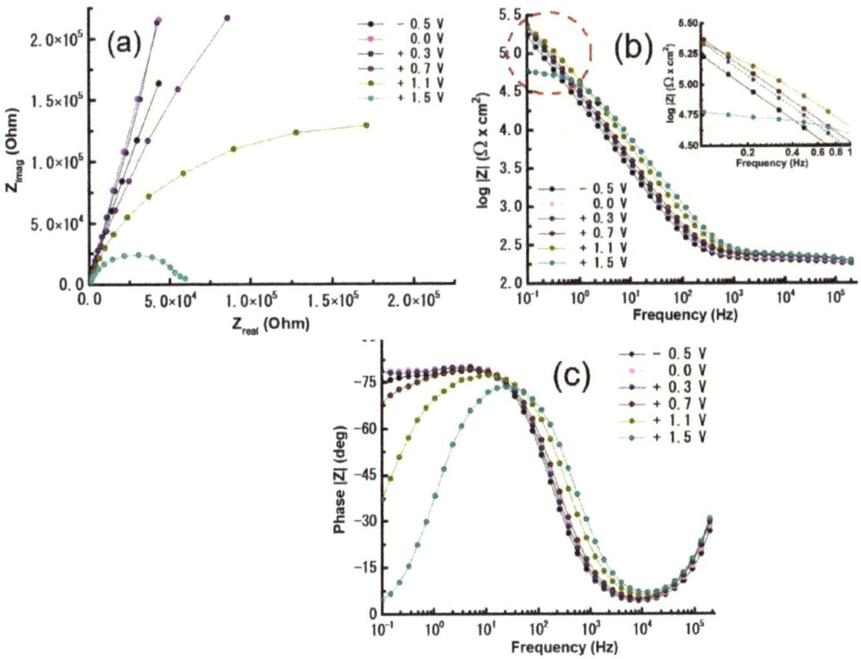

Figure 6. EIS data: (**a**) Nyquist diagram; (**b**) Bode log IZI; (**c**) Bode phase of Ti-20Zr alloy tested in simulated body fluid.

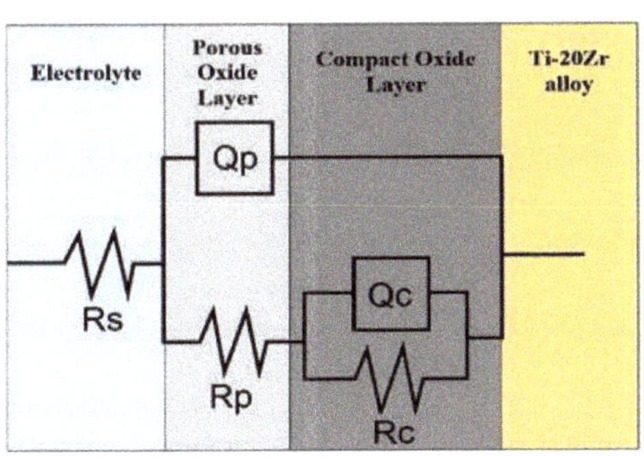

Figure 7. The electrical circuit used to fit the experimental data.

It can be observed from Table 2 that the corrosion resistance of the alloy across the potential range is more than double the reported values [13] in the absence of heat treatment.

Table 2. Electrochemical parameters of equivalent circuits.

Potential	$Y_p{}^0$ [$S\cdot cm^{-2}\cdot s^n$]	n_p	R_p [$\Omega\cdot cm^2$]	$Y_c{}^0$ [$S\cdot cm^{-2}\cdot s^n$]	n_p	R_c [$\Omega\cdot cm^2$]	χ^2
−0.5 V	4.1×10^{-9}	0.95	200.1	8.3×10^{-6}	0.89	1.3×10^6	9.7×10^{-4}
0 V	5.1×10^{-9}	0.93	213.9	6.5×10^{-6}	0.91	2.8×10^6	8.3×10^{-4}
0.3 V	5.9×10^{-9}	0.92	224.5	6.5×10^{-6}	0.91	2.7×10^6	9.0×10^{-4}
0.7 V	6.1×10^{-9}	0.92	228.3	5.7×10^{-6}	0.89	9.1×10^6	9.3×10^{-4}
1.1 V	5.4×10^{-9}	0.93	232	4.1×10^{-6}	0.89	2.9×10^6	6.6×10^{-4}
1.5 V	5.9×10^{-9}	0.92	233.6	3.0×10^{-6}	0.89	5.7×10^6	5.1×10^{-4}

3.4. Scanning Electron Microscopy Observations after Corrosion Tests

Figure 5 shows the topographical surface of the Ti-20Zr alloy, illustrating the impact of simulated body fluid on the sample's surface. One can see that the passive film has a porous microstructure (Figure 8a) that is suitable for osteoinduction. The EDS spectra (Figure 8b) and quantification (Figure 5c) show the presence of increased oxygen content which, according to the quantification, is about four times higher compared to the samples investigated before the electrochemical tests [23]. Altogether, these prove with the EIS results that the passive film formed on the sample's surface is an oxide layer, which enhances the corrosion resistance of the alloy.

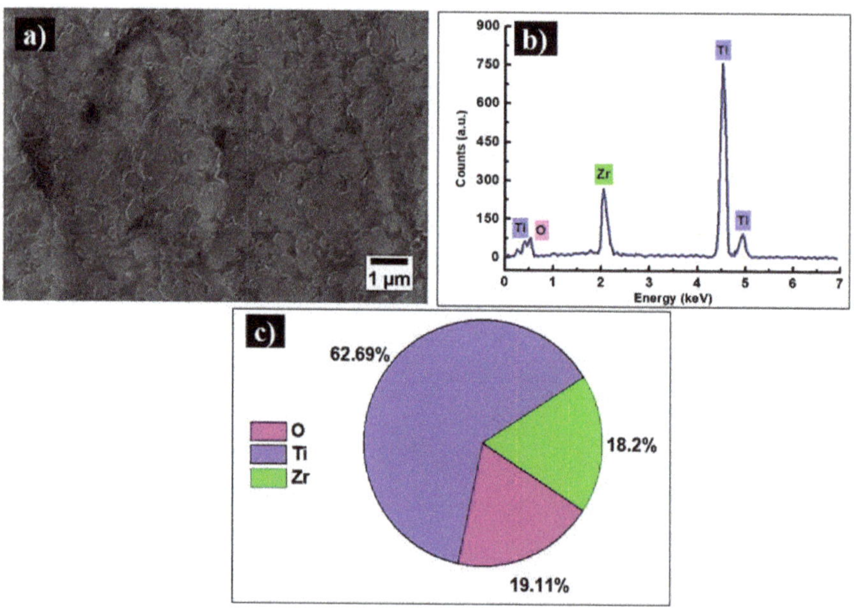

Figure 8. Microstructure of Ti-20Zr sample after corrosion test: (**a**) surface topography collected using SE detector at 10.000x; (**b**) associated EDX spectra; and (**c**) quantification in wt%.

4. Conclusions

In the present study, the influence of heat treatment at 1000 °C, followed by air-cooling, on the microstructure, mechanical properties, and corrosion behavior of a Ti-20Zr dental alloy were investigated. From the obtained results, the following conclusions can be summarized:

- The heat-treated Ti-20Zr alloy exhibited a fully lamellar α and β microstructure with α-rich grain boundaries. A fine needle-like microstructure was observed in the vicinity of the grain boundaries. The needle microstructure is a typical α'martensitic phase of Ti which is attributed to the substitutional Zr added to the alloy's composition, as well as to the heat treatment followed by fast cooling in air.

- In our study, a reversed ISE dominated the hardness measurements when loads up to 20 gf were employed; moreover, using a low range of loads, the α and β phases could be studied in more detail, and it was found out that the α phase is softer by about 30% in comparison to the β phase.
- The Ti-20Zr alloy showed better mechanical properties in terms of hardness and estimated tensile strength than cp-Ti and pure Zr. The estimated tensile strength calculated based on the measured hardness values was around 825 MPa, indicating an increased value in comparison to the non-treated Ti-20Zr.
- The heat treatment applied to Ti-20Zr increased the corrosion resistance across the potential range compared to Ti Cp.
- Surface analysis by SEM and EDS methods showed that the surface topography is characterized by a porous external film, suitable for the osteoinductive growth of bone. The high concentration of oxygen, compared to the bulk material prior to testing, indicates the formation of a protective oxide layer which enhances the electrochemical properties of the alloy in simulated body fluid.

Author Contributions: Conceptualization, I.H. and J.C.M.-R.; methodology, I.H.; software, S.B.-G.; validation, I.-D.U. and J.C.M.-R.; formal analysis, I.-D.U. and A.V.-V.; investigation, I.H. and S.B.-G.; resources, I.-D.U.; data curation, J.C.M.-R., writing—original draft preparation, I.H.; writing—review and editing, I.H. and J.C.M.-R.; visualization, S.B.-G. and A.V.-V.; supervision, I.-D.U. All authors have read and agreed to the published version of the manuscript.

Funding: The research was supported by Gran Canaria Cabildo, project number CABINFR2019-07, and the Spanish Ministry of Universities and European Union Maria Zambrano, project number SI-1821.

Data Availability Statement: The original contributions presented in the study are included in the article, further inquiries can be directed to the corresponding author.

Conflicts of Interest: The authors declare no conflicts of interest.

References

1. Elani, H.W.; Starr, J.R.; Da Silva, J.D.; Gallucci, G.O. Trends in Dental Implant Use in the U.S., 1999–2016, and Projections to 2026. *J. Dent. Res.* **2018**, *97*, 1424–1430. [CrossRef] [PubMed]
2. Popa, M.V.; Vasilescu, E.; Drob, P.; Anghel, M.; Vasilescu, C.; Mirza-Rosca, I.; Lopez, A.S. Anodic passivity of some titanium base alloys in aggressive environments. *Mater. Corros.* **2002**, *53*, 51–55. [CrossRef]
3. Kandaswamy, E.; Harsha, M.; Joshi, V.M. Titanium corrosion products from dental implants and their effect on cells and cytokine release: A review. *J. Trace Elem. Med. Biol.* **2024**, *84*, 127464. [CrossRef] [PubMed]
4. Liu, X.; Chen, S.; Tsoi, J.K.; Matinlinna, J.P. Binary titanium alloys as dental implant materials—A review. *Regen. Biomater.* **2017**, *4*, 315–323. [CrossRef] [PubMed]
5. Naji, F.A.A.; Murtaza, Q.; Niranjan, M. Challenges and opportunities in nano finishing of titanium alloys for biomedical applications: A review. *Precis. Eng.* **2024**, *88*, 81–99. [CrossRef]
6. Vasilescu, E.; Drob, P.; Popa, M.; Anghel, M.; Lopez, A.S.; Mirza-Rosca, I. Characterisation of anodic oxide films formed on titanium and two ternary titanium alloys in hydrochloric acid solutions. *Werkst. Und Korros.* **2000**, *51*, 413–417. [CrossRef]
7. Wachi, T.; Shuto, T.; Shinohara, Y.; Matono, Y.; Makihira, S. Release of titanium ions from an implant surface and their effect on cytokine production related to alveolar bone resorption. *Toxicology* **2015**, *327*, 1–9. [CrossRef]
8. Brunette, D.M.; Tengvall, P.; Textor, M.; Thomsen, P. *Titanium in Medicine*; Springer: Berlin, Germany, 2001; ISBN 1868-1212.
9. Exley, C.; Clarkson, E. Aluminium in human brain tissue from donors without neurodegenerative disease: A comparison with Alzheimer's disease, multiple sclerosis and autism. *Sci. Rep.* **2020**, *10*, 7770. [CrossRef]
10. Mirza, A.; King, A.; Troakes, C.; Exley, C. Aluminium in brain tissue in familial Alzheimer's disease. *J. Trace Elements Med. Biol.* **2017**, *40*, 30–36. [CrossRef]
11. Altuna, P.; Lucas-Taulé, E.; Gargallo-Albiol, J.; Figueras-Álvarez, O.; Hernández-Alfaro, F.; Nart, J. Clinical evidence on titanium–zirconium dental implants: A systematic review and meta-analysis. *Int. J. Oral Maxillofac. Surg.* **2016**, *45*, 842–850. [CrossRef]
12. Bao, X.; Li, X.; Ding, J.; Liu, X.; Meng, M.; Zhang, T. Exploring the limits of mechanical properties of Ti-Zr binary alloys. *Mater. Lett.* **2022**, *318*, 132091. [CrossRef]
13. Takahashi, M.; Kikuchi, M.; Okuno, O. Grindability of Dental Cast Ti-Zr Alloys. *Mater. Trans.* **2009**, *50*, 859–863. [CrossRef]
14. Dima, C.; Agop-Forna, D.; Moldoveanu, S.B.; Forna, C.N. Influence of Ti-6Al-4V and Ti-15Zr Dental Implants on the Stress in Mandibular Bone A finite element analysis study. *Mater. Plast.* **2019**, *56*, 469–473. [CrossRef]

15. Grandin, H.M.; Berner, S.; Dard, M. A Review of Titanium Zirconium (TiZr) Alloys for Use in Endosseous Dental Implants. *Materials* **2012**, *5*, 1348–1360. [CrossRef]
16. Izquierdo, J.; Mareci, D.; Bolat, G.; Santana, J.J.; Rodríguez-Raposo, R.; Fernández-Mérida, L.C.; Burtan, L.; Trincă, L.C.; Souto, R.M. Improvement of the Corrosion Resistance of Biomedical Zr-Ti Alloys Using a Thermal Oxidation Treatment. *Metals* **2020**, *10*, 166. [CrossRef]
17. Fukuo, M.; Kariya, S.; Umeda, J.; Kondoh, K.; Yoshiya, M. Strengthening Mechanisms of Powder Metallurgy Extruded CP Titanium Materials with Zirconium and Oxygen Solid Solution via Decomposition of ZrO_2 Additives in Sintering. *J. Jpn. Soc. Powder Powder Met.* **2018**, *65*, 746–755. [CrossRef]
18. Moreno, J.M.C.; Vasilescu, E.; Drob, P.; Osiceanu, P.; Vasilescu, C.; Drob, S.I.; Popa, M. Surface analysis and electrochemical behavior of Ti–20Zr alloy in simulated physiological fluids. *Mater. Sci. Eng. B Solid-State Mater. Adv. Technol.* **2013**, *178*, 1195–1204. [CrossRef]
19. Zhou, H.; Hou, R.; Yang, J.; Sheng, Y.; Li, Z.; Chen, L.; Li, W.; Wang, X. Influence of Zirconium (Zr) on the microstructure, mechanical properties and corrosion behavior of biodegradable zinc-magnesium alloys. *J. Alloys Compd.* **2020**, *840*, 155792. [CrossRef]
20. Correa, D.; Vicente, F.; Donato, T.; Arana-Chavez, V.; Buzalaf, M.; Grandini, C. The effect of the solute on the structure, selected mechanical properties, and biocompatibility of Ti–Zr system alloys for dental applications. *Mater. Sci. Eng. C* **2014**, *34*, 354–359. [CrossRef]
21. Wang, B.; Ruan, W.; Liu, J.; Zhang, T.; Yang, H.; Ruan, J. Microstructure, mechanical properties, and preliminary biocompatibility evaluation of binary Ti–Zr alloys for dental application. *J. Biomater. Appl.* **2019**, *33*, 766–775. [CrossRef]
22. Ou, P.; Hao, C.; Liu, J.; He, R.; Wang, B.; Ruan, J. Cytocompatibility of Ti-xZr alloys as dental implant materials. *J. Mater. Sci. Mater. Med.* **2021**, *32*, 50. [CrossRef] [PubMed]
23. Florido-Suarez, N.; Hulka, I.; Mirza-Rosca, J.; Saceleanu, A. Stability Range of Ti-Zr Alloy for Dental Implants. *Microsc. Microanal.* **2022**, *28*, 1034–1039. [CrossRef]
24. Issariyapat, A.; Huang, J.; Teramae, T.; Kariya, S.; Bahador, A.; Visuttipitukul, P.; Umeda, J.; Alhazaa, A.; Kondo, K. Microstructure refinement and strengthening mechanisms of additively manufactured Ti-Zr alloys prepared from pre-mixed feedstock. *Addit. Manuf.* **2023**, *73*, 103649. [CrossRef]
25. *ASM Handbook: Metallography and Microstructures*; Voort, G.F.V. (Ed.) ASM International: Novelty, OH, USA, 1992; Volume 9.
26. Salsi, E.; Chiumenti, M.; Cervera, M. Modeling of Microstructure Evolution of Ti6Al4V for Additive Manufacturing. *Metals* **2018**, *8*, 633. [CrossRef]
27. Moreno, J.M.C.; Popa, M.; Ivanescu, S.; Vasilescu, C.; Drob, S.I.; Neacsu, E.I.; Popa, M.V. Microstructure, mechanical properties, and corrosion resistance of Ti-20Zr alloy in undoped and NaF doped artificial saliva. *Met. Mater. Int.* **2014**, *20*, 177–187. [CrossRef]
28. Chávez, J.; Jimenez, O.; Diaz-Luna, J.; Bravo-Barcenas, D.; Alvarado-Hernández, F.; Flores, M.; Suárez-Martínez, R. Microstructure and corrosion characterization of a Ti-30Zr alloy with Ta additions processed by arc-melting for biomedical applications. *Mater. Lett.* **2021**, *284*, 129041. [CrossRef]
29. Min, X.H.; Zhang, L.; Sekido, K.; Ohmura, T.; Emura, S.; Tsuchiya, K.; Tsuzaki, K. Strength evaluation of α and β phases by nanoindentation in Ti–15Mo alloys with Fe and Al addition. *Mater. Sci. Technol.* **2012**, *28*, 342–347. [CrossRef]
30. Hulka, I.; Florido-Suarez, N.R.; Mirza-Rosca, J.C.; Saceleanu, A. Ti–Ta dental alloys and a way to improve gingival aesthethic in contact with the implant. *Mater. Chem. Phys.* **2022**, *287*, 126343. [CrossRef]
31. Petrík, J.; Palfy, P. The Influence of the Load on the Hardness. *Metrol. Meas. Syst.* **2011**, *18*, 223–234. [CrossRef]
32. Jiang, J.; Zhou, C.; Zhao, Y.; He, F.; Wang, X. Development and properties of dental Ti-Zr binary alloys. *J. Mech. Behav. Biomed. Mater.* **2020**, *112*, 104048. [CrossRef]
33. Froes, F.H.; Qian, M. *Titanium in Medical and Dental Applications*; Elsevier: Amsterdam, The Netherlands, 2018; ISBN 9780128124567.
34. Wang, X.; Zhan, W.; Gui, B. Effect of cerium nitrate and salicylic acid on the titanium-zirconium chemical conversion coating of 6061 aluminum alloy. *Anti-Corros. Methods Mater.* **2020**, *67*, 205–213. [CrossRef]
35. Zhang, Y.; Davenport, A.J.; Burke, B.; Vyas, N.; Addison, O. Effect of Zr Addition on the Corrosion of Ti in Acidic and Reactive Oxygen Species (ROS)-Containing Environments. *ACS Biomater. Sci. Eng.* **2018**, *4*, 1103–1111. [CrossRef] [PubMed]
36. Boukamp, B.A. A Nonlinear Least Squares Fit procedure for analysis of immittance data of electrochemical systems. *Solid State Ionics* **1986**, *20*, 31–44. [CrossRef]

crystals MDPI

Article

The Analysis of Changes in the Crystal Structure of Near-Beta Titanium Alloy in the Solution-Treated and Aged Conditions after Static Tensile Testing

Janusz Krawczyk *, Łukasz Frocisz, Marcin Goły, Sylwia Tomasik and Tomasz Śleboda

Faculty of Metals Engineering and Industrial Computer Science, AGH University of Science and Technology, A. Mickiewicza 30, 30-059 Krakow, Poland; lfrocisz@agh.edu.pl (Ł.F.); marcing@agh.edu.pl (M.G.); stomasik0@gmail.com (S.T.); sleboda@agh.edu.pl (T.Ś.)
* Correspondence: jkrawcz@agh.edu.pl

Abstract: Titanium alloys are characterized by insufficient ductility. One of the parameters affecting their ductility is their crystal structure and texture. The present study characterizes the changes in the crystallographic texture of the Ti-3Al-8V-6Cr-4Zr-4Mo alloy in solution-treated and aged conditions on the basis of texture intensity indices and pole figures. Analysis of crystal structure changes was performed before and after tensile testing. The investigated alloy in the solution-treated condition showed a single-phase β-solution structure with a body-centered cubic (BCC) crystal structure. The process of β phase aging affected the result of the tensile test, affecting the parameters of the texture of the β phase. The analysis of the texture intensity indices for each set of planes (hkl) related to the intensity for the plane (110) indicated that the highest texture intensity occurs for β titanium alloy aged at 550 °C both before and after tensile test. After plastic deformation, the largest difference with respect to the benchmark value was observed for the (220) and (310) planes. The least amount of texture intensity occurred after aging at 450 °C. The most varied values of diffraction peak intensity in relation to the benchmark were obtained for the alloy aged at 450 °C for the (310), and (200) and (211) planes, indicating the dominance of the (211) orientation, where an elongation of 10.4% was achieved. For the highest elongation of 14.2%, achieved for the sample solution-treated at 550 °C, the diffraction peak intensities were intermediate with the dominance of peaks from the planes (200) and (310).

Keywords: near-β titanium alloy; supersaturation; aging; tensile test; crystallographic texture

1. Introduction

Titanium alloys, due to their unique properties, have found applications in various industries. They are used in the aerospace industry [1], for components for the automotive industry, structural elements in the chemical and petrochemical industries, and for the manufacturing of fasteners [2]. In particular, β titanium alloys compete with the widely used α + β alloys due to the possibility to control their properties resulting in their higher strength properties at a lower Young's modulus [3]. Titanium alloys with a body-centered cubic (BCC) crystal structure are characterized by numerous variants of deformation mechanisms. Both dislocation slip, twinning, strain-induced phase transformations and numerous combinations of the above mechanisms are observed. Deformation at the room temperature of a β titanium alloy confirms formation of $\{-1011\}\langle 101\text{-}2\rangle\alpha$ twins. Twinning may occur either through atomic displacement in (1-101) planes along the twinning direction [1-10-2] or due to stress concentration at the α/β interface. Additionally, the formation of the $\{112\}\langle 111\rangle\beta$ and stress-induced α'' martensitic transformation can be observed. Formation of the twins both in α and β phases can result in the formation of high dislocation density areas, which during further deformation can develop into new grain boundary, promoting the dynamic recrystallization processes. In the case of the quasi-static deformation, the TWIP mechanism can be observed in the metastable titanium alloys. On the other hand, the increase of strain rate promotes the TRIP effect, essential to maintain the sufficient ductility of titanium alloys [4–12]. Deformation of the near-beta titanium alloys due to their low ductility is mostly performed by hot deformation and the control of the microstructure by continuous recrystallization mechanisms. The continuous formation of new grains during the deformation enables obtaining more complex shapes at higher strain rates during processing. Hot deformation processes of metastable beta titanium alloys at different equivalent strain ranges reveal the strong role of this parameter in the strain hardening of these alloys. High strain rates result in significant strain instability influencing the microstructure. Dynamic recrystallisation is observed in areas of stable deformation, with shear bands mainly observed in areas of higher strain instability. Additionally, dynamic recrystallisation is mostly located at the grain boundaries of the deformed phase, due to the strong accumulation of strain in this area of the material. The energy accumulated in the material provides the driving force for the process of annihilation of linear defects. Continuous dynamic recrystallisation is realized by progressive subgrain rotation, while discontinuous recrystallisation only results in a change in the curvature of the grain boundaries. Continuous recrystallisation processes mainly occur in the material deformed under high strain rates. These processes are related to the dissipation of energy in the material, indicating its correlation to the strain parameters [13–16]. Some researchers point out the main role of dynamic recrystallisation as a factor promoting grain fragmentation in β phase titanium alloys. Dynamic recrystallisation is observed in the areas near the grain boundaries of the β phase. Weakening of the texture intensity in the material associated with this phenomenon was also observed [17–23]. Deformation of titanium alloys below the recrystallization temperature is characterized by the recovery mechanism, while a temperature increase favoring greater mobility of grain boundaries promotes material recrystallisation. The advance of recrystallisation processes also depends on the size of the primary grains in the material before plastic deformation. This is determined by the small number of areas of large change in the crystallographic orientation between grains characteristic of areas of strain inhomogeneity, which is favored by a small amount of grain boundaries. In addition, in the case of the rolling process under larger strains, recrystallisation occurs both at grain boundaries, interfacial boundaries, or inside the grains [24–29]. A non-negligible aspect of the deformation of titanium alloys is the relationship between plastic deformation processes and crystallographic structure. In the case of titanium alloys, it is possible to encounter local perturbations of crystallographic orientation when obtaining a structure that is considered to be mono-dimensional. Certain areas of the material characterized by a distinct crystallographic texture mostly correspond to the orientation of the β phase from which they separated. They preserve the Burgers orientation (BOR) between the grains of the α phase and the β phase in the case of two-phase alloys [30–32]. The occurrence of areas of clearly extreme crystallographic orientation with respect to the rest of the material significantly lowers the material's resistance to fatigue wear [7,33,34]. In titanium alloys, the main crystallographic studies are focused on the study of crystallographic relationships in multiphase systems. The study of deformation of two-phase alloys has indicated that as the strain intensity increases, areas with homogeneous crystallographic orientation are fragmented. In addition, this is positively influenced by intense cooling during heat treatment of these alloys, affecting fragmentation of the original structure before deformation [32]. In contrast, the formation of "<001> and <111> double fiber texture" is observed during the deformation of β titanium alloys [33]. The analysis of changes in the crystallographic structure of near-β titanium alloys is an issue that is still new and requires further research. Hence, this paper focuses on the analysis of changes in the crystallographic structure of near-β titanium alloy under tensile stress and strain during cold deformation. In addition, the effect of heat treatment involving solution treatment and aging on changes in the crystallographic structure of the material, both before and after deformation, was analyzed.

2. Materials and Methods

The test material was a beta C titanium alloy Ti-3Al-8V-6Cr-4Zr-4Mo (designated in the work as TBe). The chemical composition of the alloy is shown in Table 1.

Table 1. The chemical composition of β titanium alloy (wt.%).

Elements	Al	V	Cr	Zr	Mo
Nominal content	3.0–4.0	7.5–8.5	5.5–6.5	3.5–4.5	3.5–4.5
Actual content	3.46	8.19	6.41	4.15	4.17

The tested alloy was subjected to solution treatment at 950 °C for 1 h, and then rapidly cooled in water. Due to the heat treatment, a single-phase microstructure (β phase) was obtained due to the dissolution of the non-equilibrium α phase. The alloy was then subjected to aging at different temperatures. The designations of the test specimens with the corresponding heat treatment parameters are listed in Table 2. The heat treatment was carried out in an FCF 8 M chamber furnace produced by CZYLOK Company (Poland).

Table 2. Heat treatment parameters and the sample notation.

TBe Alloy Sample	Solution Treatment	Aging
TBe-S	950 °C/1 h	-
TBe-A450	950 °C/1 h	450 °C/2 h
TBe-A500	950 °C/1 h	500 °C/2 h
TBe-A550	950 °C/1 h	550 °C/2 h

The mechanical tests were carried out on 5-fold specimens having rectangular cross-section. The test material was cut by EDM on a wire EDM machine (AG200L, Sodick, Soditronk, Warsaw, Poland) providing precise cutting of any shape. The prepared samples were ground on sandpaper. Cutting the machined material and grinding it allowed to remove microstructural changes in the surface layer of the sample. Static tensile testing was conducted on a Z250 hydraulic press (Zwick/Roell, Wroclaw, Poland). Macroscopic observations of the fracture surfaces were carried out using a Leica Stereozoom S9i macroscope (Leica, PIK–Instruments, Piaseczno, Poland).

Diffraction studies were performed on a D8 Advance diffractometer from Bruker corporation using filtered cobalt anode lamp radiation ($\lambda_{CoK\alpha}$ = 0.179 nm). Measurements were performed in the angular range 2θ = [40–130°] in step mode with a step of Δ2θ = 0.04° and an exposure time per step of 10 s. The phase analysis was performed using PDF (powder diffraction file) patterns from the ICDD (International Center for Diffraction Data) database. A line profile analysis was also carried out. The obtained parameters made it possible to determine texture coefficient (TC), given in Equation (1), which was calculated based on specific intensities of individual diffraction peaks [35]:

$$TC(hkl) = \frac{I(hkl)}{I_0(hkl)}\left\{\frac{1}{n}\sum\frac{I(hkl)}{I_0(hkl)}\right\}^{-1} \tag{1}$$

where:

I—relative intensity measured for the diffraction line measurement for a given (hkl),

I_0—benchmark relative intensity for the diffraction line for a given (hkl) for the non-textured state.

3. Research Results

The static tensile test conducted allowed the determination of parameters such as tensile strength (TS), yield strength (YS) for equal values of elongation, elongation at break (ε), yield strength ratio (YS/TS), reduction of area (RA), Young's modulus (E), as well as elastic admissible strain (δ) for individual specimens, all of which are summarized in

Table 3. The elastic admissible strain is defined as the yield strength to Young's modulus ratio (δ = YS/E)

Table 3. Tensile test results for the investigated samples.

TBe Alloy	TS (MPa)	YS (MPa)	YS/TS (%)	E (GPa)	δ (%)	EL (%)	RA (%)
TBe-S	853	849	99.5	182	0.47	9.9	60.2
TBe-A450	879	873	99.3	168	0.52	10.4	53.3
TBe-A500	869	865	99.5	171	0.51	9.6	43.7
TBe-A550	865	864	99.9	163	0.53	14.2	52.9

Based on the data shown in Table 3, it can be observed that the tensile strength of the tested materials changes slightly with the aging process of the samples. A slight increase in the value of this parameter can result from the aging process, relative to the solution treated material, with an increase in the aging temperature causing a decrease in the value of tensile strength. A similar relationship can be observed for the yield stress. The yield strength ratio expressed by the YS/TS ratio remains extremely unfavorable. In every case, the value of the yield strength is practically equal to the tensile strength of the material. The aging process of the samples reduced the Young modulus. For β titanium alloy that was not aged, the Young modulus was 181 GPa, while for β titanium alloy aged at 500 °C the Young modulus was 171 GPa, and for β titanium alloy aged at 450 °C the Young modulus value was 168 GPa. The lowest value of 163 GPa was noticed for β titanium alloy aged at 550 °C. A slight increase in the elongation of the sample after the test was also observed. As a consequence of aging, the highest elongation value was obtained for the sample aged at the highest temperature. The values for the solution-treated samples and the samples aged at lower temperatures were similar. The tensile curves obtained during the tests are shown in Figure 1. Three stages of the sample deformation can be observed. Initially, the samples deformed elastically, then after exceeding the yield point, a stage of homogeneous plastic deformation occurred. The last stage involved deformation accompanied by a decrease in stress until rupture. Similar to the data presented in Table 3, it can be observed that the first three samples exhibited a similar character of the flow curve. However, in the case of the sample aged at the highest temperature, the curve showed significantly greater ductility.

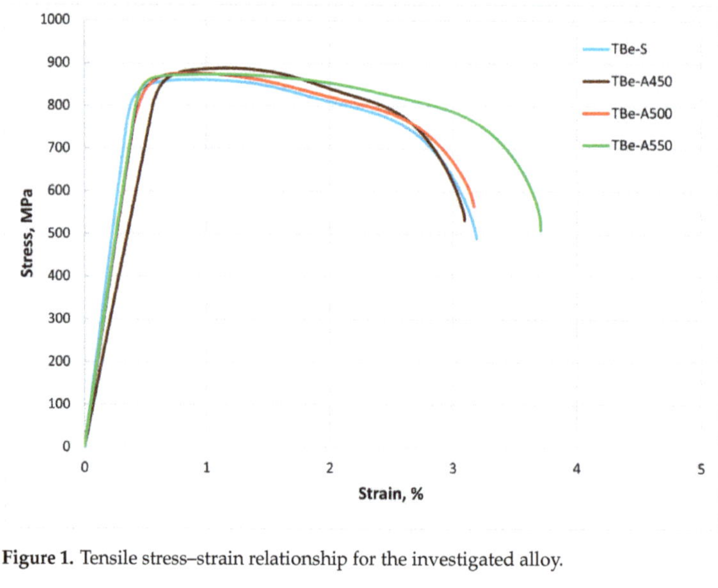

Figure 1. Tensile stress–strain relationship for the investigated alloy.

The next stage of the study was the macroscopic examination of the fractured samples (Figure 2).

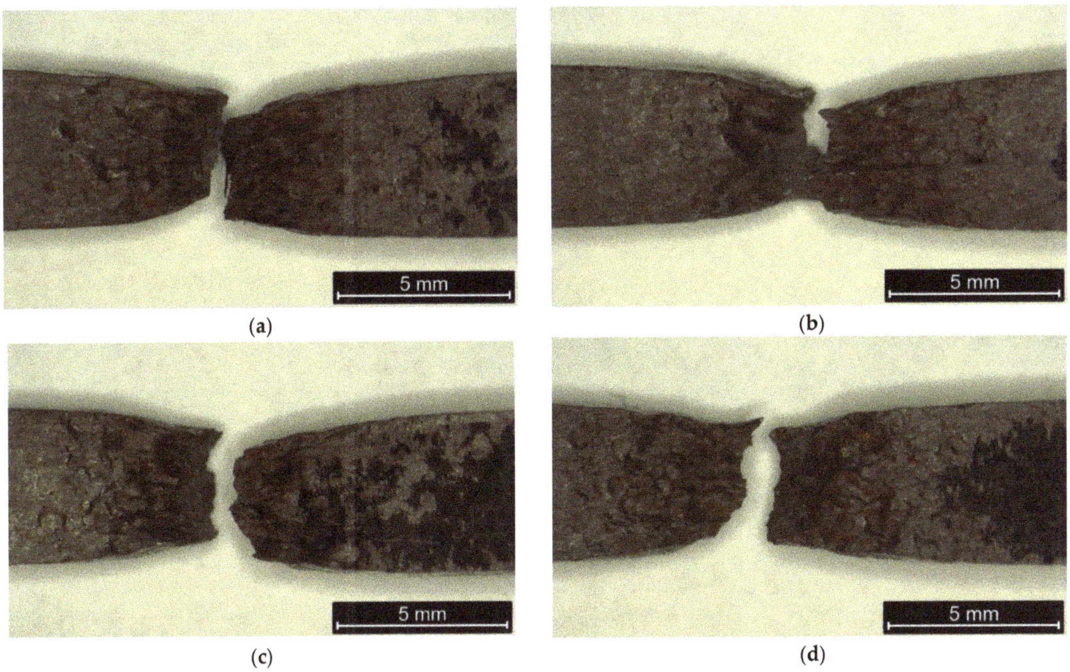

Figure 2. Macroscopic images of the fractured Ti3Al8V6Cr4Zr4Mo alloy samples, (**a**) after solution treatment at 950 °C/1 h, (**b**) after solution treatment at 950 °C/1 h and aging at 450 °C/2 h, (**c**) after solution treatment at 950 °C/1 h and aging at 500 °C/2 h, (**d**) after solution treatment at 950 °C/1 h and aging at 550 °C/2 h.

It can be observed that there are no significant differences in the fracture surfaces. It can be seen that the sample cross-section reduction and fracture occur fairly parallel on both sides of the samples. Macro images show relief on the surface of the specimen especially in the area of the neck forming. After tensile tests, diffraction studies were performed. The results of X-ray phase analysis of the tested samples are shown in Figure 3.

A slight increase in the intensity of the line (211) for the β phase can be observed as the aging temperature increases. For aging at 550 °C, the peaks from the (220) and (310) planes decrease noticeably. No reflections from the α phase were recorded in any of the analyzed variants. For the reflections from the crystallographic planes observed on the main diffraction lines, the values of TC were determined based on Equation (1). The results of these analyses are shown in Figure 4.

If the TC value significantly deviates from the benchmark value, this allows us to determine which crystallographic planes play a significant role in the deformation process. We can see that the smallest texture was exhibited by β titanium alloy after aging at 450 °C, as indicated by the TC values that were closest to the benchmark value. The largest differences from the benchmark value for β titanium alloy aged at 550 °C (TBe-A550) were noticed. The dominant plane is the (200) plane, with all sets of planes having values greater than the benchmark sample. In the case of the sample aged at 500 °C (TBe-A500), only the reflection from planes (310) showed the highest TC value. On the other hand, as for the reflections from the other planes, they remain at a much lower level than in the case of the benchmark sample (Figure 3a).

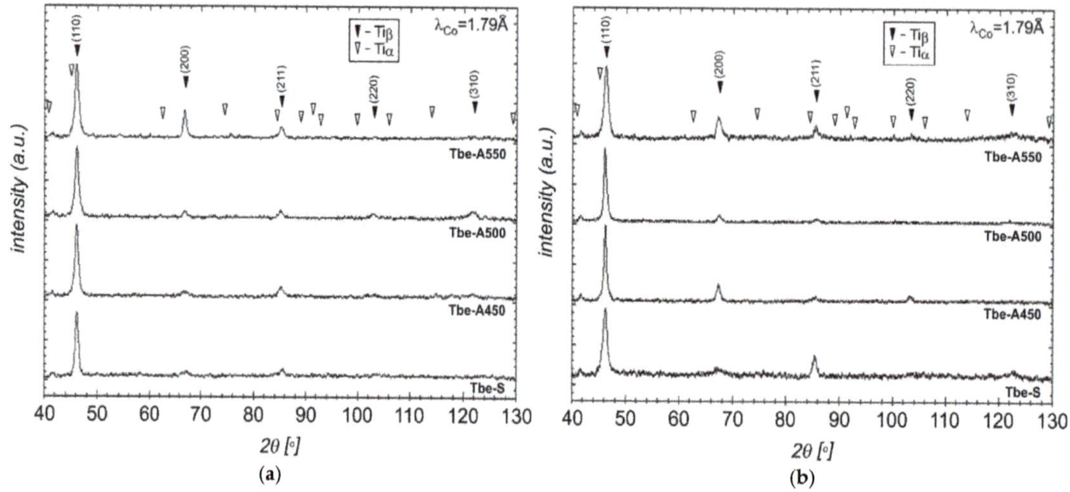

Figure 3. Results of the X-ray diffraction analysis. Phase diffraction patterns for the investigated material after heat treatment: (**a**) before tensile test, (**b**) after tensile test.

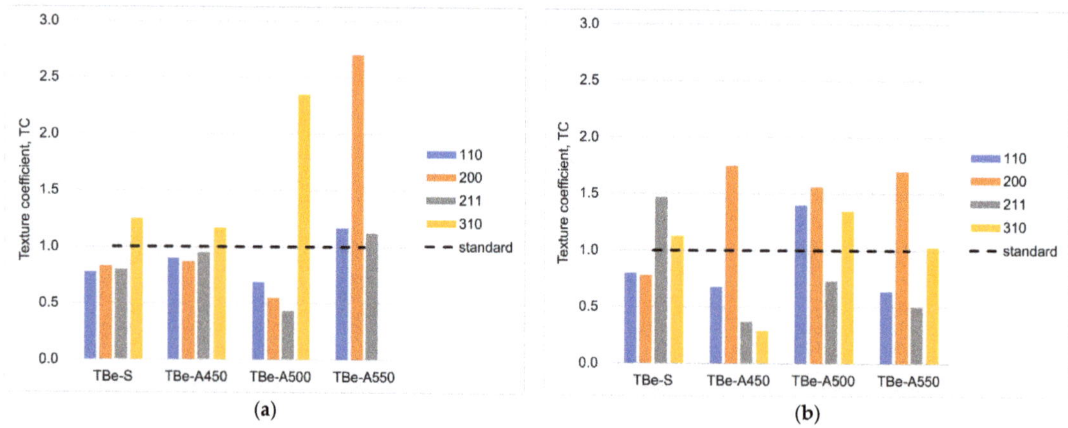

Figure 4. Variation of TC for different diffraction peaks with varied aging temperature (black line signifies TC value for nontextured material): (**a**) after heat treatment, (**b**) after heat treatment and tensile test.

After tensile test, an increase in TC values for planes (211) was observed in the case of solution-treated samples. Other values of the TC parameter were similar to those determined before deformation. An increase in the degree of texture intensity was observed for the aged samples. For the lowest aging temperature, a decrease in the degree of texture intensity was observed. Only reflections from planes (200) showed a higher degree of texture intensity. For the aging temperature of 500 °C, the degree of texture intensity increased for all sets of planes. However, for the sample aged at the highest temperature, the texture intensity for the reflections from all planes decreased. Yet, a reflection from a set of planes (310) appeared, which was not observed for the sample before deformation.

In order to illustrate changes in the crystallographic structure of the material as a consequence of deformation processes, it was decided to make polar figures showing changes in the crystallographic texture of the material. The results of this analysis are shown in Table 4.

Table 4. Measured and calculated polar figures for reflection (110) for the test material, rotated by 20°.

Material Condition	Before Plastic Deformation		After Plastic Deformation	
Pole Figures	Measured	Calculated	Measured	Calculated
Supersaturation 950 °C/1 h	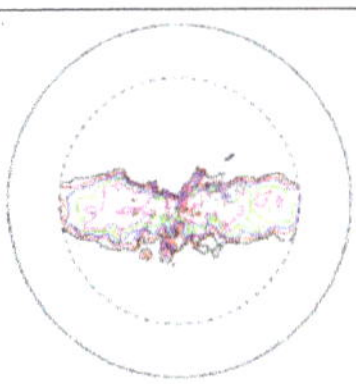	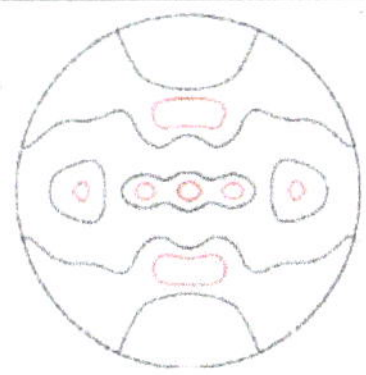	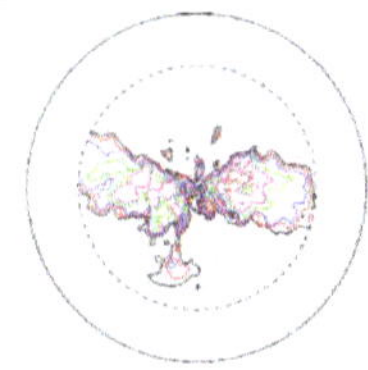	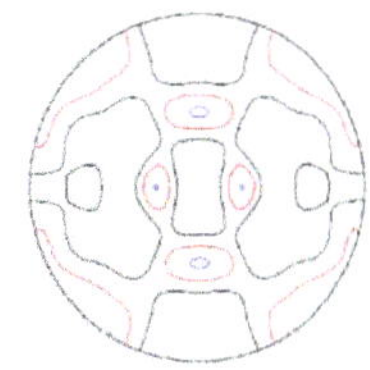
Supersaturation 950 °C/1 h Aging 450 °C/2 h	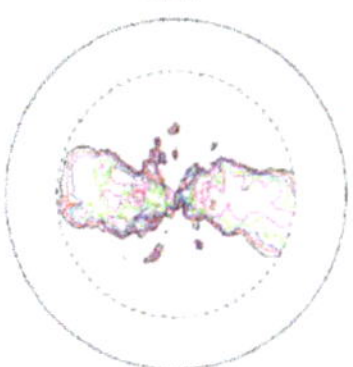	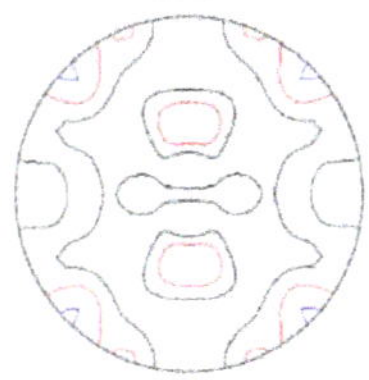	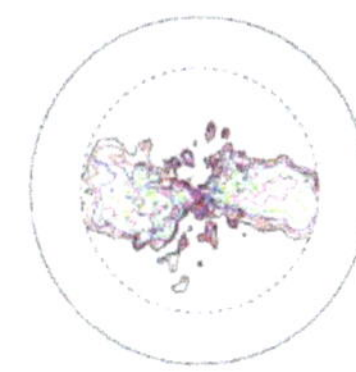	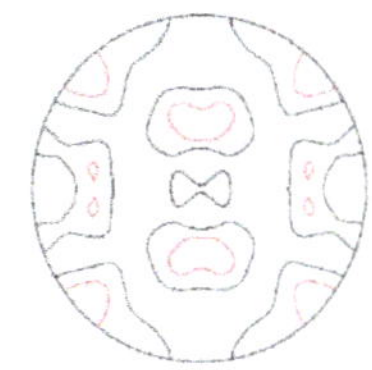
Supersaturation 950 °C/1 h Aging 500 °C/2 h	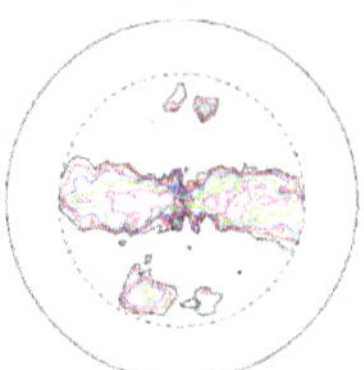	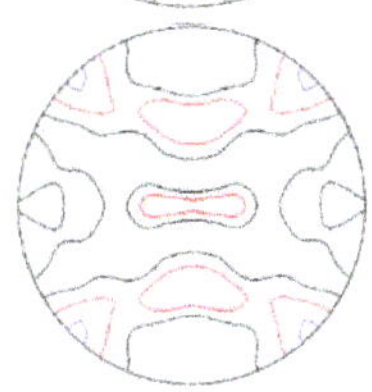	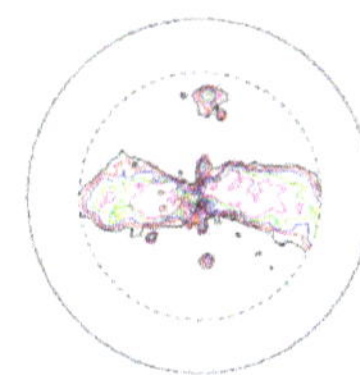	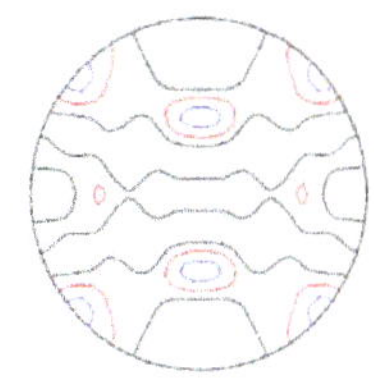
Supersaturation 950 °C/1 h Aging 550 °C/2 h	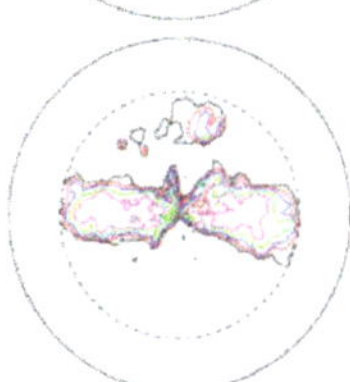	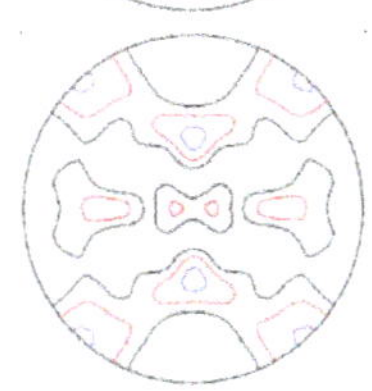	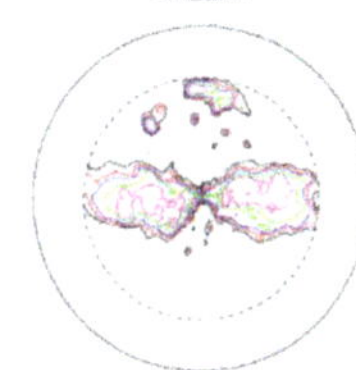	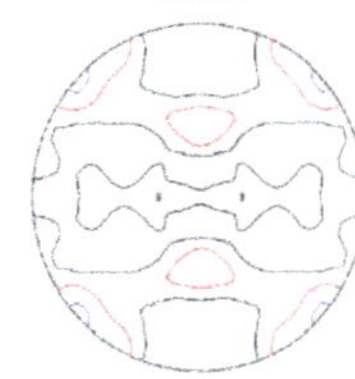

Polar figures obtained during the test show that the material was only slightly textured after the heat treatment process. The appearance of texture components for the reflection plane (110) was observed especially in the case of the samples aged at the temperatures of 500 °C and 550 °C.

The analysis of the calculated polar figures allowed to indicate greater differences in the texture components. For the (110) plane, a new component enhanced by the aging process became apparent. A similar component remained after deformation of the aged samples. Stronger textures were obtained for the samples aged at 500 °C and 550 °C. Plastic deformation of the solution-treated sample resulted in texture components similar to those obtained for samples aged, undeformed samples.

4. Discussion

The results obtained established that the tested material during cold deformation under tensile stresses was characterized by a very low ratio of yield strength to tensile strength. It can be concluded that the material did not show a range of steady-state flow, where gradual slippage in the material's crystalline structure could occur. In addition, heat treatment showed no significant effect on the strength properties. In the case of this material, the heat treatment did not result in significant changes in the phase composition, and hence the activation of deformation mechanisms associated with sliding along interfacial boundaries or deformation within α phase could be expected [34]. The heat treatment, however, had a significant effect in terms of elastic properties, affecting the reduction of

Young's modulus values. As the aging temperature increased, a gradual decrease in this parameter was observed. This is probably related to the gradual disappearance of the supersaturation of the β phase with alloying elements and the relaxation of stresses in the material, which makes it possible to achieve higher amount of strain in the elastic range. Another important aspect is the increase in sample elongation at the highest aging temperature. It can be supposed that aging at 550 °C resulted in insignificant local increase in grain size. This conjecture coincides with the results of diffraction analysis, where the tested materials for both solution-treated material and the aged material did not show clear texture. However, in the case of aging at 500 °C and 550 °C, higher texture intensity was observed for the (310) and (200) planes, respectively (Figure 3a). Yet, the high values of crystallographic indices for these planes were not conducive to obtaining high strain for this material. The (110)<110> systems supported by slippage along interfacial boundaries are conducive to the high strain values in titanium alloys [36]. No texture oriented in this system was observed in the studied material. On the other hand, after deformation, it was observed that the degree of texture of the material decreased, and the TC value increased mainly for reflections from planes with low texture indices (Figure 3b). However, the small amount of deformation only caused the appearance of new components for planes (110), which confirmed the deformation by sliding along these planes (Figure 4). Nonetheless, high values of tensile stress at low ductility resulted in the activation of deformation mechanisms also in the area of high crystallographic indices, resulting in the appearance of reflections from planes (310) for the material aged at 550 °C also, where previously these reflections were not observed. Strong deformation and the activation of multiple slip systems can also be inferred from the nature of the samples after fracture, where pronounced relief indicates strong deformation and flow of the material, but only in a small volume of the material as the activation of high-index slip systems is associated with strong strain accumulation, without significant deformation of other parts of the material.

Author Contributions: Conceptualization, J.K. and T.Ś.; methodology, M.G., Ł.F. and J.K.; validation, T.Ś. and S.T.; formal analysis, J.K. and Ł.F.; investigation, M.G. and Ł.F.; writing—original draft preparation, Ł.F., T.Ś. and S.T.; writing—review and editing, J.K. and T.Ś. All authors have read and agreed to the published version of the manuscript.

Funding: The research was financed by the Ministry of Education and Science (AGH University of Krakow, the Faculty of Metals Engineering and Industrial Computer Science, research subsidy No. 16.16.110.663, Task No. 18: "Formation of microstructure and properties of beta phase in titanium alloy").

Data Availability Statement: Not applicable.

Acknowledgments: The authors also thank Angelika Dubiel for her support in this research.

Conflicts of Interest: The authors declare no conflict of interest.

References

1. Banerjee, D.; Williams, J.C. Perspectives on Titanium Science and Technology. *Acta Mater.* **2013**, *61*, 844–879. [CrossRef]
2. Okazaki, Y.; Gotoh, E. Implant Applications of Highly Corrosion-Resistant Ti-15Zr-4Nb-4Ta Alloy. *Mater. Trans.* **2002**, *43*, 2943–2948. [CrossRef]
3. Wen, M.; Wen, C.; Hodgson, P.; Li, P. Improvement of the biomedical properties of titanium using SMAT and thermal oxidation. *Colloids Surf. B Biointerfaces* **2014**, *116*, 658–665. [CrossRef]
4. Ahmed, M.; Pereloma, E.V. Observation of simultaneous operation of deformation twins in both α and β phases in metastable β titanium alloy. *J. Alloys Compd.* **2022**, *910*, 164794. [CrossRef]
5. Zhu, X.; Fan, Q.; Wang, D.; Gong, H.; Gao, Y.; Yuan, J.; Chen, K.; Qian, F. Influence of twins found in adiabatic shear bands on dynamic recrystallization of a near β Ti-5.5Mo-7.2Al-4.5Zr-2.6Sn-2.1Cr alloy. *Mater. Sci. Eng. A* **2022**, *842*, 143084. [CrossRef]
6. Zhao, Q.; Bolzoni, L.; Chen, Y.; Xu, Y.; Torrens, R.; Yang, F. Processing of metastable beta titanium alloy: Comprehensive study on deformation behaviour and exceptional microstructure variation mechanisms. *J. Mater. Sci. Technol.* **2022**, *126*, 22–43. [CrossRef]
7. Zhao, J.; Zhong, J.; Yan, F.; Chai, F.; Dargusch, M. Deformation behaviour and mechanisms during hot compression at supertransus temperatures in Ti-10V-2Fe-3Al. *J. Alloys Compd.* **2017**, *710*, 616–627. [CrossRef]

8. Sinyakova, E.; Panin, A.; Perevalova, O.; Shugurov, A.; Kalashnikov, M.; Teresov, D. The effect of phase transformations on the recovery of pulsed electron beam irradiated Ti-6Al-4V titanium alloy during scratching. *J. Alloys Compd.* **2019**, *795*, 275–283. [CrossRef]
9. Ellyson, B.; Fezzaa, K.; Sun, T.; Parab, N.; Saville, A.; Finfrock, C.; Rietema, C.; Smith, D.; Copley, J.; Johnson, C.; et al. Transformation and twinning induced plasticity in metastable Ti-Mo alloys under high strain rate deformation. *Mater. Sci. Eng. A* **2022**, *857*, 143716. [CrossRef]
10. Wu, Z.; Kou, H.; Chen, N.; Zhang, Z.; Qiang, F.; Fan, J.; Tang, B.; Li, J. The effect of cubic-texture on fatigue cracking in a metastable β titanium alloy subjected to high-cycle fatigue. *Int. J. Fatigue* **2020**, *141*, 105872. [CrossRef]
11. Chen, K.; Fan, Q.; Yang, L.; Yao, J.; Xu, S.; Lei, W.; Gao, Y. Deciphering the microstructural evolution and adiabatic shearing behavior of the titanium alloy with stress-induced ω phase transformation during dynamic compression. *Mater. Des.* **2022**, *221*, 110939. [CrossRef]
12. Zhang, J.; Qian, B.; Lin, W.; Zhang, P.; Wu, Y.; Fu, Y.; Fan, Y.; Chen, Z.; Cheng, J.; Li, J.; et al. Compressive deformation-induced hierarchical microstructure in a TWIP β Ti-alloy. *J. Mater. Sci. Technol.* **2022**, *112*, 130–137. [CrossRef]
13. Li, C.; Huang, L.; Zhao, M.; Guo, S.; Li, J. Hot deformation behavior and mechanism of a new metastable β titanium alloy Ti–6Cr–5Mo–5V–4Al in single phase region. *Mater. Sci. Eng. A* **2021**, *814*, 141231. [CrossRef]
14. Fan, X.; Zhang, Y.; Gao, P.; Lei, Z.; Zhan, M. Deformation behavior and microstructure evolution during hot working of a coarse-grained Ti-5Al-5Mo-5V-3Cr-1Zr titanium alloy in beta phase field. *Mater. Sci. Eng. A* **2017**, *694*, 24–32. [CrossRef]
15. Chuan, W.; Liang, H. Hot deformation and dynamic recrystallization of a near-beta titanium alloy in the β single phase region. *Vacuum* **2018**, *156*, 384–401. [CrossRef]
16. Fan, J.; Kou, H.; Lai, M.; Tang, B.; Chang, H.; Li, J. Hot deformation mechanism and microstructure evolution of a new near β titanium alloy. *Mater. Sci. Eng. A* **2013**, *584*, 121–132. [CrossRef]
17. Lu, T.; Dan, Z.-H.; Li, T.-J.; Dai, G.-Q.; Sun, Y.-Y.; Guo, Y.-H.; Li, K.; Yi, D.-Q.; Chang, H.; Zhou, L. Flow softening and microstructural evolution of near β titanium alloy Ti-35421 during hot compression deformation in the α+β region. *J. Mater. Res. Technol.* **2022**, *19*, 2257–2274. [CrossRef]
18. Matsumoto, H.; Kitamura, M.; Li, Y.; Koizumi, Y.; Chiba, A. Hot forging characteristic of Ti–5Al–5V–5Mo–3Cr alloy with single metastable β microstructure. *Mater. Sci. Eng. A* **2014**, *611*, 337–344. [CrossRef]
19. Chen, Z.-Q.; Xu, L.-J.; Cao, S.-Z.; Yang, J.-K.; Zheng, Y.-F.; Xiao, S.-L.; Tian, J.; Chen, Y.-Y. Characterization of hot deformation and microstructure evolution of a new metastable β titanium alloy. *Trans. Nonferrous Met. Soc. China* **2022**, *32*, 1513–1529. [CrossRef]
20. Chen, J.; Li, J.; Tang, B.; Chen, Y.; Kou, H. Microstructure and texture evolution of a near β titanium alloy Ti-7333 during continuous cooling hot deformation. *Prog. Nat. Sci.* **2019**, *29*, 50–56. [CrossRef]
21. Chen, Y.; Li, J.; Tang, B.; Kou, H.; Xue, X.; Cui, Y. Texture evolution and dynamic recrystallization in a beta titanium alloy during hot-rolling process. *J. Alloys Compd.* **2015**, *618*, 146–152. [CrossRef]
22. Wang, J.; Wang, K.; Lu, S.; Li, X.; OuYang, D.; Qiu, Q. Softening mechanism and process parameters optimization of Ti-4.2Al-0.005B titanium alloy during hot deformation. *J. Mater. Res. Technol.* **2022**, *17*, 1842–1851. [CrossRef]
23. Chen, W.; Wang, H.; Lin, Y.C.; Zhang, X.; Chen, C.; Lv, Y.; Zhou, K. The dynamic responses of lamellar and equiaxed near β-Ti alloys subjected to multi-pass cross rolling. *J. Mater. Sci. Technol.* **2020**, *43*, 220–229. [CrossRef]
24. Chen, W.; Yang, S.; Lin, Y.; Chen, C.; Zhang, X.; Zhou, K. Multi-scale characterization of deformation features and precipitation behavior in a near β-Ti alloy. *Mater. Charact.* **2020**, *169*, 110637. [CrossRef]
25. Zhu, X.; Fan, Q.; Wang, D.; Gong, H.; Gao, Y.; Qian, F.; Jin, S.; Sha, G. A new dynamic recrystallization mechanism in adiabatic shear band of an α/β dual phase titanium alloy: Composition redistribution. *Scr. Mater.* **2022**, *206*, 114229. [CrossRef]
26. Furuhara, T.; Poorganji, B.; Abe, H.; Maki, T. Dynamic recovery and recrystallization in titanium alloys by hot deformation. *JOM* **2007**, *59*, 64–67. [CrossRef]
27. Ning, Y.; Luo, X.; Liang, H.; Guo, H.; Zhang, J.; Tan, K. Competition between dynamic recovery and recrystallization during hot deformation for TC18 titanium alloy. *Mater. Sci. Eng. A* **2015**, *635*, 77–85. [CrossRef]
28. Wang, H.; Ge, J.; Zhang, X.; Chen, C.; Zhou, K. Investigation of the Dynamic Recovery and Recrystallization of Near-β Titanium Alloy Ti-55511 during Two-Pass Hot Compression. *Metals* **2021**, *11*, 359. [CrossRef]
29. Li, J.; Dong, R.; Kou, H.; Fan, J.; Zhu, B.; Tang, B. Texture evolution and the recrystallization behavior in a near β titanium alloy Ti-7333 during the hot-rolling process. *Mater. Charact.* **2020**, *159*, 109999. [CrossRef]
30. Gu, B.; Chekhonin, P.; Schaarschuch, R.; Oertel, C.-G.; Xin, S.; Ma, C.; Zhou, L.; Gan, W.; Skrotzki, W. Microstructure, texture and hardness of a metastable β-titanium alloy after bar-rolling and annealing. *J. Alloys Compd.* **2020**, *825*, 154082. [CrossRef]
31. Krawczyk, J.; Tokarski, T.; Łukaszek-Sołek, A.; Dąbrowski, R.; Śleboda, T.; Lypchanskyi, O. Dynamic Recrystallization in Titanium Alloys. *Key Eng. Mater.* **2016**, *687*, 47–54. [CrossRef]
32. Lypchanskyi, O.; Śleboda, T.; Wojtaszek, M.; Muszka, K.; Łukaszek-Sołek, A.; Stanik, R.; Gude, M. The analysis of flow behavior of Ti-6Al-2Sn-4Zr-6Mo alloy based on the processing maps. *Int. J. Mater. Form.* **2020**, *14*, 523–532. [CrossRef]
33. Zhang, C.; Jiang, X.; Han, J.; Zhang, S.; Peng, P.; Feng, H.; Wang, T.; Cao, P. Probing the texture transition sequence of near β titanium matrix composites by observing the uneven hot deformation microstructure. *Mater. Charact.* **2023**, *198*, 112728. [CrossRef]

34. Nasirpouri, F.; Sanaeian, M.R.; Samardak, A.S.; Sukovatitsina, E.V.; Ognev, A.V.; Chebotkevich, L.A.; Hosseini, M.-G.; Abdolmaleki, M. An investigation on the effect of surface morphology and crystalline texture on corrosion behavior, structural and magnetic properties of electrodeposited nanocrystalline nickel films. *Appl. Surf. Sci.* **2014**, *292*, 795–805. [CrossRef]
35. Zhao, Z.B.; Zhang, B.H.; Sun, H.; Wang, Q.J.; Liu, J.R.; Yang, R. Influence of Globularization Process on Local Texture Evolution of a Near-a Titanium Alloy with a Transformed Microstructure. *Metall. Mater. Trans. A* **2014**, *54A*, 2849–2857. [CrossRef]
36. Du, Z.; Liu, J.; Jiang, S.; Xiao, S.; Kong, F.; Chen, Y. Strain rate dependence of microstructural evolution in β titanium alloy during subtransus superplastic deformation. *J. Alloy. Compd.* **2015**, *647*, 1–5. [CrossRef]

crystals

MDPI

Article

Influence of Powder Size on Pore Characteristics and Intermetallic Phase Kinetics in Porous Ti-Al Alloys †

Saif Haider Kayani [1,*], Hafiz Muhammad Salman Ajmal [2], Byung-Joo Kim [1], Nho-Kwang Park [1] and Kwangjun Euh [1,3]

1 Advanced Metals Division, Korea Institute of Materials Science (KIMS), Changwon 51508, Republic of Korea; kbj@kims.re.kr (B.-J.K.); pnk@kims.re.kr (N.-K.P.); keuh@kims.re.kr (K.E.)
2 Department of Electrical Engineering, University of Engineering and Technology (Narowal Campus), Lahore 54890, Pakistan; salman.ajmal@uet.edu.pk
3 Advanced Materials Engineering, Korea University of Science and Technology (UST), Daejeon 34113, Republic of Korea
* Correspondence: saifkayani@kims.re.kr

† This paper is dedicated to the memory of our co-author Nho-Kwang Park.

Abstract: This study investigates the impact of varying powder size on porosity, pore parameters, and intermetallic phase reaction during the reactive sintering of porous TiAl alloys. Ti52Al48 alloys were prepared using coarse (200 mesh) and fine (325 mesh) Ti powders through elemental powder metallurgy and were subsequently sintered at different temperatures, 600 and 1200 °C. Our findings reveal a consistent pore morphology and intermetallic phase microstructure across both alloys. However, samples containing fine Ti powder exhibited a higher number density of small pores compared to those incorporating coarse Ti powders. Additionally, alloys prepared with fine Ti powders demonstrated a higher porosity than those prepared with coarse powders. Consequently, fine Ti powder promoted enhanced diffusion between Ti and Al during sintering, as reflected by the lower onset temperature and enthalpy of intermetallic reaction during sintering.

Keywords: TiAl alloys; powder size; sintering; intermetallic compounds

1. Introduction

Porous intermetallic alloys have been considered valuable engineering materials across various industrial sectors, particularly in exhaust gas adsorption and filtration applications [1–4]. However, their utilization in high-temperature exhaust gas filtration has been limited due to inherent susceptibility to corrosion, burning, and oxidation [5–10]. Recently, there has been growing interest in porous TiAl alloys as promising candidates for filtering dust particles from exhaust gases at high temperatures [11,12]. These alloys exhibit exceptional corrosion resistance to both acids and alkalis, along with robust oxidation resistance, even at temperatures exceeding 750 °C [13,14].

Elemental powder metallurgy (EPM) remains a cost-effective method for producing porous TiAl alloys due to its simple sintering process and the use of inexpensive raw materials, i.e., elemental powders [15]. Numerous studies have elucidated the reaction behavior and diffusional phenomenon involved in the formation of porous TiAl alloys [16–18]. For instance, the Kirkendall effect plays a role in the formation of TiAl-based intermetallic compounds (IMCs) [19]. However, the formation of IMCs varies depending on the processing conditions due to disparities in the diffusion rates of Ti and Al [20]. Al particles, with their lower melting point, exhibit rapid diffusion at lower sintering temperatures, migrating towards Ti-rich skeletons and resulting in the formation of Al-rich IMCs [21]. As sintering temperatures increase, the formation of IMCs transition from Al-rich to Ti-rich phases due to the accelerated diffusion of Ti [22]. Consequently, controlling the pore size of the alloy presents a significant challenge owing to the different diffusion behaviors of Ti and Al, leading to the formation of large pores [23,24]. The presence of these diffusion-controlled coarse pores significantly diminishes the filtration efficacy of porous TiAl alloys, particularly in capturing extremely fine particles. Thus, reducing the pore size of porous TiAl alloys is crucial to enhance their suitability for high-temperature filtration applications.

Cutting-edge manufacturing processes, such as hot-isostatic pressing and spark plasma sintering using alloy powders, have reported the effective development of TiAl alloys with finer pore sizes [25,26]. However, these advanced techniques often lead to a significant increase in production costs, rendering them unsuitable for widespread commercial applications [27,28]. Alternatively, a viable approach to effectively reduce pore size involves decreasing the initial powder size utilized in the EPM process. Previous studies reported that alloys containing smaller powder sizes produce finer pores after the sintering process [29,30]. For instance, Yang et al. [30] demonstrated the fabrication of porous TiAlNb alloys using a combination of powders with varying particle sizes. However, incorporating 2 at. % of Nb yielded a similar pore size as the $Ti_{52}Al_{48}$ (at. %) alloy owing to its lower solubility in Ti [31]. In contrast, Li et al. [32] prepared $TiAl_3$ IMCs of fine pore sizes using fine Al and Ti powders. However, their study was limited to a low-temperature sintering process, specifically at 700 °C. It is well understood that complete IMC formation may not occur during reaction sintering at low temperatures, resulting in a typical skeleton microstructure with Ti-rich cores and $TiAl_3$ formed at the surfaces. Therefore, achieving a well-bonded porous structure necessitates sintering at elevated temperatures. Nevertheless, controlling the melting of fine Al particles during high-temperature sintering presents an utmost challenge. Hence, instead of employing fine Al and Ti powders, in the present study, we aimed to control the pore size of TiAl by utilizing only fine Ti powder. Furthermore, we conducted a comprehensive analysis of the microstructure, the nature of IMCs, and pore parameters of the fabricated alloys after sintering at both low (600 °C) and elevated (1200 °C) temperatures.

2. Materials and Methods

Porous TiAl alloys were fabricated with a nominal composition of 52 at. % Ti and 48 at. % Al using 99.9% Al and hydride–dehydride (HDH) Ti powders obtained from ALCO Engineering, UK, and Sejong Materials Limited, respectively. The Al powder has particle sizes of 600 mesh, while Ti powders of two different mesh sizes (200 and 325) were used. The particle size of powders was conducted using a laser diffraction particle size analyzer (ENSOL instruments, LS I3 320). The powders were dry-blended in a tumbler ball mill for 2 h at a rotational speed of 200 rpm/min under a pure (99.9%) Ar atmosphere. Afterward, the mixed powders were cold-pressed into green compacts with dimensions of 25 mm in width and 5.8 mm in height using a mechanical press (SAMWON-80 ton) under a pressure of 370 MPa. Zinc stearate (1:1 ratio) was utilized as a lubricant for the die walls during compaction. Subsequently, the green compacts were subjected to sintering at the temperatures of 600 and 1200 °C, with a heating rate of 10 °C/min. All sintering processes were conducted under a vacuum of 1×10^{-5} bar, and the samples were maintained at the sintering temperature for 2 h.

Porosity and pore parameters were determined using Mercury inclusion testing (Micromeritics, Autopore IV 9510, Norcross, GA, USA), employing an evacuation pressure of 50 mm/Hg and a 10 s intrusion time. Microstructural analysis was conducted using optical microscopy (OM; Olympus GX50, Tokyo, Japan) and scanning electron microscopy (SEM; Jeol JSM 6610LV, Tokyo, Japan). All microscopy samples underwent grinding and polishing following standard metallography procedures. Intermetallic phase analysis was carried out using an X-ray diffractometer (XRD; Rigaku Dmax 2500, Tokyo, Japan) and energy-dispersive X-ray spectroscopy (EDS) detector integrated with the SEM. The reaction mechanism of intermetallic phases was evaluated using a differential thermal analyzer (DTA; SDT Q6000, New Castle, DE, USA), operated at a heating rate of 10 °C/min. Part of the Ti 200 mesh powder sample results used in this study were taken from the author's previous work [31].

3. Results

3.1. Effect of Powder Size on Microstructure and Bulk Volume

The average particle diameter distribution of the Al 600 mesh and Ti 200 and 325 mesh powders used in this study is depicted in Figure 1a–c, respectively. The analysis revealed an average particle diameter of 20.4 μm for Al 600 mesh, 42.8 μm for Ti 200 mesh, and 26.6 μm for Ti 325 mesh powders.

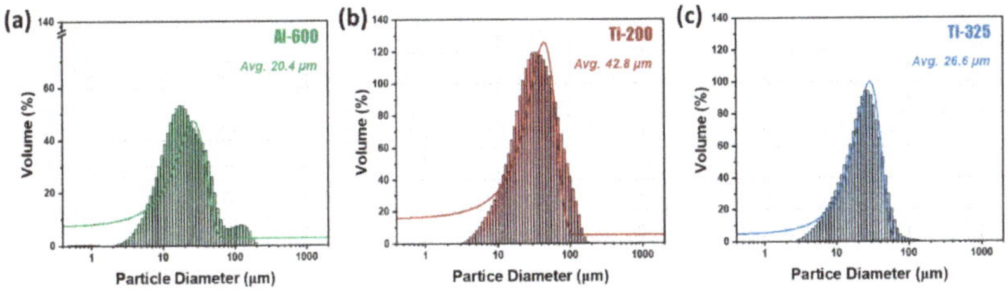

Figure 1. Particle size distribution of powders utilized in the fabrication of $Ti_{52}Al_{48}$ alloys: (**a**) Al 600 mesh, and Ti of (**b**) 200 and (**c**) 325 mesh.

Figure 2 shows secondary electron (SE) images of $Ti_{52}Al_{48}$ alloy specimens prepared to employ (a,c,e) 200 and (b,d,f) 325 mesh Ti powders. Notably, the green compacts present agglomerated powders with small pores resulting from the compaction process. Based on the SE image contrast, dark grey agglomerates were identified as Al, while light grey agglomerates were attributed to Ti. It is discernible that the Ti agglomerates in the 200 mesh sample exhibited a coarser size distribution compared to their 325 mesh counterpart. Consequently, the 200 mesh sample exhibited large inner titanium pores (ITP) as well as inter-particle pores (IPP). Subsequent to sintering at 600 °C, the contrast differentiation within the agglomerates persisted, alongside an escalation in the sizes of ITPs and IPPs. Notably, the 325 mesh Ti powders-containing sample displayed a higher density of smaller pores relative to the 200 mesh specimen. Upon sintering at 1200 °C, both alloys showed a skeleton-type microstructure characterized by large IPPs. The contrast disparity between agglomerates dissipated, indicative of the complete transformation of Ti-rich skeletons into IMCs. However, the IPPs in the 200 mesh sample appeared coarser in comparison to those observed in the 325 mesh specimen.

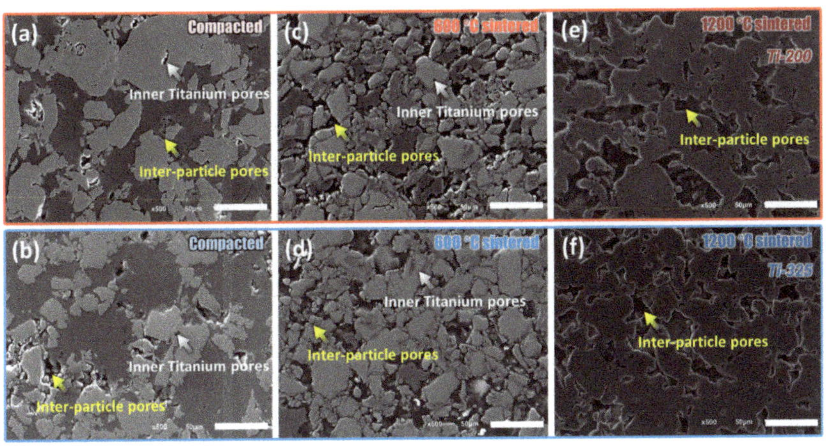

Figure 2. Secondary electron images showing $Ti_{52}Al_{48}$ alloy specimens prepared using Ti powders of varying mesh sizes: (**a**,**c**,**e**) 200 mesh and (**b**,**d**,**f**) 325 mesh samples. (**a**,**b**) Compacted (**c**,**d**) 600 °C, and (**e**,**f**) 1200 °C samples.

3.2. Effect of Powder Size on Pore Size and Porosity

Figure 3 presents the quantitative analysis of pore size and porosity of $Ti_{52}Al_{48}$ alloys, investigated via mercury intrusion tests. Figure 3a,b illustrate the variation in pore diameter for the 200 and 325 mesh samples, respectively. Notably, the pore diameter of the green compacts increased sequentially after sintering, as depicted in Figure 3c for the media pore diameter comparison. According to Table 1, it is evident that the median pore size of the compacted 200 mesh sample increased from 0.70 μm to 3.68 μm and further to 15.28 μm following sintering at 600 and 1200 °C, respectively. Conversely, the 325 mesh sample exhibited a smaller initial pore size of 0.6 μm, which increased to 1.71 μm at 600 °C and to 8.90 μm at 1200 °C, although it remained smaller than the 200 mesh specimen. Figure 3d,e display the cumulative pore volumes of the 200 and 325 mesh alloys, respectively. The corresponding porosity analysis shown in Figure 3f reveals an increase in the porosity of green compacts after sintering. Notably, the porosity of the 325 mesh sample surpassed that of the 200 mesh sample in both compacted and sintered (600 and 1200 °C) states, as presented in Table 1. This increment is attributed to a higher number and density of fine pores in the 325 mesh sample.

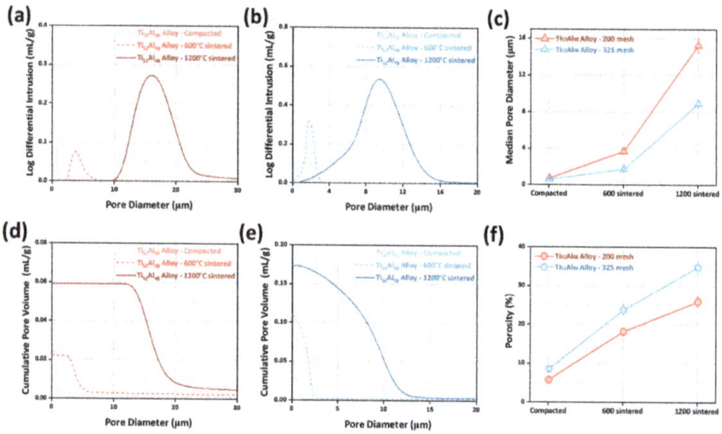

Figure 3. (**a**,**b**) The average pore diameter distributions and (**c**) median pore comparison of $Ti_{52}Al_{48}$ alloy specimens fabricated using Ti powders of different sizes under as-compacted and sintered (600 and 1200 °C) conditions. (**d**,**e**) Cumulative pore volume and (**f**) porosity of alloys for the comparison of $Ti_{52}Al_{48}$ alloy specimens prepared using Ti powders of different sizes under as-compacted and sintered (600 and 1200 °C) conditions. (Part of the 200 mesh sample results were taken from the author's previous work [31]).

Table 1. Pore size parameters of compacted and sintered (600 and 1200 °C) $Ti_{52}Al_{48}$ alloys prepared using Ti powders of different mesh sizes. (Part of the 200 mesh sample results was taken from the author's previous work [31]).

Pore Parameters	200 Mesh			325 Mesh		
	As-Compacted	600 °C Sintered	1200 °C Sintered	As-Compacted	600 °C Sintered	1200 °C Sintered
Median Pore Size (μm)	0.70	3.68	15.28	0.60	1.71	8.90
Porosity (%)	5.82	18.23	25.91	8.51	23.71	34.84

3.3. Effect of Powder Size on Intermetallic Phases

Figure 4a,b present the XRD analysis results for $Ti_{52}Al_{48}$ alloys with Ti powder sizes of 200 and 325 mesh, subjected to compaction and sintering at 600 and 1200 °C. In the compacted state, both samples displayed distinct peaks corresponding to Ti and Al. To differentiate between the powder sizes of the 200 and 325 mesh specimens, the crystallite size was determined using Scherrer's equation [33]:

$$D = \frac{K\lambda}{\beta\cos\theta} \quad (1)$$

where D represents the crystallite size in nm, K denotes Scherrer's constant (0.9) [33], λ is the wavelength of the X-ray source (0.154 nm), β indicates the full-width at half maximum (in rad), and θ denotes the peak position (in rad). The Ti peak analysis of the as-compacted samples yielded a crystallite size of 0.13 and 0.55 µm for the 200 and 325 mesh samples, respectively. These results suggest that the finer Ti powder particles in the 325 mesh specimen produced smaller agglomerates. After sintering, both samples exhibited no noticeable differences in terms of IMC phase formation. For instance, Al-rich IMCs, $TiAl_3$, and $TiAl_2$ were observed in specimens sintered at 600 °C alongside elemental Ti and Al peaks. The elemental Ti and Al peaks completely disappeared after sintering at 1200 °C, with only Ti-rich IMC peaks observed, i.e., TiAl and Ti_3Al. However, the intensity of the IMC phase peaks in the 325 mesh specimen was significantly higher compared to the 200 mesh specimen. This observation suggests that the diffusion rate is faster in the 325 mesh sample, facilitating more pronounced IMC phase reactions.

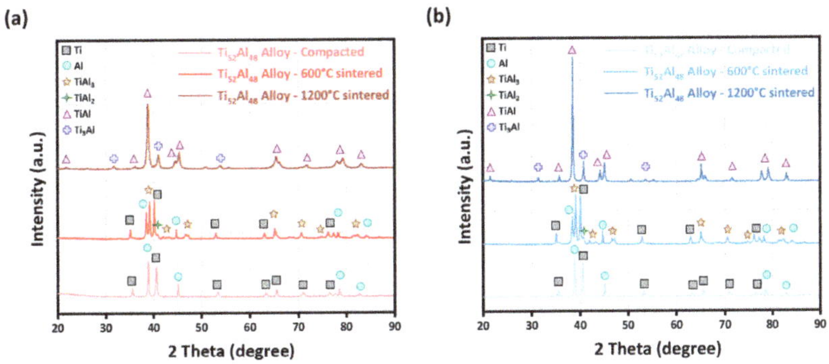

Figure 4. X-ray diffraction analysis of $Ti_{52}Al_{48}$ alloys prepared using Ti powders of varying mesh sizes, (**a**) 200 mesh and (**b**) 325 mesh, under compacted and sintered (600 and 1200 °C) conditions. (Parts of the 200 mesh sample results were taken from the author's previous work [31]).

Furthermore, we conducted backscattered electron (BSE) imaging combined with the EDS analysis of $Ti_{52}Al_{48}$ alloys. Figure 5a–c depicts the BSE images of the 325 mesh specimen in both compacted and sintered states (600 and 1200 °C), respectively. As seen in Figure 5a, the as-compacted sample reveals the presence of Al and Ti agglomerates. Upon sintering at 600 °C, Al-rich IMC Al_3Ti formed on the surface of Ti-rich agglomerates while their core remained rich in Ti. Subsequently, sintering at 1200 °C resulted in the complete transformation of the Ti-rich agglomerates into IMCs. This transformation was evidenced by the presence of the skeleton structure containing the TiAl phase on the surface and the Ti_3Al phase in their cores. The corresponding elemental point analysis results are presented in Table 2.

Table 2. EDS analysis results obtained from the locations shown in Figure 5.

Specimen	As-Compacted		600 °C Sintered		1200 °C Sintered	
Location	Ti	Al	Ti	Al	Ti	Al
i (at. %)	0.70	99.30	99.61	0.39	69.88	30.12
ii (at. %)	99.14	0.86	71.43	28.57	49.53	50.47

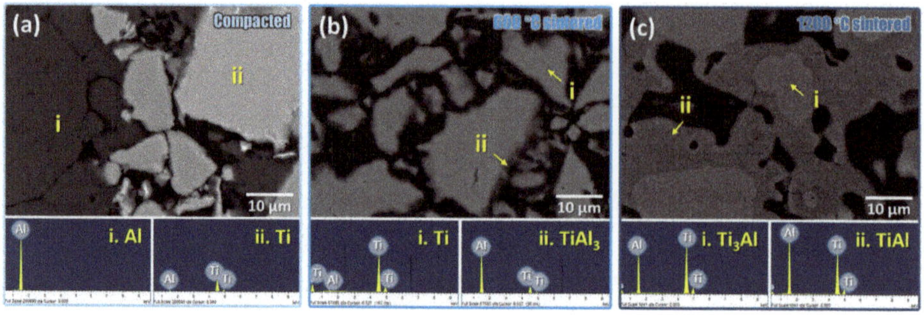

Figure 5. Backscattered electron images of $Ti_{52}Al_{48}$ alloy specimens prepared utilizing 325 mesh Ti powders under (**a**) compacted and sintered (**b**) 600 °C and (**c**) 1200 °C conditions.

4. Discussion

The sintering process of porous TiAl alloys often triggers the formation of IMCs, characterized by an exothermic reaction that induces volume expansion and pore formation. Specifically, due to the higher diffusivity of Al at lower temperatures, a solid–solid diffusion reaction precedes Al melting [34], resulting in the formation of the Al-rich Al_3Ti phase on the surface of Ti agglomerates [35]. This process generates the formation of Kirkendall pores at prior Al sites, which has been explained in detail previously [15]. As the sintering temperature increases, additional intermetallic phase transformations occur via the diffusion of Ti, such as the reactions $TiAl_3 + 2Ti \rightarrow TiAl$ and $TiAl + 2Ti \rightarrow Ti_3Al$. The activation energy for Ti diffusion in the intermetallic structure is significantly higher than that for Al at elevated temperatures [21,22]. Consequently, the presence of excess Ti enhances the thermodynamic stability of Ti-rich IMCs. Moreover, the stoichiometry of the IMCs and the size of the pores may vary depending on the sintering duration at higher temperatures [36,37].

To understand the intermetallic phase reactions in the $Ti_{52}Al_{48}$ alloy, we conducted the DSC analysis on 200 and 325 mesh samples in a compacted state. The DSC results depicted in Figure 6a,b revealed two exothermic peaks and one endothermic peak for both samples. The initial exothermic peak corresponds to the formation of the Al_3Ti phase, succeeded by the endothermic reaction of Al melting [38,39]. Notably, the onset temperature of the first exothermic peak was observed to be lower for the 325 mesh sample (630.1 °C) compared to the 200 mesh sample (638.7 °C). The XRD results presented in Figure 4 corroborate these findings. Furthermore, we calculated the enthalpy of the exothermic reaction for both samples by integrating the area under the peak in the heat flow vs. time plots, as shown in Figure S1. The resulting values of 175.68 kJ/g and 129.77 kJ/g were obtained for the 200 and 325 mesh samples, respectively. Previously, it has been suggested that coarser powders result in less intimate contact between Ti and Al agglomerates, leading to lower interfacial energy and necessitating a stronger diffusion driving force, typically denoted by a higher reaction onset temperature [32,39]. Consequently, the formation of Al_3Ti at elevated temperatures in the samples containing coarser powder releases a larger amount of exothermic heat [32]. The second exothermic peak in the heat flow curves indicates the formation of Ti-rich IMCs following Al melting. Similarly, the peak onset temperature of 325 mesh (806.2 °C) was lower compared to the 200 mesh specimen (810.3 °C). SE images in Figure 2 revealed that Ti agglomerates in the 200 mesh specimen were coarser than those in the 325 mesh sample. Thus, the lower wettability of coarse agglomerates by liquid Al requires prolonged diffusion times for Ti-rich IMC formation, resulting in an increase in the exothermic reaction temperature [40,41].

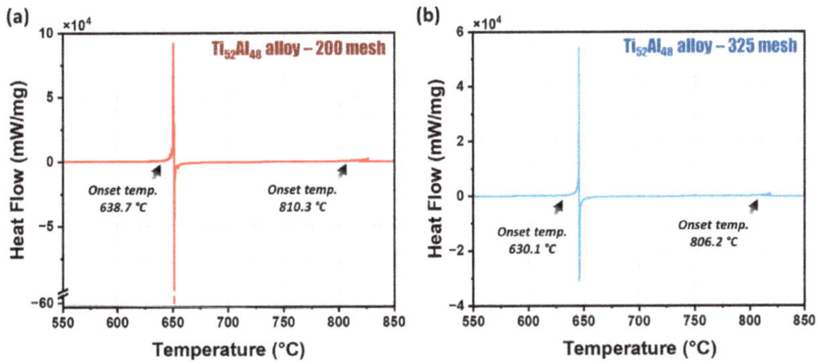

Figure 6. Heat flow curves of $Ti_{52}Al_{48}$ alloys prepared using Ti powders of varying mesh sizes, (**a**) 200 mesh and (**b**) 325 mesh, under compacted and sintered (600 and 1200 °C) conditions.

The driving force for the diffusion of liquid Al is defined by the pressure difference (P) existing between pores originating from compaction and those formed during sintering. Assuming partial melting of all Al agglomerates, the pressure exerted by the liquid Al spreading towards Ti agglomerates can be quantified as follows [32]:

$$P = \frac{2\gamma}{r} \tag{2}$$

where γ represents the surface tension, and r denotes the particle size. Equation (2) elucidates that larger Ti particles result in lower P. Consequently, reaction speeds diminish during sintering in the 200 mesh sample, thereby prolonging diffusion time and facilitating pore enlargement [42]. Conversely, Al diffusion into fine Ti agglomerates is easier, leading to the formation of IMCs with reduced heat flow.

Despite the higher diffusion rates and faster IMC phase reactions observed, the 325 mesh alloy exhibited greater porosity compared to the 200 mesh sample. This increased porosity is attributed to the finer Ti powder particles' higher surface energy [32], which promotes agglomeration. Thus, voids can become trapped within these agglomerates, leading to a higher number and density of pores compared to the coarser Ti powder. This is corroborated by the cumulative pore volume data in Figure 3d,e and the SE images in Figure 2a,b. Additionally, the Kirkendall effect, which involves differential diffusion rates between Ti and Al, can contribute to porosity through the formation of vacancies during sintering [19]. In fine powders, the higher surface area to volume ratio exacerbates this effect, potentially resulting in more pronounced porosity despite the theoretically higher densification rate. Consequently, the 325 mesh samples exhibited a higher density of fine pores with elevated porosity, rendering them favorable for exhaust gas filtration applications.

5. Conclusions

The present study investigated the effect of varying powder sizes on the porosity, pore characteristics, and intermetallic phase formation in porous TiAl alloys. $Ti_{52}Al_{48}$ alloys were fabricated using both fine (325 mesh) and coarse (200 mesh) Ti powders, revealing a skeleton-like IMC microstructure in both cases. At lower sintering temperatures, an Al-rich Al_3Ti phase was formed on the surface of Ti-rich skeletons while the core remained Ti-rich. Upon sintering at 1200 °C, the skeleton cores transformed into the Ti_3Al phase, while the surface exhibited the TiAl phase. Notably, the intermetallic reaction remained consistent across the alloys, with a lower reaction onset temperature and enthalpy observed for the 325 mesh sample. This resulted in the accelerated diffusion of Al into the Ti skeletons at lower temperatures, leading to a high number and density of fine pores and increased porosity in the 325 mesh specimen.

Supplementary Materials: The following supporting information can be downloaded at: https://www.mdpi.com/article/10.3390/cryst14060559/s1, Figure S1: Heat flow vs. time plots of Ti52Al48 alloys prepared using Ti powders of varying mesh sizes; (a) 200 mesh and (b) 325 mesh, under compacted and sintered (600 and 1200 °C) conditions.

Author Contributions: Conceptualization, S.H.K.; methodology, S.H.K.; investigation, S.H.K.; resources, S.H.K.; data curation, S.H.K.; writing—original draft preparation, H.M.S.A. and B.-J.K.; writing—review and editing, H.M.S.A. and B.-J.K.; visualization, N.-K.P.; project administration, supervision, K.E. All authors have read and agreed to the published version of the manuscript.

Funding: This work was financially supported by prime research program of the Korea Institute of materials science (KIMS), and an alliance academic program of Korea University of science and technology (UST), South Korea.

Data Availability Statement: The raw data supporting the conclusions of this article will be made available by the authors on request.

Conflicts of Interest: The authors declare no conflicts of interest.

References

1. Li, X.; Guan, B.Y.; Gao, S.; Lou, X.W.D. A general dual-templating approach to biomass-derived hierarchically porous heteroatom-doped carbon materials for enhanced electrocatalytic oxygen reduction. *Energy Environ. Sci.* **2019**, *12*, 648–655. [CrossRef]
2. Yang, X.-Y.; Chen, L.-H.; Li, Y.; Rooke, J.C.; Sanchez, C.; Su, B.-L. Hierarchically porous materials: Synthesis strategies and structure design. *Chem. Soc. Rev.* **2017**, *46*, 481–558. [CrossRef]
3. Sina, H.; Iyengar, S. Reactive synthesis and characterization of titanium aluminides produced from elemental powder mixtures. *J. Therm. Anal. Calorim.* **2015**, *122*, 689–698. [CrossRef]
4. Jiang, Y.; Deng, C.; He, Y.; Zhao, Y.; Xu, N.; Zou, J.; Huang, B.; Liu, C.T. Reactive synthesis of microporous titanium-aluminide membranes. *Mater. Lett.* **2009**, *63*, 22–24. [CrossRef]
5. Shi, Q.; Qin, B.; Feng, P.; Ran, H.; Song, B.; Wang, J.; Ge, Y. Synthesis, microstructure and properties of Ti–Al porous intermetallic compounds prepared by a thermal explosion reaction. *RSC Adv.* **2015**, *5*, 46339–46347. [CrossRef]
6. Sun, M.-H.; Huang, S.-Z.; Chen, L.-H.; Li, Y.; Yang, X.-Y.; Yuan, Z.-Y.; Su, B.-L. Applications of hierarchically structured porous materials from energy storage and conversion, catalysis, photocatalysis, adsorption, separation, and sensing to biomedicine. *Chem. Soc. Rev.* **2016**, *45*, 3479–3563. [CrossRef]
7. Zhang, G.-H.; Zhu, Q.-H.; Zhang, L.; Yong, F.; Zhang, Z.; Wang, S.-L.; Wang, Y.; He, L.; Tao, G.-H. High-performance particulate matter including nanoscale particle removal by a self-powered air filter. *Nat. Commun.* **2020**, *11*, 1653. [CrossRef]
8. Sharma, P.; Yadav, P.; Ghosh, C.; Singh, B. Heavy metal capture from the suspended particulate matter by Morus alba and evidence of foliar uptake and translocation of PM associated zinc using radiotracer (65Zn). *Chemosphere* **2020**, *254*, 126863. [CrossRef] [PubMed]
9. Petriev, I.; Pushankina, P.; Bolotin, S.; Lutsenko, I.; Kukueva, E.; Baryshev, M. The influence of modifying nanoflower and nanostar type Pd coatings on low temperature hydrogen permeability through Pd-containing membranes. *J. Membr. Sci.* **2021**, *620*, 118894. [CrossRef]
10. Aydoğmuş, T.; Palani, D.K.H.; Kelen, F. Processing of porous β-type Ti74Nb26 alloys for biomedical applications. *J. Alloys Compd.* **2021**, *872*, 159737. [CrossRef]
11. Cobbinah, P.V.; Matizamhuka, W.; Machaka, R.; Shongwe, M.B.; Yamabe-Mitarai, Y. The effect of Ta additions on the oxidation resistance of SPS-produced TiAl alloys. *Int. J. Adv. Manuf. Technol.* **2020**, *106*, 3203–3215. [CrossRef]
12. Cai, X.; Bo, X.; Feng, P.; Ren, X.; Kang, X.; Xu, C.; Zhang, P. Porous NbAl3/$TiAl_3$ intermetallic composites with controllable porosity and pore morphology prepared by two-step thermal explosion. *J. Mater. Res. Technol.* **2019**, *8*, 3188–3197. [CrossRef]
13. Kothari, K.; Radhakrishnan, R.; Wereley, N.M. Advances in gamma titanium aluminides and their manufacturing techniques. *Prog. Aerosp. Sci.* **2012**, *55*, 1–16. [CrossRef]
14. Jiang, Y.; He, Y.H.; Xu, N.P.; Zou, J.; Huang, B.Y.; Liu, C.T. Effects of the Al content on pore structures of porous Ti–Al alloys. *Intermetallics* **2008**, *16*, 327–332. [CrossRef]
15. Kayani, S.H.; Cui, M.; Ahmed, R.T.M.; Cho, Y.-H.; Lee, J.-M.; Park, N.-K.; Ajmal, H.M.S.; Euh, K. Pore formation mechanism and intermetallic phase transformation in Ti–Al alloy during reactive sintering. *J. Mater. Res. Technol.* **2023**, *22*, 1878–1887. [CrossRef]
16. Wang, Y.H.; Lin, J.P.; He, Y.H.; Wang, Y.L.; Chen, G.L. Effect of Nb on pore structure and tensile property of Ti–48Al cellular alloy. *J. Alloys Compd.* **2008**, *456*, 297–303. [CrossRef]
17. Meng, X.L.; Cai, W.; Lau, K.T.; Zhao, L.C.; Zhou, L.M. Phase transformation and microstructure of quaternary TiNiHfCu high temperature shape memory alloys. *Intermetallics* **2005**, *13*, 197–201. [CrossRef]
18. Ma, Y.; Fan, Q.; Zhang, J.; Shi, J.; Xiao, G.; Gu, M. Microstructural evolution during self-propagating high-temperature synthesis of Ti-Al system. *J. Wuhan Univ. Technol.-Mater Sci. Ed.* **2008**, *23*, 381–385. [CrossRef]
19. He, Y.H.; Jiang, Y.; Xu, N.P.; Zou, J.; Huang, B.Y.; Liu, C.T.; Liaw, P.K. Fabrication of Ti–Al Micro/ Nanometer-Sized Porous Alloys through the Kirkendall Effect. *Adv. Mater.* **2007**, *19*, 2102–2106. [CrossRef]

20. Mishin, Y.; Herzig, C. Diffusion in the Ti–Al system. *Acta Mater.* **2000**, *48*, 589–623. [CrossRef]
21. Gupta, R.K.; Pant, B.; Agarwala, V.; Sinha, P.P. Differential scanning calorimetry and reaction kinetics studies of $\gamma + \alpha 2$ Ti aluminide. *Mater. Chem. Phys.* **2012**, *137*, 483–492. [CrossRef]
22. Herzig, C.; Przeorski, T.; Mishin, Y. Self-diffusion in γ-TiAl: An experimental study and atomistic calculations. *Intermetallics* **1999**, *7*, 389–404. [CrossRef]
23. Mackowiak, J.; Shreir, L.L. The nature and growth of interaction layers formed during the reaction between solid titanium and liquid aluminium. *J. Less Common Met.* **1959**, *1*, 456–466. [CrossRef]
24. Sohn, H.Y.; Wang, X. Mathematical and experimental investigation of the self-propagating high-temperature synthesis (SHS) of $TiAl_3$ and Ni3Al intermetallic compounds. *J. Mater. Sci.* **1996**, *31*, 3281–3288. [CrossRef]
25. Mphahlele, M.R.; Olubambi, P.A.; Olevsky, E.A. Advances in Sintering of Titanium Aluminide: A Review. *JOM* **2023**, *75*, 2877–2896. [CrossRef]
26. Karunanithi, R.; Prashanth, M.; Kamaraj, M.; Sivasankaran, S.; Kumaraswamidhas, L.A.; Alhomidan, A.A. Synthesis, characterization, and mechanical behavior of ultra-fine-grained Ti-6Al-5V alloy prepared by mechanical alloying and spark plasma sintering. *Mater. Today Commun.* **2024**, *38*, 108228. [CrossRef]
27. Gerling, R.; Clemens, H.; Schimansky, F.P. Powder Metallurgical Processing of Intermetallic Gamma Titanium Aluminides. *Adv. Eng. Mater.* **2004**, *6*, 23–38. [CrossRef]
28. Gao, H.Y.; He, Y.H.; Shen, P.Z.; Jiang, Y.; Liu, C.T. Effect of pressure on pore structure of porous FeAl intermetallics. *Adv. Powder Technol.* **2015**, *26*, 882–886. [CrossRef]
29. Liang, Y.; Yang, F.; Zhang, L.; Lin, J.; Shang, S.; Liu, Z.-K. Reaction behavior and pore formation mechanism of TiAl–Nb porous alloys prepared by elemental powder metallurgy. *Intermetallics* **2014**, *44*, 1–7. [CrossRef]
30. Yang, F.; Zhang, L.; Lin, J.; Liang, Y.; He, Y.; Shang, S.; Liu, Z.-K. Pore structure and gas permeability of high Nb-containing TiAl porous alloys by elemental powder metallurgy for microfiltration application. *Intermetallics* **2013**, *33*, 2–7. [CrossRef]
31. Kayani, S.H.; Park, N.-K. Effect of Cr and Nb on the phase transformation and pore formation of Ti-Al base alloys. *J. Alloys Compd.* **2017**, *708*, 308–315. [CrossRef]
32. Li, K.; Zhang, T.; Zhu, Y. Reaction Behavior of Porous $TiAl_3$ Intermetallics Fabricated by Thermal Explosion with Different Particle Sizes. *Materials* **2021**, *14*, 7417. [CrossRef] [PubMed]
33. Liu, Y.; Zhang, W.; Peng, Y.; Fan, G.; Liu, B. Effects of TiAl Alloy as a Binder on Cubic Boron Nitride Composites. *Materials* **2021**, *14*, 6335. [CrossRef]
34. Sienkiewicz, J.; Kuroda, S.; Molak, R.M.; Murakami, H.; Araki, H.; Takamori, S.; Kurzydłowski, K.J. Fabrication of TiAl intermetallic phases by heat treatment of warm sprayed metal precursors. *Intermetallics* **2014**, *49*, 57–64. [CrossRef]
35. Wang, T.; Lu, Y.X.; Zhu, M.L.; Zhang, J.S.; Ji, S.J. DSC Research on Critical Temperature in Thermal Explosion Synthesis Reaction $Ti+3Al \rightarrow TiAl_3$. *J. Therm. Anal. Calorim.* **2002**, *67*, 605–611. [CrossRef]
36. Školáková, A.; Leitner, J.; Salvetr, P.; Novák, P.; Deduytsche, D.; Kopeček, J.; Detavernier, C.; Vojtěch, D. Kinetic and thermodynamic description of intermediary phases formation in Ti-Al system during reactive sintering. *Mater. Chem. Phys.* **2019**, *230*, 122–130. [CrossRef]
37. Sujata, M.; Bhargava, S.; Sangal, S. On the formation of $TiAl_3$ during reaction between solid Ti and liquid Al. *J. Mater. Sci. Lett.* **1997**, *16*, 1175–1178. [CrossRef]
38. Jiao, X.; Wang, X.; Feng, P.; Liu, Y.; Zhang, L.; Akhtar, F. Microstructure Evolution and Pore Formation Mechanism of Porous $TiAl_3$ Intermetallics via Reactive Sintering. *Acta Metall. Sin. Engl. Lett.* **2018**, *31*, 440–448. [CrossRef]
39. Jiao, X.; Ren, X.; Feng, P. Visible Observation and Formation Mechanism of Porous $TiAl_3$ Intermetallics During the Continuous Sintering Process. *JOM* **2020**, *72*, 3652–3660. [CrossRef]
40. Ye, S.; Hao, H.; Mo, W.; Yu, K.; Liu, L.; Deng, C.; Yu, P. Effects of cold compacting pressure on the expansion behavior of Ti-48Al during sintering. *J. Alloys Compd.* **2016**, *673*, 399–404. [CrossRef]
41. Che, H.Q.; Fan, Q.C. Microstructural evolution during the ignition/quenching of pre-heated Ti/3Al powders. *J. Alloys Compd.* **2009**, *475*, 184–190. [CrossRef]
42. Jiang, Y.; He, Y.; Liu, C.T. Review of porous intermetallic compounds by reactive synthesis of elemental powders. *Intermetallics* **2018**, *93*, 217–226. [CrossRef]

crystals

MDPI

Article

Optimizing the Rolling Process of Lightweight Materials

Jessica Rawles *, Svitlana Fialkova, Kai Hubbard, Zhigang Xu, Christopher Hale and Jagannathan Sankar

Mechanical Engineering, North Carolina Agricultural and Technical State University, Greensboro, NC 27411, USA; sfialkov@ncat.edu (S.F.); kmhubbard@aggies.ncat.edu (K.H.); zhigang@ncat.edu (Z.X.); clhale@ncat.edu (C.H.); sankar@ncat.edu (J.S.)
* Correspondence: jlrawles@aggies.ncat.edu

Abstract: Conventional rolling is a plastic deformation process that uses compression between two rolls to reduce material thickness and produce sheet/plane geometries. This deformation process modifies the material structure by generating texture, reducing the grain size, and strengthening the material. The rolling process can enhance the strength and hardness of lightweight materials while still preserving their inherent lightness. Lightweight metals like magnesium alloys tend to lack mechanical strength and hardness in load-bearing applications. The general rolling process is controlled by the thickness reduction, velocity of the rolls, and temperature. When held at a constant thickness reduction, each pass through the rolls introduces an increase in strain hardening, which could ultimately result in cracking, spallation, and other defects. This study is designed to optimize the rolling process by evaluating the effects of the strain rate, rather than the thickness reduction, as a process control parameter.

Keywords: lightweight materials; conventional rolling; metal forming; warm working; manufacturing optimization

1. Introduction

Lightweight metals and alloys are highly valued in the automotive, aerospace, and medical industries for their excellent strength-to-weight ratios. These properties are ideal for applications that prioritize fuel efficiency and energy conservation. Titanium and magnesium alloys have been studied extensively for lightweight applications. Magnesium is the lightest of all structural metals, with a density that is one-quarter that of steel [1]. Although Mg-based alloys typically exhibit lower strength and ductility compared to other structural metals, the addition of alloying elements can yield materials with an excellent strength-to-weight ratio, good fatigue and impact strengths, and relatively high thermal and electrical conductivities [2]. Titanium (Ti) alloys also present favorable characteristics, including high strength, corrosion resistance, and reliable mechanical properties at elevated temperatures.

Aluminum is a common alloying element in both magnesium and titanium. Magnesium–aluminum alloys are valued for their low density, making them ideal for lightweight applications. Adding aluminum to magnesium increases hardness, strength, and formability while having a minimal impact on density [3]. When aluminum dissolves in α-Mg, it induces solid solution strengthening. However, a higher aluminum content leads to the formation of the hard and brittle intermetallic phase $Al_{12}Mg_{17}$, which is undesirable due to its hardness, brittleness, and low thermal stability [4,5]. The fragility at the Mg/$Al_{12}Mg_{17}$ interface stems from a mismatch in crystal structures: α-Mg has a hexagonal close-packed (HCP) arrangement, whereas γ-$Al_{12}Mg_{17}$ has a body-centered cubic (BCC) structure [6].

Ti-6Al-4V is a dual-phase (α + β) titanium alloy that incorporates aluminum and vanadium as substitutional elements. The α and β of this alloy have different crystal structures. The α phase Ti-6Al-4V is stable under 650 °C and has an HCP structure while the β phase occurs at temperatures above 900 °C and is of BCC structure [7]. Compared to pure titanium, Ti-6Al-4V exhibits greater hardness, yield strength, and tensile strength, with a tensile yield strength of 900 MPa [8]. This alloy has a superior strength-to-weight ratio, excellent corrosion resistance, and ease of machinability [8,9]. However, due to its high strength and limited formability, Ti-6Al-4V is typically processed at high temperatures, which can be costly.

Enhancing the strength and ductility of lightweight alloys through plastic deformation processes can improve material sustainability and durability while maintaining a low weight. Grain refinement, a common method to enhance material properties, can be achieved through severe plastic deformation techniques such as conventional rolling, cold rolling, hard-plate rolling, and ultrasonic surface rolling. Bulk forming processes like rolling produce deformation textures influenced by the strain path during processing. Rolling processes typically result in a texture similar to plain strain tension, with compression being the primary stress state.

The cold rolling of magnesium alloys leads to a high density of dislocations and deformation twins, significantly contributing to strain hardening. This process can adversely affect the anisotropic behavior due to the development of a strong basal texture [10,11]. However, the strain hardening from cold rolling also increases the yield strength and tensile strength of the Mg alloys. Additionally, cold rolling increases the hardness of Mg alloys, making them more susceptible to brittle cracking. Ti alloys processed through cold rolling experience significant deformation as well as a high density of dislocations. The process of strain hardening the alpha phase in titanium alloys results in a more refined microstructure, which in turn leads to a rise in dislocation density. Similar to magnesium, titanium alloys sacrifice ductility through cold rolling in exchange for increased strength and hardness. Hard plate rolling, being a cold rolling process, exhibits similar effects on both magnesium and titanium alloys to those observed in traditional cold rolling techniques.

Ultrasonic surface rolling (USRP) is similar to conventional rolling but incorporates ultrasonic vibration. This technique aims to enhance the fatigue performance of metallic materials by improving their surface properties. The plastic deformation induced by USRP results in grain refinement and increased dislocation density. In magnesium alloys, USRP creates ultra-fine grain structures on the surface, leading to a more uniform microstructure. This process improves the surface hardness by up to 50% in some Mg alloys and by more than 50% in some Ti alloys [12]. The enhancements in fatigue performance are due to the compressive residual stress induced by USRP. In titanium alloys, USRP promotes the formation of a nanocrystalline surface layer, resulting in reduced surface roughness. The improvement in the surface hardness and fatigue life of titanium alloys after USRP can be attributed to the compressive residual stresses introduced by the process. These stresses make titanium alloys less susceptible to crack initiation and propagation under cyclic loading [13].

Conventional rolling is a plastic deformation process in which the dislocation of polycrystalline material structure is initiated through the stress applied by two compressive rolls [14,15]. In plastic deformation processes, continuous dislocation induces changes in the grain morphology, ultimately leading to an augmentation in the overall grain boundary area. The increase in dislocations can also be attributed to internal structures within the grains such as secondary phases [16]. These secondary phases deform less readily and sometimes not at all and tend to be more brittle in nature. The crystal orientation tends to correspond to the direction of the applied stress establishing the preferred orientation and generating texture. The energy stored from dislocation and increased grain boundary area is what causes the strengthening of material through plastic deformation [17]. For this study, conventional rolling was the preferred deformation process because it minimizes the critical shear stress and is less likely to initiate stress-induced phase transformation.

Deformation mechanisms in metallic structures are often linked to the strain rate sensitivity, defined as the variation of flow stress with increasing strain rate. Key parameters influencing the strain rate sensitivity include the deformation temperature, texture or grain size, and deformation velocity. As the deformation temperature rises, the strain rate sensitivity increases more rapidly compared to room temperature conditions. Dislocations, which cause the reorientation of crystals within a metallic structure, generate texture in plastically deformed alloys. The crystallographic orientation, deformation modes, and slip systems significantly influence the strain. Complex slip systems limit dislocations, and fewer dislocation sites increase the likelihood of concentrated strain in plastically deformed regions [18].

During severe plastic deformation, dislocations tend to move along the slip plane's direction, and this movement also relies on the direction in which the stress is applied. The direction of slip typically corresponds with the most densely packed plane. Typically, under ambient conditions, both magnesium and titanium alloys exhibit a hexagonal close-packed (HCP) crystal structure. HCP polycrystalline materials have limited deformation systems, which has a negative impact on material ductility. HCP structures are sensitive to the crystallographic texture because of their low symmetry and anisotropic behavior [19,20]. This results in a significant effect on texture evolution for HCP material during plastic deformation. Material texture generated for rolling HCP crystals is heavily dictated by the level of impurities, the effect of the c/a ratio of the crystal cell, and the initial texture.

The HCP crystal structure has a limited number of slip systems, which leads to reduced ductility when subjected to severe plastic deformation. For material like Ti that have a lower than ideal (1.6333) c/a ratio, the preferred slip system during deformation is along the prismatic plane. Whereas for Mg alloys, basal slip (0001) is the preferred direction of deformation because of its high c/a ratio [16,21,22]. During plastic deformation, mechanical twinning of Mg occurs on the $\{10\bar{1}2\}$ planes in the $\langle 10\bar{1}1\rangle$ direction, specifically observed under c-axis tension. The formability of Mg is limited by its insufficient slip systems necessary to meet the criteria for homogeneous plastic deformation. However, as the temperature rises above 225 °C, the formability of Mg increases due to a decrease in the critical resolved shear stress [23].

The texture formed during the rolling process offers valuable insights into the material's behavior and mechanical properties. This research aims to examine these characteristics using electron backscatter diffraction (EBSD) and X-ray diffraction (XRD). XRD has been widely used in various studies to analyze the lattice behavior of crystallographic materials and to determine the phase composition through diffraction intensity. In this study, XRD will help to identify whether a preferred orientation is achieved through conventional rolling and verify the presence of stress-induced secondary phases in the rolled samples as a result of plastic deformation. EBSD is an excellent tool for providing detailed crystallographic data and orientation. Within the parameters of this research, we will use EBSD to observe changes in the material's texture and orientation induced by the rolling process. This combination of XRD and EBSD techniques will allow for a comprehensive analysis of the material's behavior and texture following rolling.

As for the present investigation, the aim is to evaluate the effects of the rolling parameters on the structure and strain rate of lightweight materials, with the objective of developing a process map that prioritizes strain rate control over the traditionally used reduction percentage. By focusing on the strain rate control, this study seeks to preserve the ductility of lightweight materials, mitigate strain hardening effects, and promote texture development while enhancing the strength properties. The findings of this research could improve material performance in various applications, particularly in industries in which weight reduction and strength are crucial. Advancing the understanding of strain rate control addresses a specific need in materials science and engineering.

The novelty of this research lies in its approach to understanding the texture and mechanical properties of lightweight alloys, specifically Mg-Al and Ti-6Al-4V, using advanced analytical techniques like electron backscatter diffraction (EBSD) and X-ray diffraction (XRD). These techniques provide detailed insights into crystallographic texture and phase composition changes induced by the rolling process, which are critical for optimizing the material performance. By focusing on Mg-Al and Ti-6Al-4V, this research addresses materials that are important for the automotive, aerospace, and medical industries. Understanding how rolling processes affect these lightweight alloys could lead to enhanced performance in applications in which weight and strength are paramount.

Controlling the rolling temperature is crucial to avoid the formation of undesirable phases, such as the brittle γ-$Al_{12}Mg_{17}$ phase in Mg-Al alloys and the martensitic transformation in Ti-6Al-4V. Careful temperature management is essential for preserving material integrity while investigating deformation mechanisms. Enhancing the strength and ductility of lightweight alloys through plastic deformation processes can improve material sustainability and durability while maintaining a low weight. By addressing the gap in the current literature regarding the optimization of rolling processes for lightweight materials, this study contributes valuable insights to the field, offering a pathway to optimizing these alloys for high-performance applications in which weight reduction is critical.

2. Materials and Methods

This study utilized three different lightweight alloys with varying compositions. T5 commercial-grade titanium (Ti-6Al-4V) was cut from a commercially procured rolled sheet into strips measuring 25.4 × 152.4 mm with a thickness of 1.6 mm. The other two specimen types were Mg-Al alloys with aluminum contents of 6 and 9 wt%. Both Mg-6Al and Mg-9Al were cast as slabs in-house, with their chemical compositions listed in Table 1. To avoid the formation of the brittle γ-$Al_{12}Mg_{17}$ phase, the Mg-Al alloys were fully solutionized. Four strips measuring 21.6 × 119.4 mm were cut from Mg-6Al, and four strips measuring 19.1 × 106.7 mm were cut from Mg-9Al, both with an approximate thickness of 3 mm. All the samples were roughly polished to remove surface roughness and sharp edges from the cutting process.

Table 1. Rolling parameters.

Materials	Speed Ratio	Preheating Temperature (°C)	Initial Thickness (mm)	Thickness Reduction per Pass	Total Thickness Reduction	Number of Passes
Mg-6Al	1:1	315	2.30	15%	56.5%	5
Mg-9Al	1:1	400	2.40	15%	58.3%	5
Ti-6Al-4V	1:1	650	1.60	5%	37.5%	9

After polishing, measurements of each specimen were recorded and listed in Table 1. Each specimen underwent warm working through conventional rolling, using work rolls with a diameter of 50 mm, a rolling speed of 1.0 m/min, and a 1:1 speed ratio between the rolls. The rolling parameters were preset for each specimen based on the initial thickness, reduction percentage per pass, and material properties. The target final thickness for each specimen was 1 mm. The initial thicknesses were 2.3 mm for Mg-6Al, 2.4 mm for Mg-9Al, and 1.6 mm for Ti-6Al-4V. The reduction percentages per pass were 15% for Mg-Al alloys and 5% for Ti-6Al-4V, set to prevent defects such as cracking or spallation [24,25]. These parameters determined the number of passes needed for each specimen: 5 passes for Mg-Al alloys and 9 for the Ti alloy.

Prior to each pass, the specimens were preheated for 5 min to maintain the rolling temperature. Mg-Al specimens were solution-treated by heating above the γ phase region to avoid undesirable phases, with preheating temperatures of 315 °C for Mg-6Al and 400 °C for Mg-9Al, to prevent the $Al_{12}Mg_{17}$ phase [2]. The Ti-6Al-4V specimens were preheated at 650 °C to prevent martensitic phase transformation [9]. The conventional roller used in this experiment had a rolling temperature limit of 300 °C. Consequently, the temperature during the rolling process was maintained at a constant 300 °C.

By controlling the rolling temperature, the formation of undesirable phases, such as the brittle γ-$Al_{12}Mg_{17}$ phase in Mg-Al alloys and martensitic transformation in Ti-6Al-4V, are avoided. This careful temperature management is essential for preserving the material integrity while investigating deformation mechanisms.

One of the four Mg-6Al samples was used as a dummy sample to calibrate, preheat, and prep the roller for experimentation, leaving the remaining three samples for data analysis. The sample thickness of each specimen type was measured after each pass and averaged. The final pass for each sample was a double pass in which the thickness was reduced to the final size, and then the specimen was flipped and passed through again without additional thickness reduction. After rolling, each specimen was quenched in room-temperature water for 5 s to preserve the material's microstructure. It is important to note that there was no significant change in the chemical composition of each sample after rolling. Table 2 shows the chemical composition of each alloy before and after rolling.

Table 2. Material composition.

Materials	Element	Pre-Rolled Weight %	Post-Rolled Weight %
Mg-6Al	Mg Al	92.69 7.31	92.51 7.49
Mg-9Al	Mg Al	89.25 10.75	89.25 10.75
Ti-6Al-4V	Ti Al V	85.34 6.39 6.19	86.92 6.55 6.53

The post-rolled samples were prepared for microscopic and spectroscopic analysis. This involved a series of grinding and polishing steps, conducted using a MultiPrep System polisher (Allied High Tech Products, Inc, Compton, CA, USA). For etching, the Mg-Al samples were treated with a 2.5% nitric acid solution for 30 s, while the Ti samples were etched using a 10% HF solution for 10 s. The rolling normal direction of each sample was then examined using a Zeiss Imager M2 optical microscope.

To observe the phase composition, a Hitachi SU8000 scanning electron microscope (SEM, Tokyo, Japan) equipped with an Oxford Instruments Aztec electron backscattering diffraction (EBSD, Abingdon, UK) CMOS EBSD detector was used. EBSD data analysis was carried out with Oxford Instruments Aztec software (version 6.0).

For the analysis of the crystal direction and orientation, a Bruker D8 Discover X-ray diffraction (XRD, Billerica, MA, USA) machine was employed. The XRD measurements were taken over a range of 30° to 80° with a scanning speed of 2 s per step and a step size of 0.0172°. The resulting XRD data were processed using Bruker Diffrac Eva software (version 5.2) to accurately determine the crystal structures present in the samples.

3. Results

The final average thicknesses achieved for the rolled Mg-6Al, Mg-9Al, and Ti-6Al-4V samples were 0.98 mm, 0.95 mm, and 0.95 mm, respectively, corresponding to total thickness reductions of 57.6%, 60.4%, and 40.6%. Throughout the rolling process, each specimen remained intact and exhibited no visible defects. However, the strain rate increased with each pass. Figure 1 presents the true strain rate for each pass, where the numbers on each point indicate the draft per pass (in mm), defined as the difference between the initial and final thickness.

According to the figure, there is a noticeable peak in the strain rate during the final pass for Mg-9Al and Ti-6Al-4V. This peak is likely attributed to the double pass applied during the final rolling reduction. Reorienting the sample and passing it twice without reducing the thickness provides an opportunity for dislocations to move in a different direction compared to the previous rolling path, thus contributing to the observed strain rate increase. This study highlights the importance of controlling the strain rate during the rolling process to maintain the material's ductility and manage strain hardening. By understanding the relationship between draft per pass and strain rate, the process parameters can be optimized to enhance the mechanical properties of the rolled alloys. This detailed analysis of the strain rate's behavior and its implications on the material properties highlights the significance of precise control in the rolling of lightweight materials.

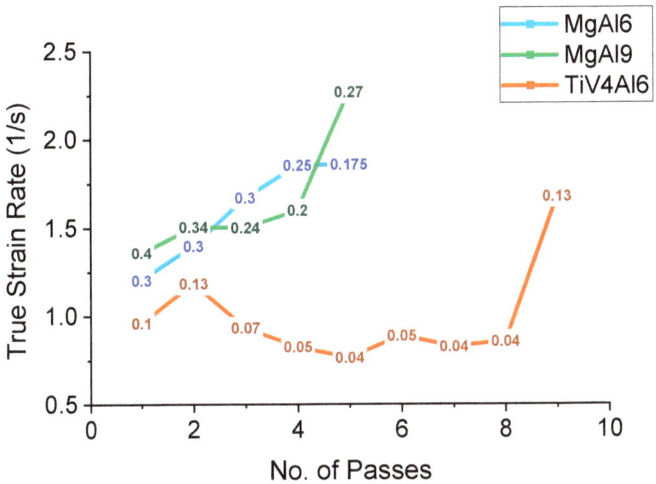

Figure 1. The true strain rate of each specimen per pass with draft values.

In Figure 1, it is observed that the Mg-Al samples experienced an increase in the strain rate per pass with fluctuating draft values. For the Ti-6Al-4V sample, the strain rate was more constant with each pass. This can be credited to two factors: firstly, the 5% thickness reduction per pass for Ti-6Al-4V, contrasting with the 15% reduction applied to the Mg-Al samples; secondly, the greater hardness of Ti-6Al-4V. From this result, it can be concluded that reducing the draft decreases the strain rate in the Ti-6Al-4V samples. Reducing the draft in Mg-Al presents more of a challenge considering it is a much softer metal allowing for more atomic dislocations. This finding underscores the complex interplay between the material properties and the process parameters in determining the strain rate during rolling.

3.1. Microstructure Evolution

Figure 2 shows the microstructure of all the samples before and after rolling. For the Mg-Al alloys, the average grain sizes of the solution-treated specimens were larger at about 53.1 μm for Mg-6Al and 64.4 μm for Mg-9Al. The increased amount of Al content increases the grain size in the Mg-Al alloys. After rolling, the average grain size of Mg-6Al was about 2.5 μm and that for Mg-9Al was about 4.4 μm. For the Mg-Al alloys, the reduction in grain size was greater than 90%. This was not the case for the Ti-6Al-4V specimen. After rolling the Ti alloys, the average grain size reduced by less than 5% with a value, prior to rolling, of about 2.7 μm and 2.6 μm after rolling. This reduction in grain size is considered to be negligible.

Through the EBSD analysis, shown in Figure 3, it was observed that the preferred orientation of crystals was achieved in all rolled specimens. Prior to rolling, the solution-treated Mg-Al alloys crystals were large and randomly oriented. Appendix A contains the sample orientation that correspond to the crystal orientation of the phases present in each specimen. The rolled Mg-Al alloys have a preferred orientation of the (0001) plane, which corresponds with basal slip. There was also evidence of stressors within the larger grains of the rolled Mg-Al samples. The phase most present in the Mg-Al alloys was the magnesium phase, and a phase change did not occur as a result of the rolling process. For the Ti alloy, the preferred orientation of ($\bar{1}2\bar{1}0$) was achieved but with less concentration than that of the Mg-Al alloys. Before rolling, Ti-6Al-4V consisted primarily of the Ti hexagonal phase with strong phases of vanadium intermingled between the phases of Ti. After rolling, the vanadium phase was more disbursed through the dominating Ti hexagonal phase in smaller concentrations.

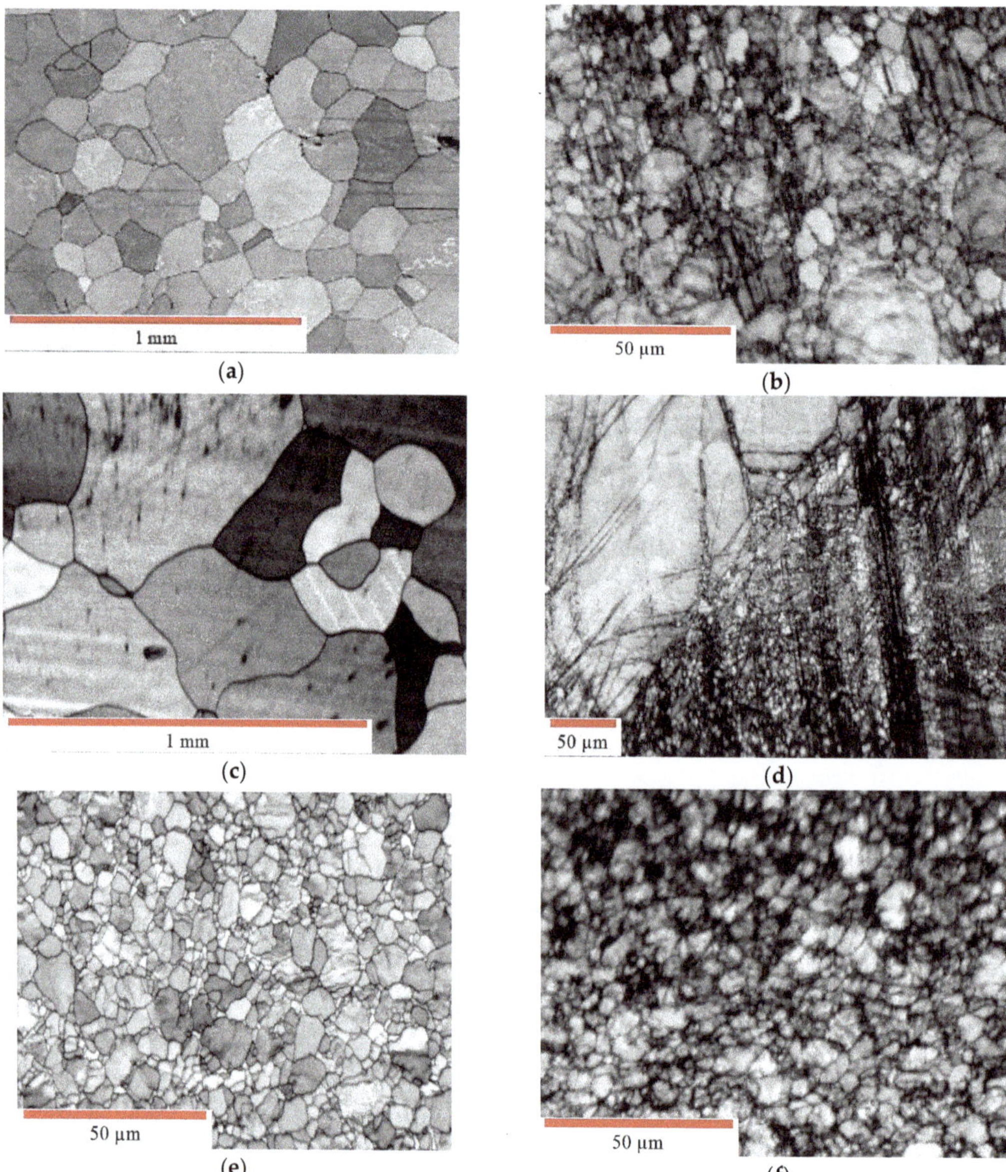

Figure 2. Grain size of material before and after rolling: (**a**) Mg-6Al, solution treated; (**b**) Mg-6Al, rolled; (**c**) Mg-9Al, solution treated; (**d**) Mg-9Al, rolled; (**e**) Ti-6Al-4V; (**f**) Ti-6Al-4V, rolled.

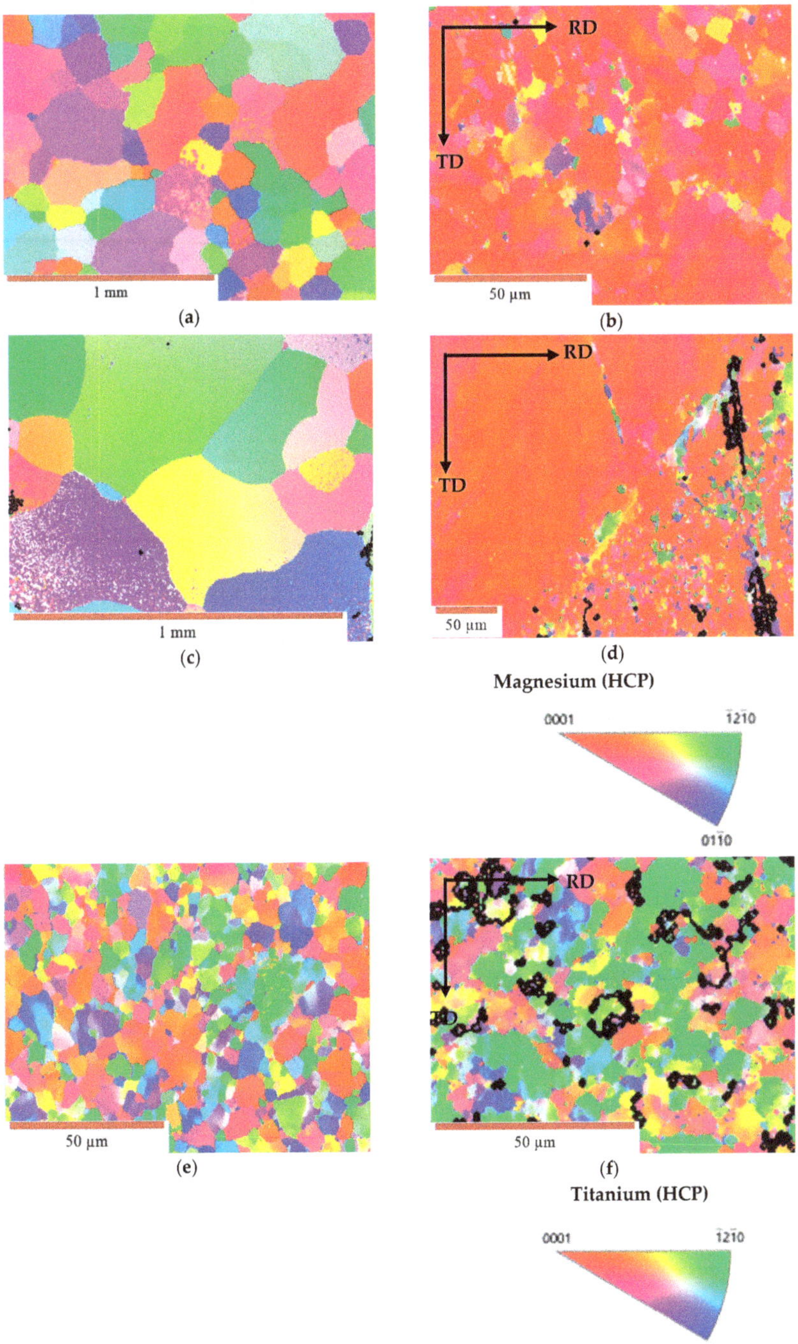

Figure 3. EBSD crystallographic orientation of the normal surface of samples before and after rolling: (**a**) Mg-6Al, solution treated; (**b**) Mg-6Al, rolled with a (0001) preferred planar orientation; (**c**) Mg-9Al, solution treated; (**d**) Mg-9Al, rolled with a (0001) preferred planar orientation; (**e**) Ti-6Al-4V; (**f**) Ti-6Al-4V, rolled with a primary preferred planar orientation of ($\bar{1}2\bar{1}0$).

3.2. XRD Analysis

Figure 4 depicts the XRD diffraction patterns of the Mg-6Al, Mg-9Al, and Ti-6Al-4V alloys before and after rolling. The peaks of each of the samples were matched to the crystallographic database for Mg and Ti patterns, respectively. Each peak corresponds to a planar direction of a certain intensity. In all the samples, the number of diffraction peaks did not significantly change. The peaks observed on the solution-treated Mg-Al alloys were generally narrow indicating large crystallite structures. After rolling Mg-6Al, the rolled sample peak intensity significantly dropped at about 32° 2θ, while the peak intensity around 35° 2θ shot up. This increased intensity at 35° 2θ suggests that there is a preferred orientation of (0002) as a result of rolling Mg-6Al. A similar phenomenon occurred in the Mg-9Al sample in that there was an intensity increase at 35° 2θ, once again suggesting a preferred orientation along the (0002) plane. After rolling, the peak intensity at 57° 2θ for solution-treated Mg-9Al decreased to a negligible level. For the Ti-6Al-4V alloys, the XRD data displayed broader peaks, indicating very small crystallite structures. The intensity of each peak broadened and significantly increased. The broadening of each peak can be attributed to the occurrence of non-uniform strain as a result of rolling. For the rolled Ti-6Al-4V sample, the peak intensity increased at about 38° and 40°, suggesting that the crystals were becoming smaller and follow a more preferred direction of (0002) and (01$\bar{1}$1). The ratios between the 38° and 40° peaks before rolling were about 3:2 and 2:1 after rolling.

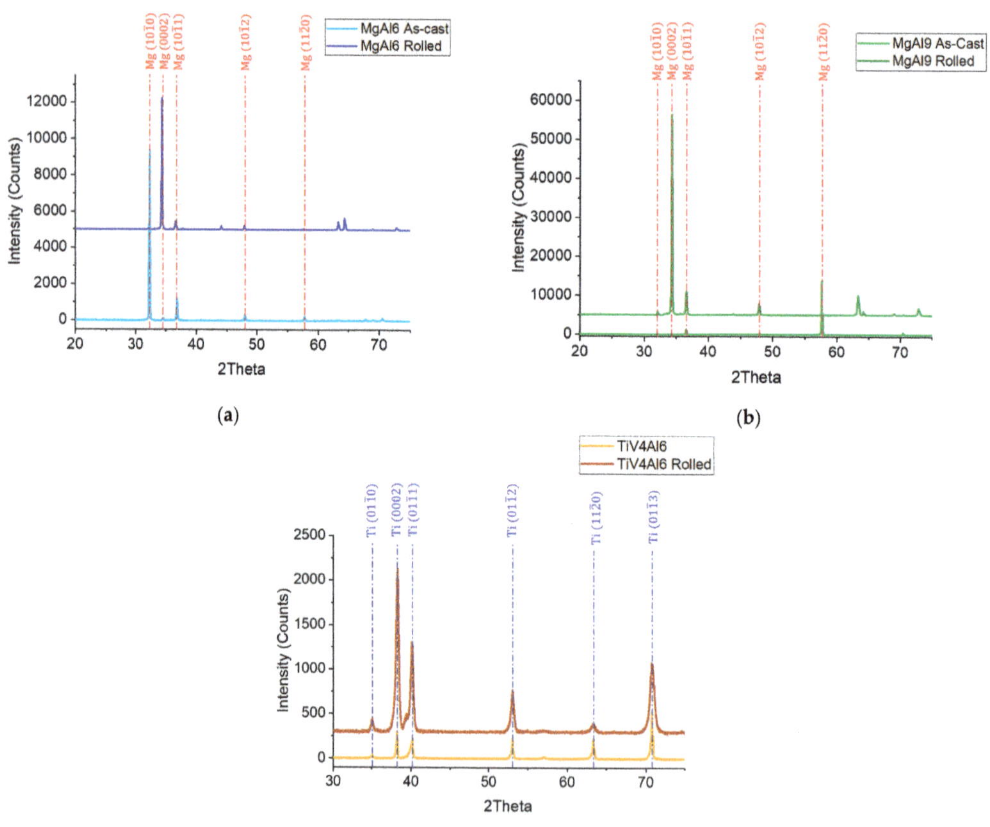

Figure 4. XRD data of samples before and after rolling: (**a**) Mg-6Al, solution treated vs. rolled; (**b**) Mg-9Al, solution treated vs. rolled; (**c**) Ti-6Al-4V as procured vs. rolled.

3.3. Formatting of Mathematical Components

Figure 5 depicts the free-body diagram of the forces applied to the rolled material by the roll. The volume of the rolled material remains the same as there is no material being added or taken away.

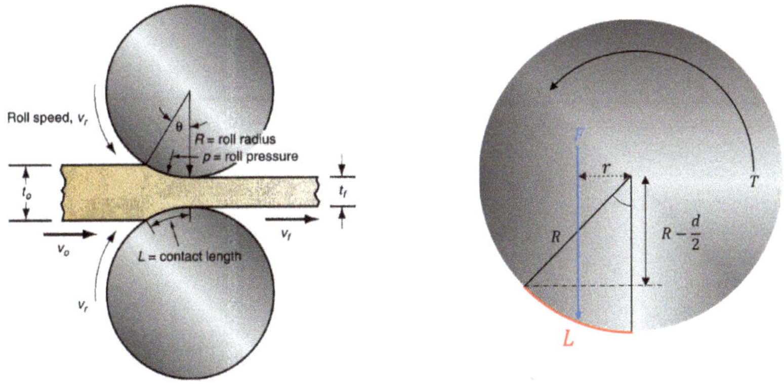

(**a**) (**b**)

Figure 5. The free-body diagram of conventional rolling: (**a**) image from Fundamentals of Modern Manufacturing: Materials, Processes, and Systems textbook [10]—forces exerted on the sample by both top and bottom rollers; (**b**) forces exerted by a single roller.

Variables that impact the true strain rate include the reduction percentage, temperature, and roll speed. The draft d of a rolled material is the difference in the thickness before and after rolling. Reduction r is often represented as a percentage and is the draft divided by the initial thickness.

$$r = \frac{d}{t_0} = \frac{t_0 - t_f}{t_0} \tag{1}$$

The experiment was recalculated so that the strain rate stayed constant. The equation for the true strain rate is as follows:

$$\dot{\epsilon} = \frac{v}{L} ln \frac{t_0}{t_f} \tag{2}$$

where v is velocity, L is contact length, and t_f and t_0 are final and initial thicknesses. The contact length can be expressed in terms of thickness as follows:

$$L = \sqrt{Rd} = \sqrt{R\left(t_o - t_f\right)} \tag{3}$$

where R is the radius of the rolls and d is the draft. In order to manipulate the strain rate, the effects of changes in the draft size or reduction percentage and velocity were analytically observed.

4. Discussion

The expectation was to keep a constant percentage of reduction for each pass; however, the reduction percentage varied and spiked on the last pass of each specimen. This anomaly can be explained by the heightened dislocation density resulting from the final pass being a double pass, leading to increased strain hardening due to a flipped specimen orientation. It was also observed that increases in the percentage of reduction with each pass were correlated with higher strain rates. With each pass, the ability to stay true to the constant reduction percentage became more of a challenge because, with each pass, the strain increases, which limits dislocation. It was also concluded through calculation that, even if a constant percentage were maintained, the strain rate would still increase with each pass. Increasing the strain rate for each material increases the likelihood of failure. From this, it can be hypothesized that, if the rolling process is controlled keeping a constant strain rate, it can ensure that no damage occurs during the process.

A process map was developed in order to control the strain rate for each pass. The variables that affect the true strain rate are the temperature, roll speed, and the initial and final thickness of the material. The temperature must remain constant in order to avoid phase transformation. Therefore, for the purposes of developing the process map, the factors that may vary are the roll velocity and the material thickness or draft. If the roll speed is too fast, there is more possibility of a fracture due to the spring back effect as well as the formation of twin structure especially for the much softer Mg-Al alloys. Because of this, the velocity domain was observed within the ranges of 1–2.5 m/min. The developed process map is depicted in Figure 6a. It can be observed that, by maintaining a constant strain rate, we can avoid failure and have more control over the rolling process. Figure 6 also shows a breakdown of the process map in terms of the daft and the reduction percentage, respectively.

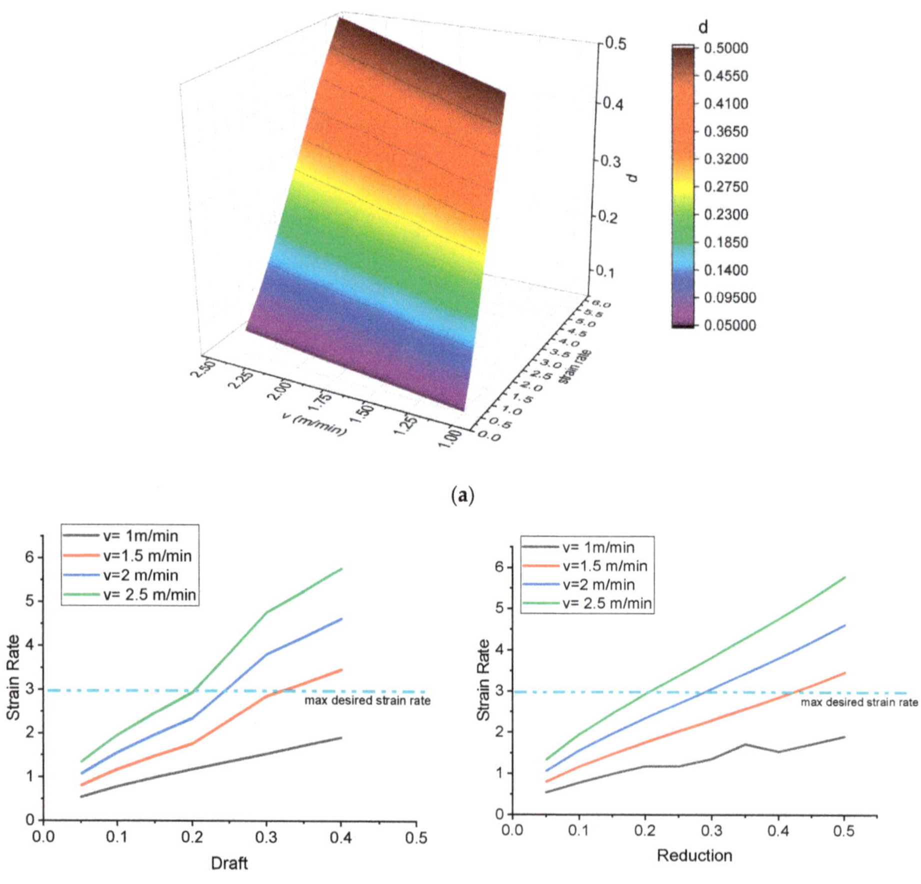

Figure 6. Controlling rolling process by strain rate. (**a**) Process map of velocity and draft effect on strain rate, (**b**) draft vs. strain rate for various velocities, (**c**) reduction percentage vs. strain rate for various velocities.

The proposed model will help control the strain rate by selecting the appropriate velocity and draft for each pass. By selecting a velocity and draft that corresponds to the desired strain rate, the strain rate can be held constant retaining more ductility and reducing the likelihood of failure during processing. This ensures that we can strengthen material through a cold working process without compromising the material's integrity.

5. Conclusions

The conventional rolling process significantly reduced the grain size in the Mg-Al system by over 90% at ~58% cold work (CW) and in the Ti-6Al-4V alloy by 4% at ~41% CW. This reduction in grain size, coupled with an increase in the dislocation density and strain after each pass, indicates that strain hardening is occurring and generating texture. Texture (preferred plane orientation) was observed in all alloys after rolling.

This study aimed to evaluate the impact of the texture on the strain rate and to identify process parameters to maintain a more consistent strain rate. This goal was achieved through the development of a process map that assesses the parameters contributing to strain values. This study specifically investigates how conventional rolling affects the texture of Mg-Al and Ti-6Al-4V alloys. By analyzing the texture formation and its implications on mechanical properties, this research provides insights that are critical for optimizing rolling processes in industrial applications. It not only investigates the microstructural changes of the Mg-Al and Ti-6Al-4V alloys but it also develops a process map that correlates rolling parameters (such as temperature, strain rate, and reduction percentage) with the resulting strain and texture. This practical tool can guide future processing techniques to achieve the desired material properties.

The proposed process map can be utilized in future studies and, given that the draft is directly associated with the reduction percentage, this process can potentially be modified and applied to other severe plastic deformation processes. This study can also be extended to include mechanical studies such as hardness and strength evaluations. Future research may investigate the effects of maintaining a constant strain rate during the rolling process on mechanical properties like strength and hardness.

Author Contributions: Methodology, J.R. and Z.X.; Formal analysis, J.R.; Investigation, J.R.; Resources, S.F., Z.X. and J.S.; Data curation, J.R. and K.H.; Writing—original draft, J.R.; Writing—review & editing, J.R., S.F., Z.X., C.H. and K.H.; Supervision, S.F., Z.X. and C.H.; Project administration, J.S. All authors have read and agreed to the published version of the manuscript.

Funding: This research was funded by the National Science Foundation under Grant No. (NSF EEC-2133630).

Data Availability Statement: The original contributions presented in the study are included in the article, further inquiries can be directed to the corresponding author.

Acknowledgments: This material is based on research sponsored by the Air Force Research Laboratory under agreement number FA8650-20-2-5853. The U.S. Government is authorized to reproduce and distribute reprints for government purposes notwithstanding any copyright notation thereon. The views and conclusions contained herein are those of the authors and should not be interpreted as necessarily representing the official policies or endorsements, either expressed or implied, of the Air Force Research Laboratory of the U.S. Government.

Conflicts of Interest: The authors declare no conflict of interest.

Appendix A

The samples were positioned on the sample mount with their normal direction perpendicular to the surface and the rolling the direction is along the x-axis as seen in Figure A1.

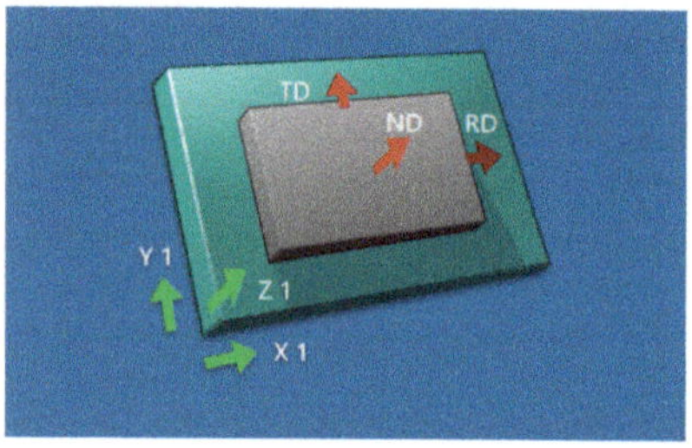

Figure A1. Sample orientation for EBSD analysis.

References

1. Gupta, M.; Sharon, N.M. *Magnesium, Magnesium Alloys, and Magnesium Composites*; John Wiley & Sons, Inc.: Hoboken, NJ, USA, 2010.
2. Mezbahul-Islam, M.; Mostafa, A.O.; Medraj, M. Essential Magnesium Alloys Binary Phase Diagrams and Their Thermochemical Data. *J. Mater.* **2014**, *2014*, 704283. [CrossRef]
3. Murugesan, R.; Venkataramana, S.H. Influence of Alloying Materials Al, Cu, and Ca on Microstructures, Mechanical Properties, And Corrosion Resistance of Mg Alloys for Industrial Applications: A Review. *ACS Omega* **2023**, *8*, 37641–37653. [CrossRef] [PubMed]
4. Zhang, L.; Cao, Y.B. Effect of Al content on the microstructures and mechanical properties of Mg–Al alloys. *Mater. Sci. Eng. A* **2008**, *508*, 129–133. [CrossRef]
5. Friedrich, H.E.; Mordike, B.L. Physical Metallurgy. In *Magnesium Technology: Metallurgy, Design Data, Apllications*, 1st ed.; Springer: Berlin/Heidelberg, Germany, 2006; pp. 63–107.
6. Sanamar, S.; Brokmeier, H.G. Formation of the intermetallic phases $Al_{12}Mg_{17}$ and Al_3Mg_2 during heating of elemental Al-Mg composites studied by high-energy X-ray diffraction. *J. Alloys Compd.* **2022**, *911*, 165114. [CrossRef]
7. Ducato, A.; Fratini, L.; La Cascia, M.; Mazzola, G. An Automated Visual Inspection System for the Classification of the Phases of Ti-6Al-4V Titanium Alloy. In Proceedings of the 15th International Conference on Computer Analysis of Images and Patterns, York, UK, 27–29 August 2013; pp. 362–369.
8. Blanco, D.; Rubio, E.M.; Marin, M.M.; Davim, J.P. Advanced materials and multi-materials applied in aeronautical and automotive fields: A systematic review approach. *Procedia CIRP* **2021**, *99*, 196–201. [CrossRef]
9. Macleod, S.G.; Errandonea, D.; Cox, G.A.; Cynn, H.; Daisenberger, D.; Finnegan, S.E.; McMahon, M.I.; Munro, K.A.; Popescu, C.; Storm, C.V. The phase diagram of Ti-6Al-4V at high-pressures and high-temperatures. *J. Phys. Condens. Matter* **2021**, *33*, 154001. [CrossRef] [PubMed]
10. Deng, B.; Dai, Y.; Lin, J.; Zhang, D. Effect of Rolling Treatment on Microstructure, Mechanical Properties, and Corrosion Properties of WE43 Alloy. *Material* **2022**, *15*, 3985. [CrossRef] [PubMed]
11. Zhu, S.Q.; Yan, H.G.; Chen, J.H.; Wu, Y.Z.; Liu, J.Z.; Tian, J. Effect of twinning and dynamic recrystallization on the high strain rate rolling process. *Scr. Mater.* **2010**, *63*, 985–988. [CrossRef]
12. John, M.; Ralls, A.M.; Dooley, S.C.; Thazhathidathil, A.K.V.; Perka, A.K.; Kuruveri, U.B.; Menezes, P.L. Ultrasonic Surface Rolling Process: Properties, Characterization, and Applications. *Appl. Sci.* **2021**, *11*, 10986. [CrossRef]
13. Liu, C.; Liu, D.; Zhang, X.; He, G.; Xu, X.; Ao, N.; Ma, A.; Liu, D. On the influence of ultrasonic surface rolling process on surface integrity and fatigue performance of Ti-6Al-4V alloy. *Surf. Coat. Technol.* **2019**, *310*, 24–34. [CrossRef]
14. Raabe, D. Deformation Processing. In *Encyclopedia of Condensed Matter Physics*; Elsevier: Amsterdam, The Netherlands, 2005; pp. 387–395.
15. Groover, M.P. *Fundamentals of Modern Manufacturing: Materials, Processes, and Systems*, 5th ed.; John Wiley & Sons, Inc.: Hoboken, NJ, USA, 2013; pp. 419–421.
16. Humphreys, F.J.; Hatherly, M. *Recyrstallization and Related Annealing Phenomena*, 2nd ed.; Elsevier: Amsterdam, The Netherlands, 2004.
17. Chapuis, A.; Liu, Q. Simulations of texture evolution for HCP metals: Influence of the main slip systems. *Comput. Mater. Sci.* **2015**, *97*, 121–126. [CrossRef]
18. Wang, H.W.; Xue, E.S.; Nan, X.L.; Yue, T.; Wang, Y.P.; Jiang, Q.C. Influence of grain size on strain rate sensitivity in rolled Mg–3Al–3Sn alloy at room temperature. *Scr. Mater.* **2013**, *68*, 229–232. [CrossRef]
19. Capolungo, I.; Beyerlein, I.J.; Wang, Z.Q. The role of elastic anisotropy on plasticity in hcp metals: A three-dimensional dislocation dynamics study. *Model. Simul. Mater. Sci. Eng.* **2010**, *18*, 16. [CrossRef]
20. Ma, Z.A.; Tang, X.Z.; Mao, Y.; Guo, Y.F. The Plastic Deformation Mechanisms of hcp Single Crystals with Different Orientations: Molecular Dynamics Simulations. *Material* **2021**, *14*, 773. [CrossRef] [PubMed]
21. Tehranchi, A.; Yin, B.; Curtin, W.A. Solute strengthening of basal slip in Mg alloys. *Acta Mater.* **2018**, *151*, 56–66. [CrossRef]
22. Yapici, G.G.; Karaman, I. Common trends in texture evolution of ultra-fine-grained hcp materials during equal channel angular extrusion. *Mater. Sci. Eng. A* **2009**, *503*, 78–81. [CrossRef]

23. Mackenzie, L.W.F.; Pekguleryuz, M. The influences of alloying additions and processing parameters on the rolling microstructures and textures of magnesium alloys. *Mater. Sci. Eng. A* **2008**, *480*, 189–197. [CrossRef]
24. Zhang, H. Study on Dynamic Recrystallization and Improving Mechanical Properties of Mg-xAl(x = 6,9) Alloy through Differential Speed Rolling. Ph.D. Thesis, North Carolina A&T State University, Greensboro, NC, USA, 2022.
25. Yu, F.; Zhang, Y.; Kong, C.; Yu, H. Microstructure and mechanical properties of Ti-6Al-4V alloy sheets via room-temperature rolling and cryorolling. *Mater. Sci. Eng. A* **2022**, *834*, 142600. [CrossRef]

crystals

MDPI

Article

Grain Refinement and Strengthening of an Aluminum Alloy Subjected to Severe Plastic Deformation through Equal-Channel Angular Pressing

Atef Korchef * and Imen Souid

College of Sciences, King Khalid University (KKU), P.O. Box 9004, Abha 61413, Saudi Arabia
* Correspondence: akorchef@kku.edu.sa

Abstract: In the present study, the microstructure, mechanical properties, and stored energy of an aluminum alloy containing iron-rich fine precipitates, subjected to severe plastic deformation through equal-channel angular pressing (ECAP), were investigated using X-ray diffraction, scanning electron microscopy, transmission electron microscopy, and atomic force microscopy. Up to four passes through ECAP resulted in significant nanometer-scale grain refinement, as well as the accumulation of lattice defects, such as dislocations and mesoscopic shear planes. This resulted in a noticeable enhancement in the Vickers microhardness and the flow stress after ECAP. Differential scanning calorimetry results showed that the ECAP'ed material exhibited two exothermal peaks at 222 ± 2 °C and 362 ± 2 °C, with total thermal effects of ΔH = 4.35 and 6.5 J/g, respectively. Slight increases in the ECAP'ed material microhardness and flow stress were observed at 200 °C. The heat release, at a relatively low temperature, and the slight improvement in the mechanical properties were attributed to the evolution of low- and high-angle misorientation, with the strain and the pinning of tangled dislocation caused by the existing fine particles. The second peak was attributed to grain growth, resulting in a significant softening of the material.

Keywords: ECAP; aluminum alloy; microstructure; mechanical properties; heat release; grain refinement

1. Introduction

In order to refine grains with moderate-to-high strain, numerous severe plastic deformation (SPD) processes have been utilized, including torsion under high pressure [1–3], ball milling [4,5], and equal-channel angular pressing (ECAP) [6–9]. During ECAP, the bulk of the sample undergoes approximately simple shear, and repetitive passes of the same sample can be achieved, since the sample cross-section remains unaltered after each pass. It was shown that the number of pressing cycles and the processing route strongly affect the microstructure and mechanical properties of ECAP'ed materials [10,11]. Metals and alloys deformed by ECAP showed heterogeneous microstructures, containing high-energy grain boundaries [12,13] or homogenous microstructures [14–16]. Indeed, the nature of the microstructure (homogeneous or heterogeneous) depends on the deformed material and the operating parameters of the ECAP. Nakashima et al. [17] conducted ECAP experiments on pure aluminum (99.99%) using protocol B (i.e., the sample was rotated by 90° between two successive passes through the die). They found that after the first pass, the microstructure consisted of elongated subgrain bands, approximately parallel to the upper and lower faces of the pressed sample. The subgrains were separated by low-angle boundaries. After two passes, the banded structure was less detectable, and the grains appeared more equiaxed. After four passes, the grains were equiaxed and separated by high-angle boundaries. The average grain size achieved was about 1.3 μm. Comparable results were found by Kawasaki et al. [18]. Indeed, they examined the microstructure of high-purity aluminum (99.99%) processed using ECAP through 1–12 passes. They found that the microstructure evolved from elongated subgrains to an essentially equiaxed array of ultrafine grains, and both the boundary misorientation angle and fraction of high-angle boundaries increased rapidly from 1 to 4 passes. From 4 to 12 passes, there was no measurable change in the average grain size. Optimum processing was achieved using ECAP through 4–8 passes. The resulting texture depends on the strain accumulated during ECAP and the strain path changes involved in the ECAP process. Skrotzki et al. [19] investigated the texture of aluminum alloy AA 5109, deformed using ECAP at room temperature through three passes, without rotation between two successive passes (so-called route A). They showed that the resulting texture was characterized by typical shear components of FCC metals that deviate from their ideal positions, and it depended on several factors, such as the starting texture, the distance from the top of the extruded billet, and changes from pass to pass.

ECAP, among other techniques of SPD, has been applied to bulk metals and alloys in order to intensively refine their structures and enhance their mechanical properties. High dislocation density and fine grains with sizes less than 100 nm can typically be achieved, and, more generally, grains with sizes in the submicrometer range were produced using ECAP [20,21]. As a result, the mechanical properties of ultrafine-grained (UFG) materials are significantly improved, compared to those of coarse-grained materials. Several UFG materials, including aluminum alloys, for instance, were found to exhibit superplastic behavior, enhanced flow stress, and high Vickers microhardness [22–24]. Mabuchi and Higashi [25] investigated the mechanical properties of a relatively low-purity Al (99.5%) and a high-purity Al (99.9999%) deformed using ECAP at room temperature. They showed that the tensile strength of Al (99.5%) was improved after ECAP, and this improvement depended on impurities and temperature. The tensile strength increased up to a total strain of 7.3, and then it remained constant for higher values. In addition, the flow stress decreased for a strain of 7.3 and remained unchanged for larger strains. For Al (99.9999%), the tensile strength was poorly improved after ECAP because of the dislocation restoration, which readily occurs below room temperature. Verlinden and Popović [26] studied the mechanical properties of free copper Al-4.4 wt% Mg and AA5182 containing 1.2 wt% Cu Alloys after four and eight ECAP passes at 200 °C. They showed that the addition of copper to the alloy increased the strength of the alloy by 20%. Post-ECAP annealing at 200 °C did not lead to further increases in hardness or strength. Indeed, the precipitates formed during ECAP at 200 °C did not directly contribute to the higher strength of the copper-rich alloy, but they slowed down the recovery.

When the temperature increases, UFG materials become softer in both their plastic and elastic regimes [27]. The changes in the elastic parameters were attributed to a change in the grain boundary, leading to important changes in the microhardness and yield stress [28]. Furthermore, UFG materials may display superplastic properties at relatively low temperatures, owing to the enhancement of grain boundary diffusion processes [29]. The models by Lifshitz [30] and Rachinger [31] describe deformation processes via grain boundary sliding (GBS) in conventional polycrystals under diffusion creep and dislocation creep, at high temperatures or superplastic conditions, respectively. Hahn et al. [28] proposed an alternative deformation model for UFG materials, based on GBS and the formation of mesoscopic shear planes. In this model, at a critical grain size, a transition from plastic deformation, dominated by dislocation, to GBS arises when the stress needed for the formation and motion of dislocation becomes higher than that needed for GBS. Consequently, the Hall–Petch relationship, which states that a material becomes harder when its grains become smaller, remains valid for some nanocrystalline materials [23,32], which is consistent with a high-angle grained structure; however, a negative slope (the inverse Hall–Petch relationship) was found in some cases of hardness measurements [33,34]. Different assumptions and models were proposed with which to explain this unusual behavior, such as the precipitation of a second phase in grain boundaries [28], a sliding of grain boundaries and formation of mesoscopic shear planes [32], and an extrinsic dislocation movement in the non-equilibrium grain boundaries when hardness measurements were taken [35].

Different techniques have been used to observe the microstructure evolution of UFG metals and alloys during the ECAP process, i.e., transmission electron microscopy (TEM), optical microscopy (OM), and X-ray diffraction (XRD). Atomic force microscopy (AFM) can be used with these techniques to give further experimental information about the microstructure and deformation mechanisms of UFG materials. However, it seems that only a few studies used AFM to investigate the microstructure of materials processed by ECAP [36,37]. AFM can be extremely useful to study the deformation mechanisms after mechanical tests of materials processed by ECAP. For example, using AFM, it was shown that the dominant deformation mechanism of nanostructured copper [38] was the dislocation-dominated plasticity and nano-void formation. In addition, AFM observations were carried out to determine the mechanism of fatigue damage in UFG copper processed by ECAP [39]. It was found that the strains initiated in UFG copper at the slip's edges caused the rotation of grains and the deformation in the adjacent nanometric grains. This resulted in the coarsening of the grains in the order of micrometers followed by the propagation of the slips. Also, an AFM investigation confirmed the transition in deformation mechanisms with decreasing strain rate and provide irrefutable evidence for the existence of GBS within the superplastic regime for Zn-22% Al eutectoid samples processed by ECAP at a temperature of 200 °C and pulled to strains in the range ~0.2–0.5 [40].

The present work aims to Investigate the microstructure and mechanical properties of aluminum that contains an iron-rich α-phase situated essentially at grain boundaries during ECAP. The material used in our work is low-purity recycled aluminum (99.1%). Unlike high-purity aluminum, it is cheap and easy to obtain at an industrial scale. Also, a major problem encountered when using UFG materials at high temperatures is the coarsening of their grains which negatively affected their mechanical properties. However, the presence of the α-phase in the aluminum matrix was expected to improve the thermal stability of the processed material. This is of utmost importance for industrial applications of the ECAP'ed aluminum alloy at relatively high temperatures. For this purpose, OM, SEM, TEM, and XRD were used for microstructure investigation. A complementary investigation on the microstructure, in particular the formation of mesoscopic shear planes, was carried out using AFM.

Tension and Vickers microhardness measurements were performed for the mechanical tests. The effect of annealing on the hardening of the alloy was then investigated using Vickers microhardness and tensile tests. Stored energy measurements were performed using differential scanning calorimetry (DSC).

2. Materials and Methods

Experiments were performed on an aluminum alloy containing 0.2 wt% Fe, 0.1 wt% Si, 0.17 wt% Mg, 0.15 wt% Zn, and 0.07% Mn, and Al was the balance. The aluminum alloy, received in the form of cast ingots, was annealed at 500 °C for 24 h and then quenched in iced water. After annealing, the microstructure of the alloy consisted of large grains (~80 μm in size) (Figure 1a). Thereafter, samples with a square section (10 × 10 mm^2) and 70 mm in length were cut for ECAP. Prior to ECAP, the samples were coated with lithium-based lubricant. The aluminum matrix contained fine and coarse precipitates (with a volume fraction of ~1%) most of which were situated at grain boundaries (Figure 1b). Analysis via X-ray emission spectrometry indicated that these precipitates contain Al, Fe, and Si (Figure 1c). The possible chemistry of this phase, usually denoted as α-AlFeSi, is $Al_{12}Fe_3Si$ or $Al_{15}Fe_3Si_2$.

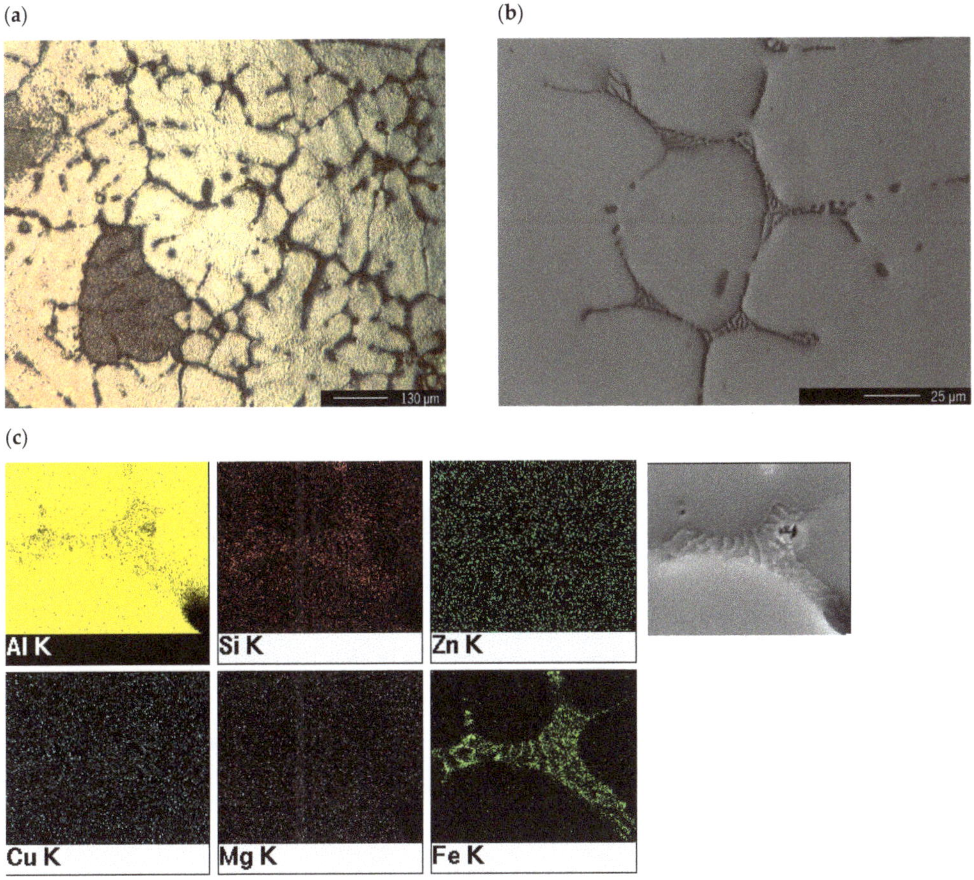

Figure 1. (**a**) Optical micrograph under polarized light of the nondeformed (N = 0) sample. (**b**) SEM micrograph and (**c**) X-ray emission spectrometry via SEM of the fine precipitates in the aluminum matrix.

The surfaces of the samples were ground mechanically and polished electrolytically for 30 min at temperatures less than 10 °C under a 30 V operating voltage in a solution of 66% $(CH_3CO)_2O$ and 34% $HClO_4$. Samples were subjected to high strains through ECAP. The ECAP die used in the present work consisted of two square channels with a cross-sectional area of 10×10 mm^2 each, which met at an angle Ø = 90°. An angle ψ = 20° was delineated by the curvature arc at the outer point of the intersection of the two square channels (Figure 2). Since the cross-section of the sample remained the same after each pass through the die during the ECAP procedure, it was possible to repeatedly press the same sample to attain high strains. Using the formula provided by Iwahashi et al. [41], the sample was subjected to an equivalent strain, ε_N, equal to ~N after N passes through the die. In the present study, the same sample was repeatedly pressed in the same direction after each pass (the so-called route A) through the die to reach an equivalent strain of up to ~4.

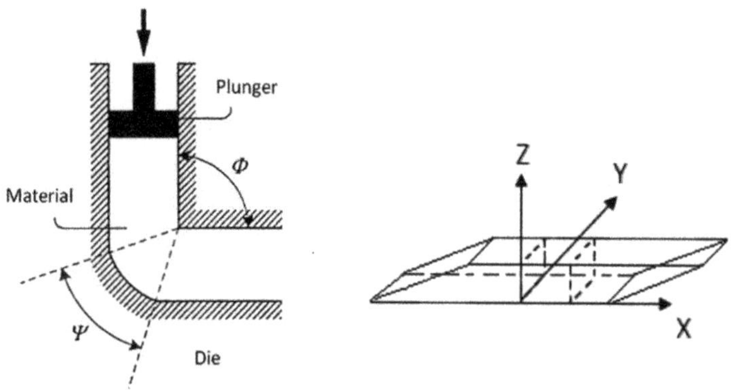

Figure 2. Schematic illustrations of the ECAP die (**left**) and the sample shape after ECAP (**right**). The *X*-axis is the pressing direction.

A Philips X'PERT PRO diffractometer (Malvern Panalytical, Malvern, United Kingdom) was used to perform XRD patterns in step scanning mode with the radiation Co-Kα. The XRD patterns were recorded at room temperature in the scanning range 2θ between 30 and 155° at a scan rate of 2×10^{-3} deg/s. The APD 1700 software (Philips, Amsterdam, Netherlands) was used to determine the full width at half maximum (FWHM) of each reflection. The following expression was used to obtain a correction for the instrumental effects' contribution to the experimental FWHM, ßexp, for each reflection:

$$ß = ([(ß^2)_{exp} - (ß^2)_{ins}]^{1/2} \times [ß_{exp} - ß_{ins}])^{1/2} \quad (1)$$

where $ß_{ins}$ is the instrumental FWHM determined using the nondeformed (N = 0) aluminum alloy as standard reference material. The Halder–Wagner (HW) method, which assumes a Voigt peak shape, was used to calculate the microstrains (lattice distortions) and the crystallite size of the ECAP'ed samples [42]:

$$(\beta^*/d^*)^2 = 1/D[\beta^*/(d^*)^2] + (\varepsilon/2)^2 \quad (2)$$

where $\beta^* = \beta \cos\theta/\lambda$ and $d^* = 2 \sin\theta/\lambda$; θ is the Bragg angle, and λ is the wavelength used. From Equation (2), the slope gives the crystallite size, and the intercept of the plot of $(\beta^*/d^*)^2$ versus $\beta^*/(d^*)^2$ gives the value of the microstrains.

Dislocations are the main defects in SPD samples, and considering that the nondeformed sample was submitted to a prolonged annealing at the high temperature of 500 °C, one can assume that the dislocation density of the nondeformed sample, ρ_i, was significantly lower than the dislocation density,]ρ, accumulated after ECAP ($\rho_i << \rho$). In that case, ρ can be estimated by the following equation [43]:

$$\rho = \frac{1}{D\left(y + \frac{b}{\varepsilon_N}\right)} \quad (3)$$

In this equation, D is the crystallite size calculated from the reflection broadening, ε_N = N is the strain introduced by ECAP, b is the magnitude of the Burgers vector of dislocation ($b = a/\sqrt{2}$ for FCC aluminum alloys, where a is the aluminum lattice parameter), and y is the annihilation distance of dislocation (y = 2 nm).

The surface topographies of samples deformed by ECAP were observed using a scanning electron microscope operating at 25 kV and a Digital instruments Nanoscope III atomic force microscope operating in the tapping mode.

A DURIMET microhardness tester, with a diamond pyramidal indenter, was used for the Vickers microhardness, H_V, measurements. Eight measurements were taken with a load of 100 g applied at randomly selected points for 15 s. Cylindric specimens with dimensions of 15 mm and 3 mm for the gage length and gage diameters, respectively, were cut along the X-axis for tensile tests. Tensile tests were performed at room temperature, using an Instron testing machine, at a crosshead displacement speed of 0.05 mm.min^{-1} (equivalent to an initial strain rate of 5.5×10^{-5} s^{-1}).

In order to investigate a possible recovery, four-pass (N = 4) specimens were annealed for 1 h at temperatures between 25 and 400 °C. Then, XRD analysis, Vickers microhardness measurements, and tensile tests were performed on each sample.

A Perkin Elmer calorimeter was used for differential scanning calorimetry (DSC) measurements. The material was heated at a rate of 40 °C·min^{-1} in closed copper crucibles. Three heating runs on the same sample were accomplished under an argon atmosphere from room temperature up to 600 °C. The two last runs were used to determine the baseline. The DSC experiments were performed on three samples in the same conditions. The obtained curves were found to be reproducible.

3. Results

3.1. Microstructure and Mechanical Properties

Figure 3a shows the XRD patterns of the nondeformed sample (N = 0) and ECAP'ed samples (N = 1 to 4). The 222 and 400 reflections of the nondeformed material (N = 0) were not detected, thus indicating a texture of the nondeformed sample. The texture was partially removed after ECAP due to the strain path changes involved in the ECAP process. After ECAP, a pronounced broadening of the X-ray reflections was observed (Figure 3b). This broadening was attributed to the grain refinement and lattice distortions of the ECAP'ed samples. In addition, the broadening of the X-ray reflections made the Kα2 reflection scarcely visible after ECAP. This reflection was well resolved for the nondeformed sample.

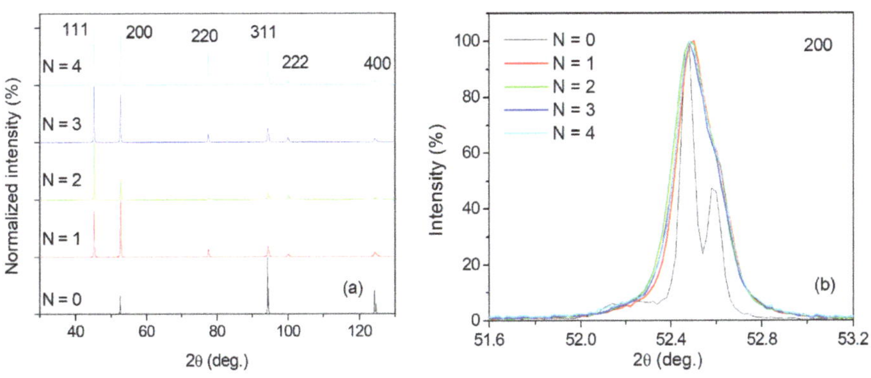

Figure 3. (**a**) XRD patterns and (**b**) the 200 reflection profile of the nondeformed (N = 0) and ECAP'ed (N = 1–4) aluminum samples.

The intensive grain refinement and accumulation of lattice defects such as dislocations during ECAP resulted in a significant strengthening of the deformed alloy, i.e., an increase in the flow stress (Figure 4), and Vickers microhardness was depicted after the ECAP process. The values of the cell size, microstrains, dislocation density, 0.2% flow stress, and Vickers microhardness obtained after the ECAP are reported in Table 1. Significant refinement of the grains was depicted after SPD through the ECAP process. Indeed, the crystallite size attained ~190 nm after the first pass and further decreased to reach ~150 nm after four passes. The lattice distortions continuously increased with the number of passes of the ECAP process. For example, the value of microstrains obtained for N = 1 was 0.08% and increased to 0.15% after four passes.

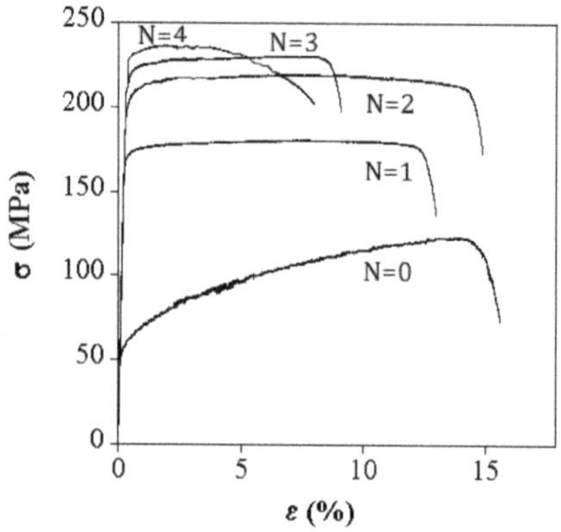

Figure 4. True stress–strain curves of the nondeformed (N = 0) and ECAP'ed (N = 1–4) samples.

Table 1. Crystallite sizes, microstrains, dislocation densities, Vickers microhardness, and 0.2% flow stress values of the aluminum alloy after ECAP.

Number of ECAP Passes N	Crystallite Size (nm)	Microstrains (%)	Dislocations Density ($\times 10^{15}$ m^{-2})	Vickers Microhardness (H_V)	$\sigma_{0.2}$ (MPa)
0	8×10^4	-	-	33	110
1	190	0.08	1.6	60	190
2	185	0.09	2.5	63	210
3	165	0.12	2.6	65	230
4	150	0.15	2.8	76	240

In addition, the ECAP process introduced an increment of dislocation density. Thus, the undeformed (N = 0) aluminum sample should contain a significantly lower dislocation density than the ECAP'ed samples since it was submitted to a prolonged annealing treatment before deformation. It is seen from Table 1 that the highest dislocation density (2.8×10^{15} m^{-2}) and the lowest crystallite size (150 nm) were obtained after four passes. TEM analysis facilitated the obtaining of direct evidence of grain refinement and the accumulation of dislocations after ECAP (Figure 5). The Vickers microhardness reached after one pass (N = 1) became twice as high as the nondeformed (N = 0) sample, and a significant increase in the flow stress at 0.2% was also depicted after the first pass. The highest values were obtained after four passes (Table 1). Ghosh et al. [44] studied the mechanical properties of Al–Zn–Mg–Cu alloys through ECAP after optimizing the outer corner angles through finite element modeling. They depicted an improvement in the hardness and yield strength, by 22% and 18%, respectively, after ECAP. This improvement was attributed to grain refinement and precipitate hardening. In the present work, the improvement of the mechanical properties (H_v and $\sigma_{0.2}$) was by more than 200% after the fourth pass. Therefore, grain refinement was strongly affected by the strain accumulated during the ECAP process. Three stages are typically involved in the intense refinement during SPD [45]: (i) the dislocations are concentrated in high-density shear bands, where the deformation is localized; (ii) these dislocations annihilate at a certain level of strain and rearrange to form small-angle grain boundaries separating the individual crystallites; (iii) the orientation of the crystallites becomes completely random with respect to their neighboring, and a deformation steady state is reached when the rate of the dislocation generation becomes equal to the rate of annihilation. According to the theory of dynamic recovery, internal stresses are reduced during mechanical deformation while also producing more internal stresses (i.e., dislocations). The density of dislocations in a material affects the kinetics of dislocation recovery. That is, the annihilation rate of dislocation increases at higher strains, which results in smaller crystallite sizes. This would explain the high strains and decreased crystallite sizes obtained in the present work after four passes.

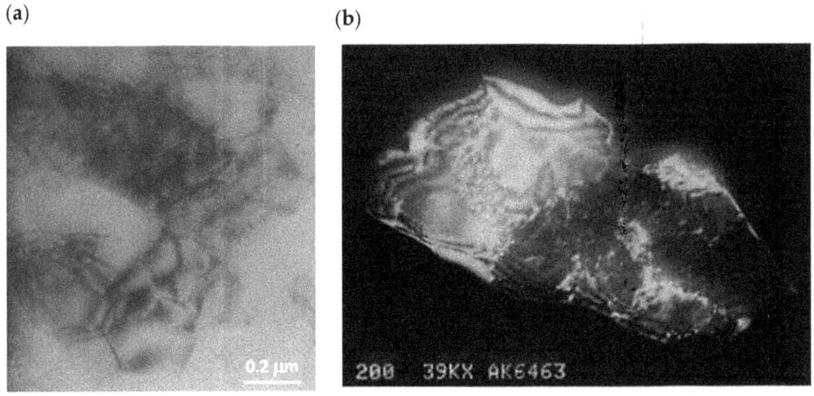

Figure 5. TEM images of the aluminum sample processed by ECAP (N = 4) showing the (**a**) grain refinement and (**b**) accumulation of dislocations after ECAP.

Typical 2D and 3D AFM images of the microstructure of the aluminum samples deformed by ECAP (N = 4) using route A are shown in Figure 6a,b, respectively. The microstructure was relatively homogeneous, and the grain size obtained was ~150 nm in size. This value was slightly higher than that obtained by XRD analysis for the same sample. During SPD, sub-grain boundaries evolve in the grain structures [46], developing coherent crystalline domains [47,48]. The distribution of dislocations may affect the size of these domains. XRD allows the sizes of these coherent domains (crystallite sizes) to be determined. However, AFM (or SEM and TEM) investigations give the grain size distribution from direct observation. This explains the difference in grain size values determined by XRD and AFM. Also, the AFM images reveal the formation of shear bands during the ECAP process. These mesoscopic bands, labeled MB in Figure 6a, are aligned with the shearing plane generated by the ECAP. Similar results were reported for ECAP'ed pure gold [36]. In addition, an examination of the AFM images (Figure 6b) revealed the development of mesoscopic grain boundary sliding (GBS) as proposed by the model of Hahn et al. [28]. In this model, planar interfaces are formed by the migration of grain boundaries. In nanostructured materials, the grain boundary width constitutes a significant fraction of the overall volume of the material, and this favors the formation of planar interfaces [40]. After that, the localized sliding shears lead to macroscopic sliding over dimensions that are significantly greater than the grain size and lead to enhanced strains and superplastic deformation. This is clearly supported by the present AFM images. The occurrence of mesoscopic GBS was established for UFG Cu and Ni with grain sizes of ~220 nm [49], pure gold with a grain size of ~500 nm [36], and pure aluminum with a grain size of 1.3 µm [40] processed by ECAP. However, mesoscopic GBS did not form in a two-phase 2% Al alloy because of the difficulties to achieve local rearrangements of the grain boundaries when two separate phases were present [40]. In the present work, the α-phase, which constitutes a low-volume fraction of the material, apparently did not affect the mesoscopic GBS formation.

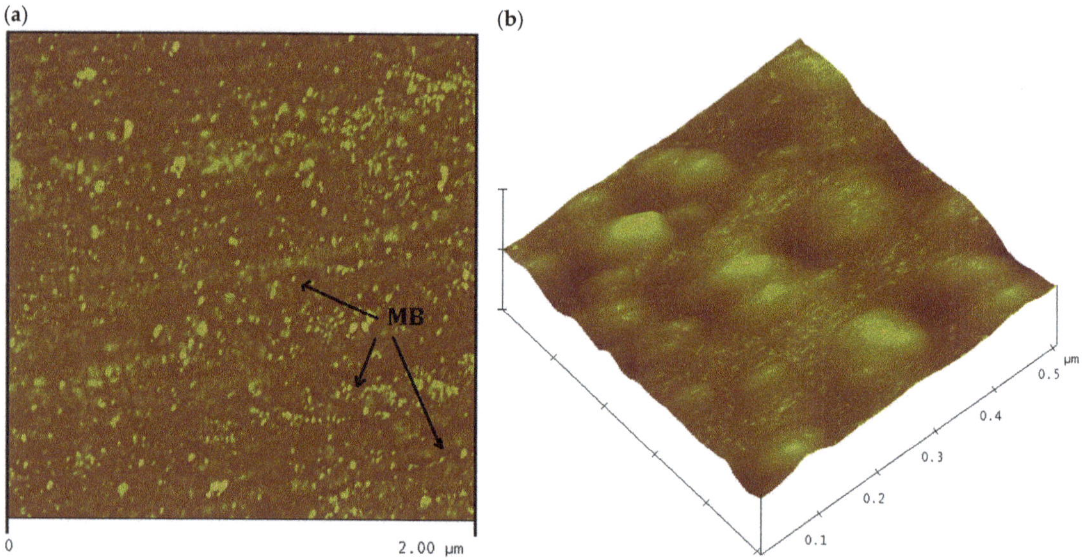

Figure 6. Typical (**a**) 2D and (**b**) 3D AFM images for the aluminum sample processed by ECAP (N = 4). MB: mesoscopic bands.

3.2. Annealing Effect

DSC results of the ECAP'ed (N = 4) material showed main two exothermal peaks at 222 ± 2 °C and 362 ± 2 °C with a total thermal effect of ΔH = 4.35 and 6.5 J/g, respectively (Figure 7). It should be noted that no transformation was detected when two successive heating runs were repeated on the same sample. This indicates irreversible transformations. Other than this, the two peaks were not observed or scarcely visible for the ECAP'ed samples up to N = 3 (results not shown herein). The values of stored energy were lower than those obtained in a cryogenic ball-milled Al-7.6 at% Mg alloy in which recovery occurred during anneals at temperatures between 100 and 230 °C, and recrystallization proceeded at higher temperatures to 370 °C [50]. A slight increase in the microhardness, H_V, and 0.2% flow stress, $\sigma_{0.2}$, of the ECAP'ed material was observed after annealing at 200 °C (Table 2). This might be explained by several microstructure features: (i) Considering the properties of coarse-grained aluminum, the thermal treatment and enhanced diffusivity facilitate the diffusion of impurities toward strained regions around dislocations. The formation of the Cottrell atmosphere decreases the dislocation's mobility, leading to higher Vickers microhardness and flow stress values. (ii) Considering the special behavior of ECAP'ed materials [47], the strengthening of the material after annealing at 200 °C can be attributed to the change of the dislocation cell walls structure from a highly disordered structure to low-angle grain boundaries more resistant than disordered walls. (iii) The pinning of tangled dislocation by the existing fine particles (Figure 8) may result in the improvement of the Vickers microhardness and flow stress.

The peak at 362 °C is in the range of recrystallization temperatures of Al-based alloys [51,52]. In a previous work [53], we demonstrated that after annealing of 1h at a temperature of 400 °C, the ECAP'ed material exhibited a fully recrystallized microstructure consisting of grains ~40 µm in size. This explains the decrease in the ECAP'ed material microhardness H_V and 0.2% flow stress, $\sigma_{0.2}$, at temperatures above 200 °C (Table 2). It should be noted that the microhardness obtained after annealing for 1 h at 400 °C was slightly greater than that of the nondeformed material (N = 0). This was attributed to the lower grain size and redistribution of the α-phase in the aluminum matrix after ECAP, which create a recrystallized structure that differs from that of the nondeformed material.

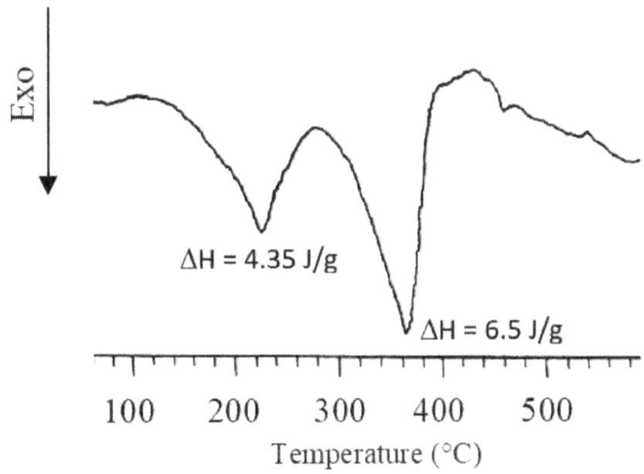

Figure 7. DSC curve of the ECAP'ed (N = 4) aluminum sample.

Table 2. Vickers microhardness and 0.2% flow stress values of the ECAP'ed (N = 1) aluminum alloy at different annealing temperatures.

Temperature (°C)	Vickers Microhardness (H_V)	$\sigma_{0.2}$ (MPa)
25	60	190
200	80	215
370	40	72

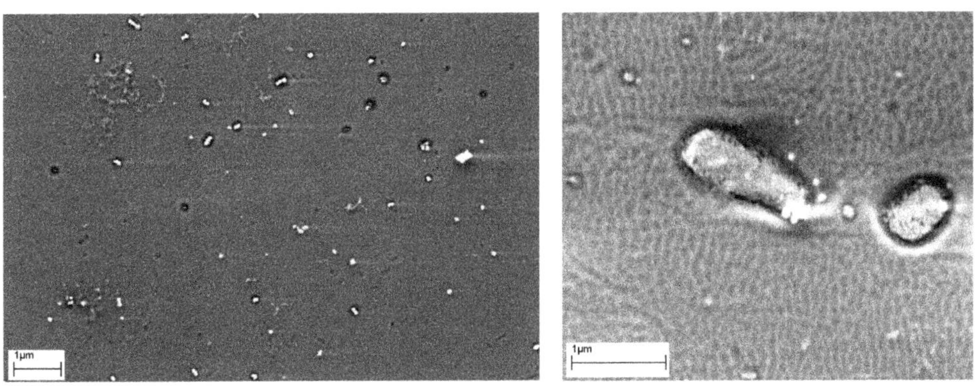

Figure 8. SEM micrographs of the fine precipitates in the ECAP'ed aluminum sample (N = 1). Precipitates can pin the tangled dislocations.

4. Conclusions

XRD, OM, SEM, TEM, and AFM were used to investigate the microstructure of an aluminum alloy containing an iron-rich α-phase deformed by ECAP. Intensive grain refinement and accumulation of strains that resulted in a significant strengthening of the aluminum alloy were observed after ECAP. After four passes, the size of the grains and the dislocation density attained ~150 nm and 2.8×10^{15} m^{-2}, respectively. AFM provided direct evidence of the formation of mesoscopic grain boundaries sliding during the ECAP process. A significant increase in the Vickers microhardness and 0.2% flow stress was depicted for the ECAP'ed samples. DSC results showed that the ECAP'ed material (N = 4) exhibited two exothermal peaks at 222 ± 2 °C and 362 ± 2 °C with a total thermal effect of ΔH = 4.35 and 6.5 J/g, respectively. The heat release at the relatively low temperature was attributed to microstructural features such as the diffusion of impurities towards the strain region around dislocation lines, the change in the dislocation cell wall structure from high disordered structure to low angle grain boundaries, and the pinning of tangled dislocations by the existing fine particles. As a result, a slight increase in the microhardness H_V and 0.2% flow stress of the aluminum alloy was observed after annealing at 200 °C. This is of utmost importance for industrial applications of aluminum alloys at relatively high temperatures since ECAP'ed materials are generally confronted with intensive softening after annealing. The second peak was attributed to grain growth. Indeed, a significant softening was observed by annealing at temperatures higher than 200 °C, which was attributed to grain growth.

Author Contributions: A.K. was involved in the methodology, experimental investigation, supervision, and writing and editing of the work. I.S. was involved in ensuring resources, data curation, and formal analysis of the work. All authors have read and agreed to the published version of the manuscript.

Funding: This research was funded by the Deanship of Scientific Research at King Khalid University, Saudi Arabia, grant number GRP/87/44.

Data Availability Statement: Not available.

Acknowledgments: The authors extend their appreciation to the Deanship of Scientific Research at King Khalid University, Saudi Arabia, for funding this work through the General Research Project under grant number GRP/87/44.

Conflicts of Interest: The authors declare no conflict of interest.

References

1. Wetscher, F.; Vorhauer, A.; Pippan, R. Strain hardening during high pressure torsion deformation. *Mater. Sci. Eng. A* **2005**, *410–411*, 213–216. [CrossRef]
2. Kawasaki, M.; Han, J.K.; Lee, D.H.; Jang, J.; Langdon, T.G. Fabrication of nanocomposites through diffusion bonding under high-pressure torsion. *J. Mater. Res.* **2018**, *33*, 2700–2710. [CrossRef]
3. Kunčická, L.; Jambor, M.; Král, P. High pressure torsion of copper; effect of processing temperature on structural features, microhardness and electric conductivity. *Materials* **2023**, *16*, 2738. [CrossRef]
4. Charfeddine, S.; Zehani, K.; Bessais, L.; Korchef, A. Microstructure characterization of an aluminium alloy processed by milling followed by spark plasma sintering. *Crys. Res. Technol.* **2018**, *53*, 1700137. [CrossRef]
5. Daniyan, I.A.; Tlhabadira, I.; Mpofu, K.; Adeodu, A.O. Process design and optimization for the milling operation of aluminum alloy (AA6063 T6). Mater. *Today Proc.* **2021**, *38*, 536–543. [CrossRef]
6. Segal, V.M.; Reznikov, V.I.; Drobyshevskii, A.E.; Kopylov, V.I. Plastic metal working by simple shear. *Izv. Akad. Nauk. SSSR Met.* **1981**, *1*, 115–123. [CrossRef]
7. Bednarczyk, W.; Wątroba, M.; Kawałko, J.; Bała, P. Can zinc alloys be strengthened by grain refinement? A critical evaluation of the processing of low-alloyed binary zinc alloys using ECAP. *Mater. Sci. Eng. A* **2019**, *748*, 357–366. [CrossRef]
8. Volokitina, I.; Bychkov, A.; Volokitin, A.; Kolesnikov, A. Natural aging of aluminum alloy 2024 after severe plastic deformation. *Metallogr. Microstruct. Anal.* **2023**, *12*, 564–566. [CrossRef]
9. Sagar, K.G.; Suresh, P.M.; Sampathkumaran, P. Tribological studies on aluminum beryl composites subjected to ECAP process. *Wear* **2023**, *523*, 204775. [CrossRef]
10. El-Garaihy, W.H.; BaQais, A.; Alateyah, A.I.; Alsharekh, M.F.; Alawad, M.O.; Shaban, M.; Alsunaydih, F.N.; Kamel, M. The impact of ECAP parameters on the structural and mechanical behavior of pure Mg: A combination of experimental and machine learning approaches. *Appl. Sci.* **2023**, *13*, 6279. [CrossRef]
11. Shin, D.H.; Park, K.T. Ultrafine grained steels processed by equal channel angular pressing. *Mater. Sci. Eng. A* **2005**, *410–411*, 299–302. [CrossRef]
12. Bodyakova, A.; Tkachev, M.; Pilipenko, A.; Belyakov, A.; Kaibyshev, R. Effect of deformation methods on microstructure, texture, and properties of a Cu–Mg alloy. *Mater. Sci. Eng. A* **2023**, *876*, 145126. [CrossRef]
13. Zhang, L.; Liu, C.Y.; Xie, H.Y. Hall-Petch relation and grain boundary slipping in Al–Mg–Sc alloys with fine equiaxed grain structure. *Mater. Charact.* **2022**, *194*, 112472. [CrossRef]
14. Terhune, S.D.; Swicher, D.L.; Ishi, K.O.; Horita, Z.; Langdon, T.G.; McNelly, T.R. An investigation of microstructure and grain-boundary evolution during ECA pressing of pure aluminum. *Metall. Mater. Trans. A* **2002**, *33*, 2173. [CrossRef]
15. Tang, L.; Peng, X.; Huang, J.; Ma, A.; Deng, Y.; Xu, G. Microstructure and mechanical properties of severely deformed Al–Mg–Sc–Zr alloy and their evolution during annealing. *Mater Sci. Eng. A* **2019**, *754*, 295–308. [CrossRef]
16. Damavandi, E.; Nourouzi, S.; Rabiee, S.M.; Jamaati, R. Effect of ECAP on microstructure and tensile properties of A390 aluminum alloy. *Trans. Nonferrous Met. Soc. China* **2019**, *29*, 931–940. [CrossRef]

17. Nakashima, K.; Horita, Z.; Nemoto, M.; Langdon, T.G. Influence of channel angle on the development of ultrafine grains in equal-channel angular pressing. *Acta Mater.* **1998**, *46*, 1589–1599. [CrossRef]
18. Kawasaki, M.; Horita, Z.; Nemoto, M.; Langdon, T.G. Microstructural evolution in high purity aluminum processed by ECAP. *Mater. Sci. Eng. A* **2009**, *524*, 143–150. [CrossRef]
19. Skrotzki, W.; Scheerbaum, N.; Oertel, C.G.; Brokmeier, H.G.; Suwas, S.; Tóth, L.S. Texture Formation during ECAP of Aluminum Alloy AA 5109. Mater. *Sci. Forum* **2006**, *503–504*, 99–106. [CrossRef]
20. Xue, D.; Wei, W.; Wen, S.; Wu, X.; Shi, W.; Zhou, X.; Gao, K.; Huang, H.; Nie, Z. Microstructural evolution of Al-Mg-Er-Zr alloy by equal channel angular extrusion at room temperature. *Mater. Lett.* **2023**, *3341*, 133759. [CrossRef]
21. Sitdikov, O.; Avtokratova, E.; Markushev, M.; Sakai, T. Structural characterization of binary Al-Cu alloy processed by equal channel angular pressing at half of the melting point. *Metall. Mater. Trans. A* **2023**, *54*, 505–525. [CrossRef]
22. Ghosh, A.; Ghosh, M.; Gudimetla, K.; Kalsar, R.; Kestens, L.A.; Kondaveeti, C.S.; Singh Pugazhendhi, B.; Ravisankar, B. Development of ultrafine grained Al–Zn–Mg–Cu alloy by equal channel angular pressing: Microstructure, texture and mechanical properties. *Arch. Civ. Mech. Eng.* **2020**, *20*, 7. [CrossRef]
23. Huang, S.J.; Subramani, M.; Borodianskiy, K.; Immanuel, P.N.; Chiang, C.C. Effect of equal channel angular pressing on the mechanical properties of homogenized hybrid AZ61 magnesium composites. *Mater. Today Commun.* **2023**, *34*, 104974. [CrossRef]
24. Dyakonov, G.S.; Raab, G.I.; Pesin, M.V.; Polyakov, A.V.; Semenova, I.P.; Valiev, R.Z. Superplastic-like behavior and enhanced strength of a two-phase titanium alloy with ultrafine grains. *Adv. Eng. Mater.* **2022**, *24*, 2101592. [CrossRef]
25. Mabuchi, M.; Higashi, K. Mechanical properties of pure aluminium processed by equal channel angular extrusion. *J. Mater. Sci. Lett.* **1998**, *17*, 215–217. [CrossRef]
26. Verlinden, B.; Popović, M. Influence of Cu on the Mechanical Properties of an Al-4.4 wt% Mg Alloy after ECAP. *Mater. Sci. Forum* **2006**, *503–504*, 107–112. [CrossRef]
27. Schiøtz, J.; Vegge, T.; Di Tolla, F.D.; Jacobsen, K.W. Atomic-scale simulations of the mechanical deformation of nanocrystalline metals. *Phys. Rev. B* **1999**, *60*, 11971–11983. [CrossRef]
28. Hahn, H.; Mondal, P.; Padmanabhan, K.A. Plastic deformation of nanocrystalline materials. *J. Nanostruct.* **1997**, *9*, 603–606. [CrossRef]
29. Valiev, R.Z.; Krasilnikov, N.A.; Tsenev, N.K. Plastic deformation of alloys with submicron-grained structure. *Mater. Sci. Eng. A* **1991**, *137*, 35–40. [CrossRef]
30. Lifshitz, I.M. On the theory of diffusion-viscous flow of polycrystalline bodies. *Sov. Phys. Jetp-Ussr* **1963**, *17*, 909–920.
31. Rachinger, W.A. Relative grain translations in the plastic flow of aluminium. *J. Inst. Met.* **1952**, *81*, 33.
32. Furukawa, M.; Horita, Z.; Nemoto, M.; Langdon, T.G. The significance of the Hall-Petch relationship in ultra-fine grained materials: Matériaux à grains ultra-fins produits par hypercorroyage. In *Annales de Chimie*; Lavoisier: Cachan, France, 1996; Volume 21, pp. 493–502.
33. Chokshi, A.H.; Rosen, A.; Karch, J.; Gleiter, H. On the validity of the hall-petch relationship in nanocrystalline materials. *Scr. Metall.* **1989**, *23*, 1679–1683. [CrossRef]
34. Nieh, T.G.; Wadsworth, J. Hall-petch relation in nanocrystalline solids. *Scr. Metall. Mater.* **1991**, *25*, 955–958. [CrossRef]
35. Valiev, R.Z.; Korznikov, A.V.; Mulyukov, R.R. Structure and properties of ultrafine-grained materials produced by severe plastic deformation. *Mater. Sci. Eng. A* **1993**, *168*, 141–148. [CrossRef]
36. Gubicza, J.; Chinh, N.Q.; Szommer, P.; Vinogradov, A.; Langdon, T.G. Microstructural characteristics of pure gold processed by equal-channel angular pressing. *Scr. Mater.* **2007**, *56*, 947–950. [CrossRef]
37. Alipour, S.; Vafaeenezhad, H.; Fesahat, M.; Yazdi, A.; Mousavi-Khoshdel, S.M.; Soltanieh, M. Electrochemical behavior of ECAP-processed Sn–5Sb alloy. *J. Mater. Res. Technol.* **2023**, *23*, 5193–5211. [CrossRef]
38. Hussainova, I.; Kommel, L.; Lohmus, R.; Volobujeva, O. Microscopic characterization of surface morphology of nanostructured copper. *Rev. Adv. Mater. Sci.* **2005**, *10*, 266–271.
39. Kimura, H.; Kojima, Y.; Akiniwa, Y.; Tanaka, K.; Ishida, T. Fatigue damage mechanism of nanocrystals in ECAP-processed copper investigated by EBSD and AFM hybrid methods. *Key Eng. Mater.* **2007**, *340*, 943–948. [CrossRef]
40. Huang, Y.; Langdon, T.G. Characterization of deformation processes in a Zn-22% Al alloy using atomic force microscopy. *J. Mater. Sci.* **2002**, *37*, 4993–4998. [CrossRef]
41. Iwahashi, Y.; Wang, J.; Horita, Z.; Nemoto, M.; Langdon, T.G. Principle of equal-channel angular pressing for the processing of ultra-fine grained materials. *Scr. Mater.* **1996**, *35*, 143–146. [CrossRef]
42. Langford, J.I. National Institute of Standards and Technology Special Publication 846. In Proceedings of the International Conference Accuracy in Powder Diffraction II, NIST, Gaithersburg, MD, USA, 26–29 May 1992; pp. 241–244.
43. Rebhi, A.; Makhlouf, T.; Njah, N.; Champion, Y.; Couzinié, J.P. Characterization of aluminium processed by equal angular extrusion: Effect of processing route. *Mater. Charact.* **2009**, *60*, 1489–1495. [CrossRef]
44. Ghosh, A.; Das, K.; Eivani, A.R.; Mohammadi, H.; Vafaeenezhad, H.; Murmu, U.K.; Jafarian, H.R.; Ghosh, M. Development of mechanical properties and microstructure for Al–Zn–Mg–Cu alloys through ECAP after optimizing the outer corner angles through FE modeling. *Arch. Civ. Mech. Eng.* **2023**, *23*, 78. [CrossRef]
45. Bate, P.S.; Humphreys, F.J.; Ridley, N.; Zhang, B. Microstructure and texture evolution in the tension of superplastic Al–6Cu–0.4 Zr. *Acta Mater.* **2005**, *53*, 3059–3069. [CrossRef]

46. Ito, Y.; Horita, Z. Microstructural evolution in pure aluminum processed by high-pressure torsion. *Mater. Sci. Eng.* **2009**, *503*, 32–36. [CrossRef]
47. Valiev, R.Z.; Islamagaliev, R.K.; Alexandrov, I.V. Bulk nanostructured materials from severe plastic deformation. *Prog. Mater. Sci.* **2000**, *45*, 103–189. [CrossRef]
48. Zangiabadi, A.; Kazeminezhad, M. Development of a novel severe plastic deformation method for tubular materials: Tube Channel Pressing (TCP). *Mater. Sci. Eng.* **2011**, *528*, 5066–5072. [CrossRef]
49. Vinogradov, A.; Hashimoto, S.; Patlan, V.; Kitagawa, K. Atomic force microscopic study on surface morphology of ultra-fine grained materials after tensile testing. *Mater. Sci. Eng. A* **2001**, *319*, 862–866. [CrossRef]
50. Zhou, F.; Liao, X.Z.; Zhu, Y.T.; Dallek, S.; Lavernia, E.J. Microstructural evolution during recovery and recrystallization of a nanocrystalline Al-Mg alloy prepared by cryogenic ball milling. *Acta Mater.* **2003**, *51*, 2777–2791. [CrossRef]
51. Zaidi, M.A.; Sheppard, T. Effect of high-temperature soak and cooling rate on recrystallization behaviour of two Al-Mg alloys (AA 5252 and AA 5454). *Met. Technol.* **1984**, *11*, 313–319. [CrossRef]
52. Koken, E.; Embury, J.D.; Ramachandran, T.R.; Malis, T. Recrystallization at shear bands in Al-Mg. *Scr. Metall.* **1988**, *22*, 99–103. [CrossRef]
53. Korchef, A.; Champion, Y.; Njah, N. X-ray diffraction analysis of aluminium containing Al_8Fe_2Si precipitates processed by equal channel angular pressing. *J. Alloys Compd.* **2007**, *427*, 176–182. [CrossRef]

crystals

MDPI

Article

Effect of Al-Cu-Fe Quasicrystal Particles on the Reinforcement of a Polymer–Matrix Composite: From Surface to Mechanical Properties

Monika Kušter [1,2,*], Zoran Samardžija [1], Matej Komelj [1], Miroslav Huskić [3], Marko Bek [4], Gaël Pierson [5], Richard Kouitat-Njiwa [5], Jean-Marie Dubois [1,2] and Sašo Šturm [1,2,6,*]

1 Jožef Stefan Institute, Department for Nanostructured Materials, Jamova cesta 39, 1000 Ljubljana, Slovenia; zoran.samardzija@ijs.si (Z.S.)
2 International Postgraduate School Jožef Stefan, Jamova cesta 39, 1000 Ljubljana, Slovenia
3 Faculty of Polymer Technology, Slovenj Gradec, Ozare 19, 2380 Slovenj Gradec, Slovenia; miroslav.huskic@ftpo.eu
4 Department of Industrial and Material Science, Chalmers University of Technology, Rännvägen 2A, SE-412 96 Gothenburg, Sweden; marko.bek@chalmers.se
5 Institut Jean Lamour, Université de Lorraine, 2 allée André Guinier, 54011 Nancy, France; richard.kouitat@univ-lorraine.fr (R.K.-N.)
6 Department of Geology, Faculty of Natural Sciences and Engineering, University of Ljubljana, Aškerčeva Cesta 12, 1000 Ljubljana, Slovenia
* Correspondence: monika.kuster@ijs.si (M.K.); saso.sturm@ijs.si (S.Š.)

Abstract: We examined the effect of $Al_{59}Cu_{25}Fe_{13}B_3$ (at.%) quasicrystalline (QC) reinforcement particles on the mechanical and surface properties of a polymer-matrix composite by applying a technical polymer polyphthalamide (PPA). The observed increase in the tensile Young's modulus ranged from 1810 MPa for the pure polymer to 4114 MPa for the composite with a QC filling of 35 vol.%. The elongation at fracture decreased with the filling fraction, being equal to 16.9% for a pure polymer and dropping to 4.8% for the composite with a QC filling of 35 vol.%. The same trend was noticeable with flexural Young's modulus, which ranged from 100 MPa for a pure polymer to 125.5 MPa for the composite with 35 vol.% of QC. It was found that the increase in the mechanical strength led to a simultaneous increase of brittleness, which was reflected in a decrease of the impact strength for a pure polymer from 98.5 kJ/m^2 to 42.4 kJ/m^2 for composites with a QC filling of 35 vol.%. In contrast, when filled with 5 vol.% of QC, the impact strength increased by 8%. The friction coefficient against 100C6 steel dropped from 0.15 for pure PPA down to 0.10 for 5 vol.% of the QC filling, followed by an increase to 0.26 for further QC fillings up to 35 vol.%. Interestingly, a local minimum of friction was achieved at filling factors between 5 to 20 vol.% of QC. Independently, a clear *surfenergy* minimum was also found for the composite material with 20 vol.% of QC filling associated with a net drop in the polar component of the *surfenergy*. *Surfenergy* refers to the surface energy related to the top of the oxide layer under ambient conditions. We hypothesise that this is related to the percolation threshold at about 13 vol.% QC, reflected in the observed behaviour of both the friction coefficient and *surfenergy*. For the pure QC annealed in air for 1 h at 500 °C significant wear tracks were observed accompanied by a wear debris formation. On the other hand, a pure polymer exhibited slightly visible wear tracks with no apparent debris formation, and for the composites with different QC filling factors, the wear traces were barely visible with negligible debris formation.

Keywords: quasicrystals; composite materials; contact angle; surface energy; mechanical tests; friction; wear

1. Introduction

Quasicrystals (QC) have revolutionized our understanding of crystal order since their discovery in 1982–1984 by Shechtman et al. [1]. These alloys represent a new class of a complex metallic alloys materials characterized by non-translationally-repeating, aperiodic patterns that exhibit a form of order not found in traditional crystals. Unlike regular crystals, quasicrystals display symmetries that were once thought to be impossible in crystal structures, such as five-fold rotational symmetry [2]. Such an atom arrangement naturally results in the creation of a profound pseudo-gap at the Fermi energy. On one hand, there are extended electronic wave functions capable of generating the pseudogap through the diffraction and interference processes with quasiperiodically-stacked atomic planes. On the other hand, there are localized electronic states arising from resonant effects involving nested atomic clusters with self-similar geometries at different. The unique physical properties observed are a direct result of the interplay between the distinct electronic and crystal structures. For instance, of the Al-Cu-Fe alloy the icosahedral (i-phase) exhibits a hardness range of 800–1000 HV. Additionally, it is distinguished by a low friction coefficient, ranging from 0.05 to 0.2, accompanied by excellent wear resistance and reduced adhesion. A comparison of the hardness of pure aluminium, which typically ranges from 25 to 45 HV, and the friction coefficient of aluminium alloys of about 0.37 reveals a superior performance of the QC alloy [3–5]. QCs already found their way into various technological applications in the form of coatings and thin films [6–8] or as reinforcement particles in metal matrices [9]. Quasicrystals used as reinforcement precipitates in maraging steel are used for razor blades, surgical tools or dental wires [9]. A close collaboration between Philips and Sandvik led to the development of a unique, QC-based, stainless-steel shaving blade marketed by Sandvik Steel in Sweden as the Sandvik Nanoflex™ [10]. QC-precipitation-strengthened steel is ductile, corrosion resistant and resilient to ageing.

Polymers exhibit a huge variety of possible chemical compositions and have relatively low cost, ease of processing, acceptable thermal and environmental resistance and recyclability [11]. A major challenge related to the polymers employed in tribological applications at high speed under heavy loads is that they are limited by a low load-carrying capacity and a short operating lifetime [12].

Nowadays, there is a lot of interest in using polymer composites for tribological applications such as gears, wheels, bearings, seals and high wear- and scratch-resistant flexible risers [13–15]. There are already some examples of tailoring polymer-composites tribological properties by using carbon fillers such as carbon nanotubes (CNTs), carbon fibres (CFs) and graphene. Some carbon fillers, such as CNTs, have drawbacks when it comes to developing CNT-reinforced polymer composites due to their high cost and resistance to dispersion in polymer matrices [16].

Based on our current knowledge of composites, we can design novel materials with enhanced properties for specific applications by combining the main features of the different materials in a given composite.

The primary challenge associated with quasicrystals is related to their brittleness around room temperature [2]. An alternative strategy to address this limitation is by using quasicrystals as reinforcing powder within the polymer matrix, where mechanical, tribological, and thermal properties were improved. Examples are epoxy resin [17–19], polyphenylene sulphide [20] and nylon polymer [21]. The most prominent QC-polymer composite was obtained by using the additive-manufacturing technique of selective laser sintering [21].

We chose a more high-end engineering polymer [22] that has never been studied in the context of QC-reinforcement, to the best of our knowledge. Hence, our work focuses on polyphthalamide (PPA) used as a polymer matrix reinforced with different volume amounts of QC powder and adhesion between the QC-powder particles and the polymer matrix with associated surface and mechanical properties.

In our previous study [23], we established that passivating the surface of QCs alters their surface energy, termed *surfenergy*. It is the surface energy of the oxidised QC's surface, typically in the range of a few tens of mJ/m^2. This is different from the surface energy of bare, unoxidised metallic alloys, which is usually within the range of several hundred mJ/m^2 or higher [24]. *Surfenergy* is crucial because it affects how these treated QCs bond with polymers, changing the overall strength and flexibility of the composite material. It also influences the wetting behaviour of the composite sample depending on the amount of QCs mixed into the polymer. As we show in the following section, the *surfenergy* of this polymer shows a significant polar contribution that is responsible for adhesion to polar liquids such as water.

In addition, we propose that the percolation effect might be an additional key factor for optimising the mechanical properties of composite materials. Percolation happens when enough isolated QC particles in the polymer matrix come together to form a continuous network. This effect likely occurs at a specific concentration of QCs. Other studies have shown that the physical properties of composite materials can peak or drop sharply at certain points. These points, where the rate of change is zero, often line up with the percolation threshold [25]. In our study, we examined how the amount of QCs in the polymer matrix affects friction. We systematically studied the relationship between friction and the concentration of QCs. In this research, we verified this hypothesis through a systematic study of the relationship between friction and the concentration of QCs in the polymer matrix.

This aspect, particularly relevant for refining the surface and mechanical properties of QC-polymer composites, has been overlooked in prior research. Our aim was to explore QC particles' behaviour within the polymer matrix and to determine to what extent the constituting species take part in mechanical and surface properties. Our conclusions are based on characterising the sample surface and mechanical testing using techniques such as scanning electron microscopy (SEM); powder X-ray diffraction (PXRD); friction coefficient (μ); tensile, bending, and toughness testing correlated with Brinell hardness; and the impact strength.

2. Experimental Details

2.1. Input Materials

For the metal part, the quasicrystalline Al-Cu-Fe icosahedral-rich powders doped with small quantities of boron were produced using a special technology by NANOCOM LLC, Moscow, Russia. It is known that small additions of boron, in the range of a few at.% to an Al-Cu-Fe alloy, modify solidification kinetics, favouring the formation of the QC primary phase and lowering the chemical segregation in the cast boron-containing alloy [26].

We characterised the powder as supplied by the producer without any additional information due to special confidentiality relating to the production protocol. The characterisation revealed that the powder was crushed from a master ingot and that it was mechanically sieved. The powder we received was stored in a glove box under argon. The powder was chosen based on several factors, including its potential low cost, low toxicity and availability. For brevity, we label this powder as QC in the subsequent text. Since commercial QC powders contain different phases and there is no corresponding crystallographic database of quasicrystalline phases, as a reference specimen, we employed an already well-studied QC sample. This high-purity reference sample, with a nominal composition of $Al_{59}Cu_{25}Fe_{13}B_3$ (at.%) and a known PXRD, purchased from the Saint Gobain Company, Courbevoie, France (labelled from here as *reference sample*), was sintered from gas atomised powder in a flow of argon using a uniaxial sintering furnace operated at 930 °C and a pressure of 100 MPa [9]. This sample contained a majority of the $Al_{62}Cu_{25}Fe_{13}$ icosahedral quasicrystal phase (i-phase), with no apparent grain texturing and the average grain size of the matrix QC phase being 20–70 micrometres. Additionally, the sample contained small amounts of minor phases, less than 4%, such as $AlFe_2B_2$ and β-AlCu(Fe) and borides AlB_{12} [23].

The polyphthalamide granules (Amodel AT-1002 HS SOLVAY), provided by ALBIS Plastic Vertriebsgesellschaft GmbH, Vienna, Austria, were chosen as the polymer matrix.

Because it is a technical polymer that is used in a wide variety of applications, especially in the automotive industry, and because it can stand relatively high temperatures, it is suitable as a polymer matrix to be reinforced with metal particles. This material is labelled PPA in the subsequent text.

2.2. Composite Fabrication

The desired amounts of polymer and quasicrystal powders were weighed and mixed using a forming process based on a co-rotating twin-screw extruder with a diameter of 20 mm and a length-to-diameter (L/D) ratio of 44 (LABTECH—LTE 20–44, Labtech Engineering Company, Ltd., Bangkok, Thailand). The process extruded the material through a die or nozzle to produce a long filament, which was additionally cut to form the specific shape of the granules. The volume fraction of QC in the composite materials ranged from 0 vol.% to 35 vol.%. Converting mass to volume involves dividing the total mass by the density to yield the volume. The density of dry PPA0 was 1.13 g/cm^3, and for QC powder it was around 4.56 ± 0.03 g/cm^3. The maximum billet temperature for the extrusion process was set to 335 °C. The final product was in the form of 1.5–2.0-mm granules in the amount of 2 kg. Furthermore, an injection-moulding machine (Arburg, Allrounder 320 C500-100 golden edition, Lossburg, Germany) was used to prepare specimens in the standard dog-bone shape, suitable for characterisation of mechanical properties by tensile, flexural and toughness tests. The melting temperature for injection moulding was fixed at 335 °C, while the mould temperature was lowered to 85 °C, with an injection speed of 40 mm/s, dropping to 10 mm/s for the last 2 mm, after being exposed to a pressure of 550 bars for 10 s, for the net PPA and PPA with 5 vol.% of QCs or 1100 bars for 15 s with a cooling time of 12 s, for the higher fillings with QCs, respectively.

The composites made of a blend of PPA polymer with a volume fraction x of quasicrystal powder (x = 0 vol.%, 5 vol.%, 20 vol.%, 30 vol.%, 35 vol.%) will be labelled PPAx in the subsequent text. For instance, PPA5 means a composite containing 5 vol.% of QC blended with 95 vol.% of PPA.

3. Characterisation Techniques

3.1. Microstructure and Crystal Structure

(a) Scanning Electron Microscopy (SEM)

A JEOL JSM-7600F (JEOL, Tokyo, Japan) field-emission-gun SEM equipped with an energy-dispersive X-ray spectrometer (EDXS) (Oxford Instruments Plc, Abingdon, UK) was used to characterise the microstructures and elemental compositions of the prepared samples. The quantitative EDXS analyses were performed using an Oxford Instruments INCA Microanalysis Suite with an X-Max 20 SDD detector (Oxford Instruments Plc, Abingdon, UK). A sample for the SEM/EDXS microstructural characterisation was prepared using standard metallographic procedures for aluminium alloys. The investigation was performed on a quasicrystalline sample to examine the phases and surface morphology. Images were taken from the central area of the sample using secondary-electron imaging (SEI) for topographic contrast and backscattered-electron imaging (BSE) for compositional Z-contrast to reveal phases with different compositions.

(b) X-ray Powder Diffraction (PXRD)

PXRD data were collected with a Malvern Panalytical Empyrean X-ray diffractometer (Malvern Panalytical, Almelo, the Netherlands) using a monochromated X-ray beam produced by a Cu-target tube ($\lambda K\alpha 1$ = 0.15406 nm and $\lambda K\alpha 2$ = 0.15444 nm). The measurements were obtained with Bragg–Brentano geometry by applying a divergence slit of 0.04 rad, in the range $10\,^\circ < 2\,\Theta < 100\,^\circ$, using a step size of 0.0131 ° and with a counting time of 1 s per step. The PXRD data were analysed using the HighScore Plus XRD Analysis Software database PDF-4+ 2023 and based on literature relating to quasicrystals [2].

3.2. Mechanical Tests for the Composite Materials

(a) Tensile test

The uniaxial tensile tests were conducted using a Shimadzu, Ag-X plus 10 KN, universal testing machine, Kyoto, Japan at 1 mm/min rates up to an elongation of 0.25% and then with 50 mm/min rates, respectively. The samples were tested according to the Standard ISO 527 on composite materials.

(b) Toughness test

The Charpy impact strength was investigated using an LY-XJJDS apparatus Liyi Environmental technology, Ltd, Dongguan, China at room temperature. The distance between the supports was 60 mm, and the initial energy assigned to the hammer was 5 J. The composite materials were tested according to ISO 179 standards on un-notched samples. The polymer sample, as well as the composite materials, bent, but they did not break during the test.

(c) Brinell Hardness

Brinell hardness measurements were carried out by applying a hardness tester INNOVATEST NEXUS 7500 (INNOVATEST Europe BV, Maastricht, The Netherlands). The tests were carried out with a load of 15.6 kg using a hard steel ball of 2.5 mm diameter.

(d) Bending test

Three-point flexure tests were carried out using Shimadzu, Ag-X plus 10 KN universal testing machine, Kyoto, Japan at a rate of 2 mm/min. Flexure tests were carried out to 7% strain; therefore, no samples were destroyed.

3.3. Surface-Characterisation Techniques

(a) Contact-angle measurements and determination of the surfenergy

We measured the contact angle and the surface energy (referred to as *surfenergy*) of pure PPA polymer and PPA mixed with different amounts of quasicrystalline pow-der. These measurements were carried out using the Theta Lite-Biolin Scientific instrument, (Biolin Scientific, Göteborg, Sweden) following the detailed methods outlined in [23]. The term *surfenergy* is used to describe the surface energy of a material that has a native oxide layer, which naturally forms on a quasicrystal when exposed to air. As a consequence, the surface energy we deal with is that of the oxidised material, not that of the naked quasicrystals. This principle also applies when the quasicrystals are incorporated into a polymer matrix.

(b) Friction test and wear traces

The friction test was implemented on two different machines: a pin-on-disk apparatus from CSM-Instruments, Peseux, Switzerland (now Anton-Paar) and a low-load tribometer (nano-tribometer) from Anton-Paar, Peseux, Switzerland.

The flat samples were prepared using the same procedure as for the *surfenergy* experiments. It is important to remember that the coefficient of friction is not a property of a single material but rather a property of the entire friction set-up, including the indenter and all its experimental parameters (hardness of the pin, roughness of both sliding surfaces, number of passes, etc.) [27]. For the pin-on-disk experiments, the ball (100C6) had a radius of 6 mm. The test was made at a normal load of 2 N. The stopping point was set at 5000 laps. For the nano-tribometer, the ball (100C6/AISI52100) had a radius of 1 mm. The test was made at a normal load of 10 mN. The length of the linear track was 1 mm. The number of cycles was 200. This test was used to assess the wear undergone by the two contacting bodies, which was expressed by the measure of the distance separating the position along the vertical direction of the indenter holder from a reference plane taken as the origin at the beginning of the test before the load was applied.

4. Results

(a) Microstructure and crystal phases of the input materials

The SEM studies investigated the morphologies and chemical compositions of the polymers and quasicrystalline powders. Figure 1a presents the typical shape and size of the B-doped $Al_{62}Cu_{25}Fe_{13}$ QC powder particles with their unique powder morphology, and Figure 1b is a representative SEM-BSE image of the phases present and identified within the QC particle. The corresponding compositions of the phases detected inside the quasicrystalline powder are presented in Table 1. For the quasicrystalline powder, we confirmed the co-existence of the matrix $Al_{62}Cu_{25}Fe_{13}$ (i-phase), the ternary ω-phase $Al_{58}Cu_{30}Fe_{12}$, the binary β-phase AlCu(Fe) and a minority of the binary Θ-$AlFe_3$ phase. This Fe-rich phase was not a part of the powder-stability region, but it appeared as a residue left in the powder batch, which implies that full mixing of the elements was not achieved upon melting. Such a situation is not unusual in the case of an industrial product. The powder's particle size distribution was bimodal with two distinct fractions, ranging from 0.5 µm to 4–5 µm and from 10 µm to a maximum of 50 µm. Figure 1c,d presents the polymer polyphthalamide resin granules with a size of 2 mm × 1 mm, alongside a BSE micrograph showing a cross-sectional view of the polymer.

Figure 1. (**a**) Overall SEM-BSE image of the as-received, B-doped $Al_{62}Cu_{25}Fe_{13}$ quasicrystalline powder, (**b**) characteristic higher-magnification BSE micrograph of the detected phases in QC powder. (**c**) Resin granules of the PPA with approximate dimensions 2 mm × 1 mm and (**d**) high-magnification BSE micrograph of the microstructure of the polyphthalamide polymer.

Table 1. Compositions detected by the EDXS with the phases assigned from the PXRD.

Phase Assigned from the PXRD	Detected Composition with EDXS (at.%)	Crystal Structure, Space Group, Number
i-phase $Al_{62}Cu_{25}Fe_{13}$	$Al_{62}Cu_{25}Fe_{13}$	Icosahedral, Fm35
ω-phase	$Al_{58}Cu_{30}Fe_{12}$	Orthorhombic, Immm, 71
β-phase AlCu(Fe)	$Al_{72}Fe_{22}Cu_6$	Cubic, Im-3m, 229
Θ-$AlFe_3$	$Al_{18}Fe_{81}Cu$	Orthorhombic, Bmmm, 65
λ-$Al_{13}Fe_4$	Undetectable by EDXS but confirmed by PXRD	Cubic, Fm-3m, 225
AlB_2	$Al_{28}B_{72}$	Hexagonal, P6/mmm, 191

The PXRD method provided information about the crystallographic structure of the phases in the microstructure of the as-received quasicrystalline powder. The diffraction peaks are indexed using the PDF-4+ 2023 database and the literature relating to quasicrystals [2]. Figure 2 presents the PXRD diffractogram obtained from the as-received QC material based on the $Al_{62}Cu_{25}Fe_{13}$ (at.%) composition, which was used to fabricate the composite materials. Six crystal phases could be confirmed. The major phase corresponds to the quasicrystal icosahedral phase (i-phase, space group Fm35), as verified by a direct correlation of the PXRD pattern obtained from a reference sample with the highest purity of quasicrystalline icosahedral phase [23]. The other minor phases are the cubic β-AlCu(Fe) phase (space group Im-3m, 229), the orthorhombic ω-phase $Al_{60}Cu_{30}Fe_{10}$ phase (space group Immm, 71), the cubic phase λ-$Al_{13}Fe_4$ (space group Fm-3m, 225), orthorhombic phase Θ-$AlFe_3$ (space group Bmmm, 65), and hexagonal phase AlB_2 (space group P6/mmm, 191).

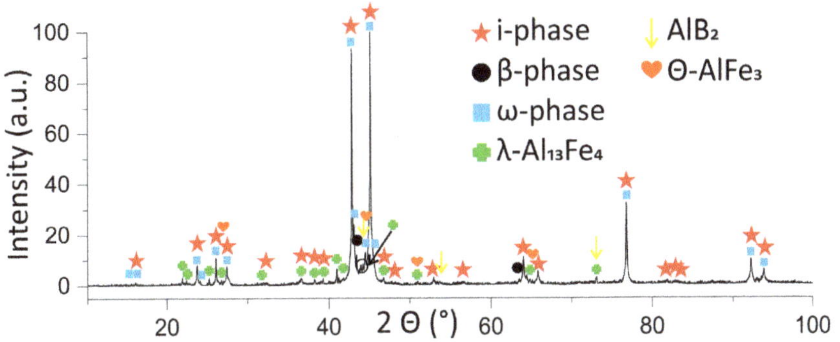

Figure 2. PXRD pattern of the as-received quasicrystalline powder with a nominal composition of $Al_{59}Cu_{25}Fe_{13}B_3$ (at.%). The highest-intensity peaks belong to the $Al_{62}Cu_{25}Fe_{13}$ i-phase. The other minority phases are the β-AlCu(Fe), the ω-$Al_{60}Cu_{30}Fe_{10}$, Θ-$AlFe_3$, λ-$Al_{13}Fe_4$ and boride AlB_2.

(b) Microstructure and structural properties of the composite materials, bonding between the particles and the matrix

Figure 3a presents the overall microstructure of the representative PPA20 fractured surface, Figure 3b presents the overall microstructure of the representative PPA35 fractured surface with visible agglomerates, Figure 3c is a higher-magnification micrograph of the composite PPA5, Figure 3d presents a higher-magnification micrograph of the composite PPA20, Figure 3e presents the overall microstructure of the PPA0 and Figure 3f presents grains of the pure PPA0 after the tensile test, where two different types of plastic deformation are visible.

An analysis of the SEM images of the PPA*x* composites with different volume fillings reveals a nearly homogenous microstructure. No defects and porosity were observed, see Figure 3a,b. Due to the atomic-number (Z) contrast, the Al-Cu-Fe-B particles appear brighter in the SEM–BSE micrographs than in the surrounding carbon-based, low-density PPA matrix. The QC powder particles were distributed uniformly in the polymer matrix. Yet, in composites PPA30 and PPA35, there were occasional cases of particle clustering. Throughout the whole composite, no microbubbles were detected. Figure 3c shows deficient adhesion between the polymer and the QC filler in PPA5, whereas Figure 3d presents good adhesion between the two materials in PPA20; this is visible as the polymer embraces the QC particles. There are no visible pores and voids formed by the ejection of the filler during the mechanical test on the fracture surface.

The PXRD of the composite material was used to verify whether the phases were preserved during the extrusion process when the materials were exposed to temperatures up to 335 °C. Figure 4 shows the PXRD diffractograms of the PPA*x* (x = 0, 5, 20 and 35) composites. The PXRD analysis confirmed the semi-crystalline nature of the polymer,

Figure 4a, which exhibits relatively broader peaks. The XRD spectrums were compared with the XRD pattern of the QC in Figure 2 and the peaks were credibly matched.

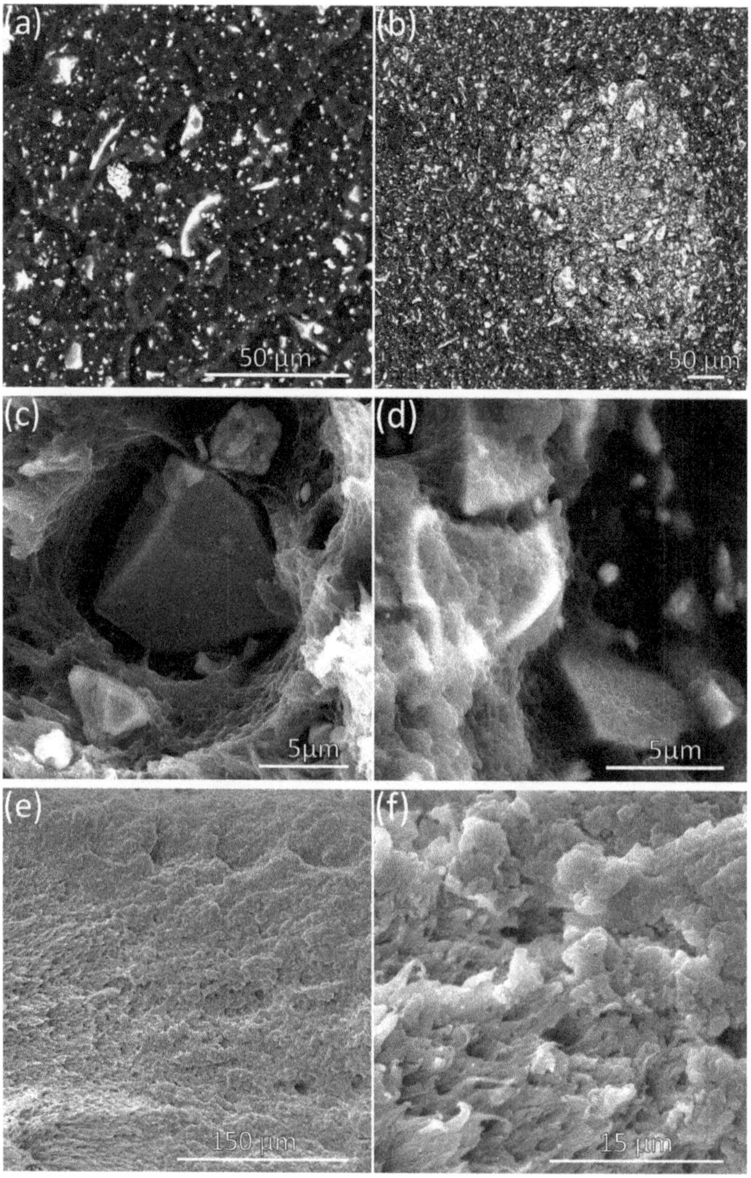

Figure 3. SEM–BSE images of the fracture surface after tensile testing for the composite material with different volumes of filling particles. Overview of the surface morphology after tensile test and fracture of the (**a**) PPA20 and (**b**) PPA35 composite materials, (**c**) characteristic higher-magnification micrograph of the PPA5 composite, (**d**) higher-magnification micrograph of the PPA20 composite, which is also representative for PPA30 and PPA35, (**e**) SEM–SEI image of the fracture surface of the PPA0 polymer after tensile test, and (**f**) characteristic higher-magnification micrograph of the fractured part.

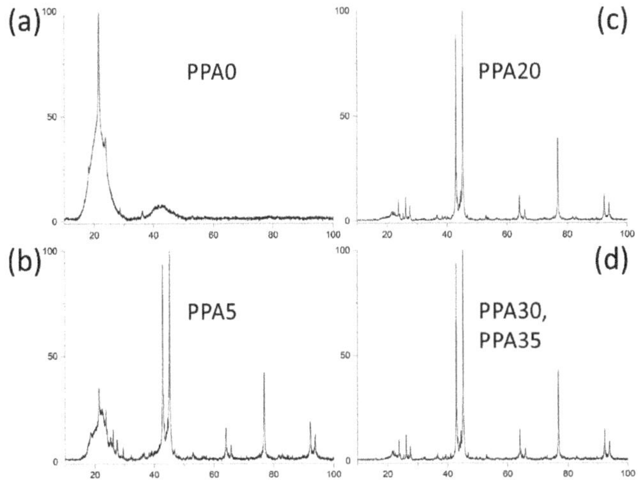

Figure 4. The PXRD patterns for (**a**) the PPA0 pure polymer, (**b**) the PPA5 composite material, (**c**) the PPA20 composite materials, and (**d**) representative diffractogram for both PPA30 and PPA35. The x-axis represents the angle 2 Θ (°), and the y-axis represents the intensity (a.u.).

(c) Mechanical properties of the composite materials

The tensile tests were performed to determine the material's ultimate tensile strength, yield strength, and ductility.

Figure 5 shows the concentration dependency of the tensile Young's modulus and elongation at fracture after tensile tests of the PPA0 and PPAx composites ($5 \leq x \leq 35$ vol.%). The Young's modulus gradually increases from 1810 MPa for the unfilled PPA to 4114 MPa for the PPA35 composite. The elongation at fracture constantly decreases with the filling fraction. It equals 16.9% for an unfilled PPA0, whereas for the PPA35, it drops to 4.8%.

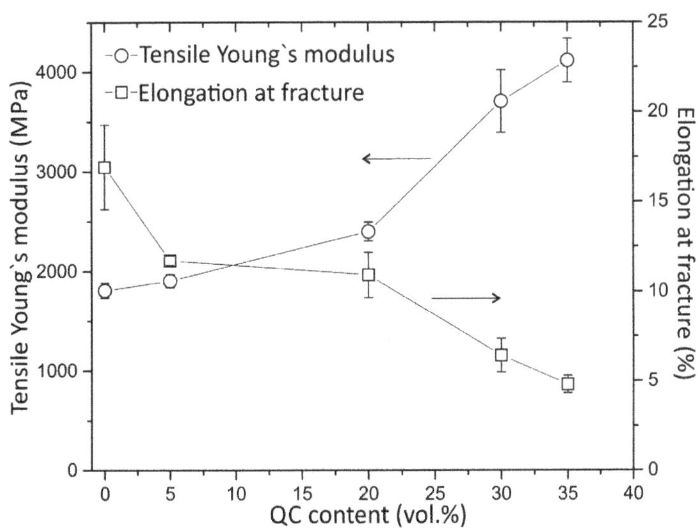

Figure 5. Young's modulus and elongation at fracture of the PPAx ($0 \leq x \leq 35$ vol.%) composites after tensile tests. The arrows specify the axes associated with the results.

Figure 6 presents the flexural strength and the flexural Young's modulus of the PPA0 and PPAx ($5 \leq x \leq 35$ vol.%) composites after three-point flexure tests. An increase of the flexural strength from 100 MPa for PPA0 to 128.5 MPa for composite PPA35 was recorded. The flexural modulus, as in the case of the uniaxial tensile test, increases with increasing the QC content from 20 MPa for PPA0 to 56 MPa for PPA35.

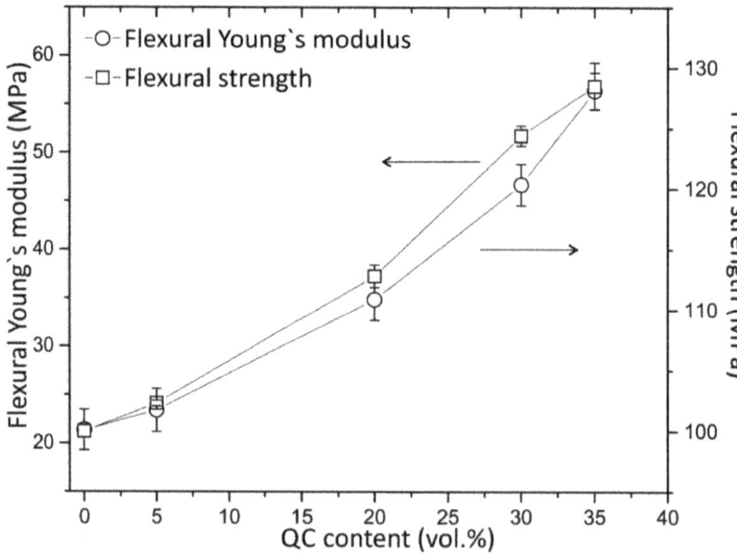

Figure 6. Results of three-point flexure tests for the PPAx ($0 \leq x \leq 35$ vol.%) composites. The arrows specify the axes associated with the results.

Polymers show different levels of brittleness under static and impact loads due to their inherent material properties, including stress-relaxation times. In static load tests, the gradual application of stress allows the material's stress-relaxation processes to play a more significant role, accommodating the applied load over a longer period. In contrast, impact loads apply stress swiftly, providing less time for these relaxation mechanisms to act, which alters the material's response. This distinction in response under different loading rates underscores the importance of impact strength as a key mechanical property in evaluating material properties [27]. Compared to other engineering thermoplastics, PPA stands out for its superior impact strength. It achieves values of 150 J/m [22], which is notably higher than that of nylon, another thermoplastic in the same group, which has an impact strength of only 60 J/m [28].

Figure 7 shows the impact strength of the PPAx ($0 \leq x \leq 35$ vol.%) samples obtained from the Charpy tests applied to our composites with different QC contents. The typical value of the impact strength for PPA0 is 98.5 kJ/m^2, whereas for PPA5, the results show an increased impact strength of up to 107 kJ/m^2. This value decreases to 69.1 kJ/m^2 for PPA20, while it amounts to 50.7 kJ/m^2 for PPA30 and 42.4 kJ/m^2 for PPA35.

The concentration dependence of the Brinell hardness is also shown in Figure 7. The hardness of the PPA0 is 19; no noticeable difference is observed at low filling percentages, but a further increase in the QC content leads to an increase in hardness to 21 for PPA20, 23 for PPA30 and 25 for PPA35.

Figure 8 shows the variation of the coefficient of friction for the PPAx ($0 \leq x \leq 35$ vol.%) composites. The coefficient of friction (μ) changed after adding QC particles, as compared to the pure PPA0, which is equal to $\mu = 0.15 \pm 0.02$. Initially, it decreased at $x = 5$ vol.% and it increased continuously for x above 20 vol.%.

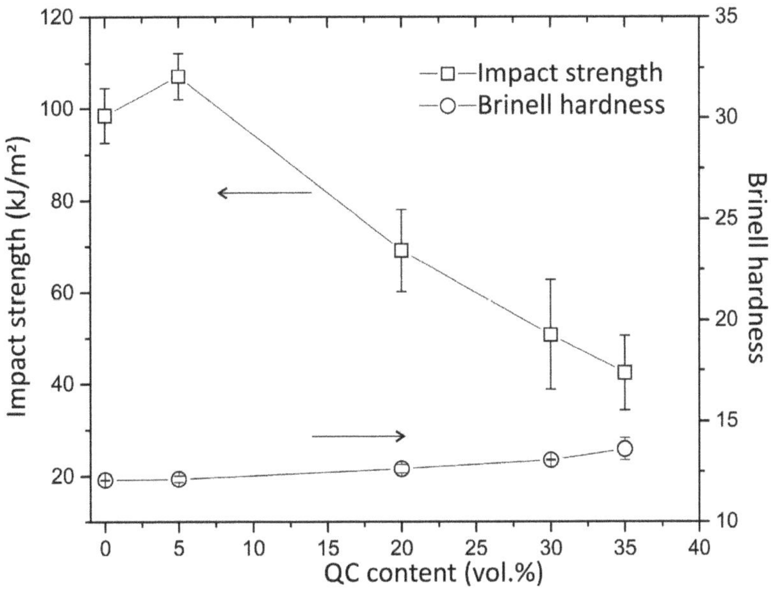

Figure 7. Impact strength and Brinell hardness of the PPAx ($0 \leq x \leq 35$ vol.%) samples. The arrows specify the axes associated with the results.

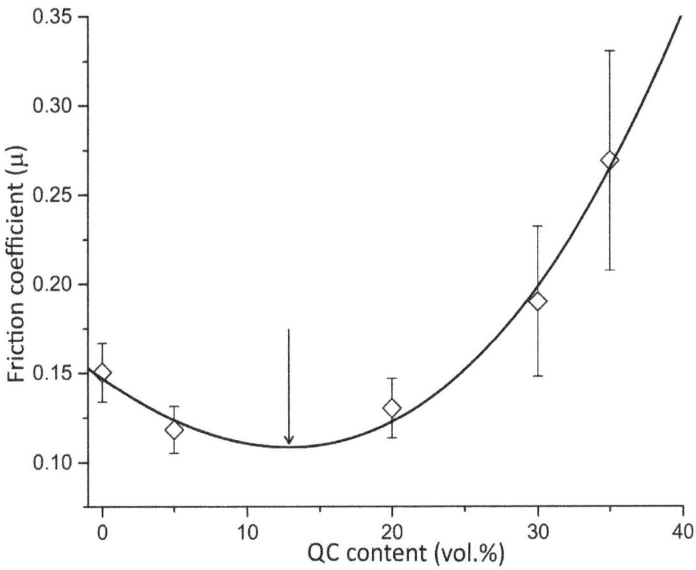

Figure 8. Coefficient of friction measured during pin-on-disk tests performed in air against a hard steel ball on our PPAx (x = 0, 5, 20, 30 and 35 vol.%) composites. The line represents a third-order polynomial fit, which is discussed in Section 4. The arrow marks the minimum of μ, see Section 4.

The SEM investigations of the wear tracks observed on our composites after the pin-on-disk friction test are presented in Figure 9. The *reference* Al-Cu-Fe-B *sample* appeared worn, leaving an exposed trail surrounded by a slightly darkened area that appears to be an Al-rich oxide. A further evaluation of the PPA0 showed a noticeable track compared to the tracks on the PPA20, PPA30 and PPA35 composites, where the tracks were almost invisible. No additional debris was detected in the wear tracks on the pure polymer and composites, and no oxide layer was found on either.

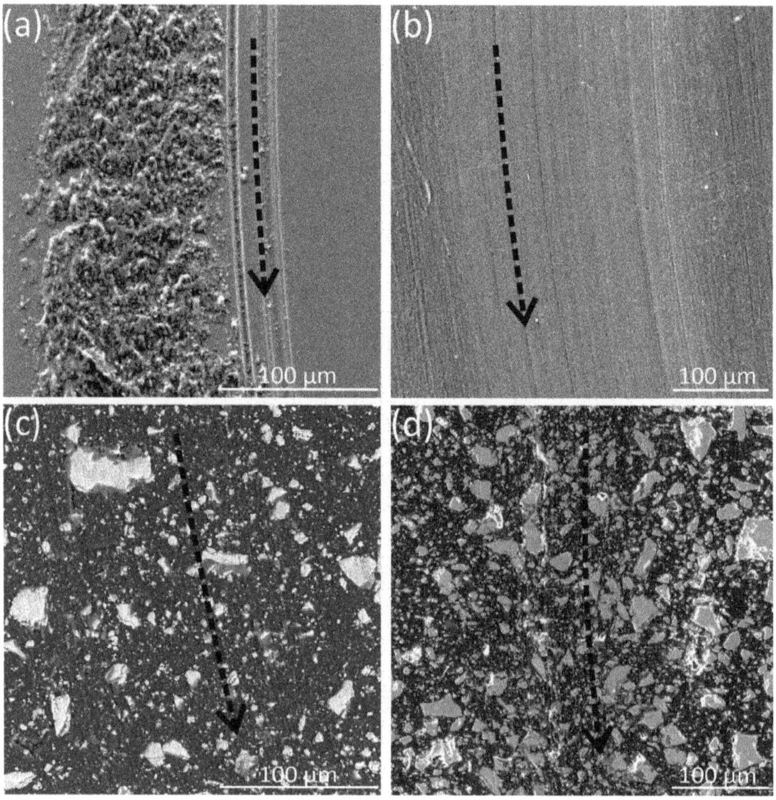

Figure 9. SEM–BSE images of the wear traces (marked by arrows) observed on (**a**) the *reference sample* annealed 1 h for 500 °C in air, (**b**) PPA0, (**c**) PPA20 and (**d**) PPA35. Arrows indicate the path of the wear tracks.

Table 2 presents the output of the contact-angle measurements performed using water and diiodomethane on the composite materials. Figure 10 presents the calculated data of the *surfenergy* and their components on the PPAx ($0 \leq x$ 35 vol.%) composites. It is noticeable that there are almost no visible changes in the dispersive component after adding up to 35 vol.% of the QC particles to the polymer matrix. The dispersive component remained in the range $\gamma_{sv}^{d} = 33 - 34 \pm 1$ mJ/m^2. A slightly different trend is visible for the polar component, which dropped from the initial value of $\gamma_{sv}^{p} = 5 \pm 2$ mJ/m^2 for the PPA0 to $\gamma_{sv}^{p} = 1.2 \pm 0.6$ mJ/m^2 for the PPA20, to $\gamma_{sv}^{p} = 4 \pm 1$ mJ/m^2 for PPA30, and to $\gamma_{sv}^{p} = 3 \pm 0.5$ mJ/m^2 for PPA35. Consequently, the *surfenergy* of the pure PPA0 was equal to $\gamma_s = 39 \pm 2$ mJ/m^2. The total value was not affected for PPA5, which resulted in $\gamma_s = 39 \pm 2$, whereas for PPA20, it resulted in a drop to $\gamma_s = 36 \pm 1$ mJ/m^2. A further measurement for PPA30 or PP35 yielded only a small change to $\gamma_s = 37 \pm 2$ mJ/m^2.

Table 2. Surface characteristics of the pure PPA and composite materials.

Material	Contact Angle Water (°)	Contact Angle Diiodomethane (°)	Dispersive Component γ_{sv}^{d} (mJ/m^2)	Polar Component γ_{sv}^{p} (mJ/m^2)	Surfenergy γ_s (mJ/m^2)
PPA0	77.1 ± 1.1	50.6 ± 0.7	33.9 ± 0.8	4.6 ± 0.9	38.5 ± 1.7
PPA5	78.8 ± 2.5	49.9 ± 0.3	34.3 ± 0.3	3.9 ± 1.7	38.2 ± 2
PPA20	88.0 ± 2.5	49.1 ± 0.6	34.8 ± 0.4	1.2 ± 0.6	36 ± 1
PPA30	79.2 ± 1.8	52.6 ± 0.8	32.8 ± 0.7	4.1 ± 1	36.9 ± 1.7
PPA35	80.5 ± 0.7	50.0 ± 0.4	34.3 ± 0.3	3.3 ± 0.5	37.6 ± 0.8

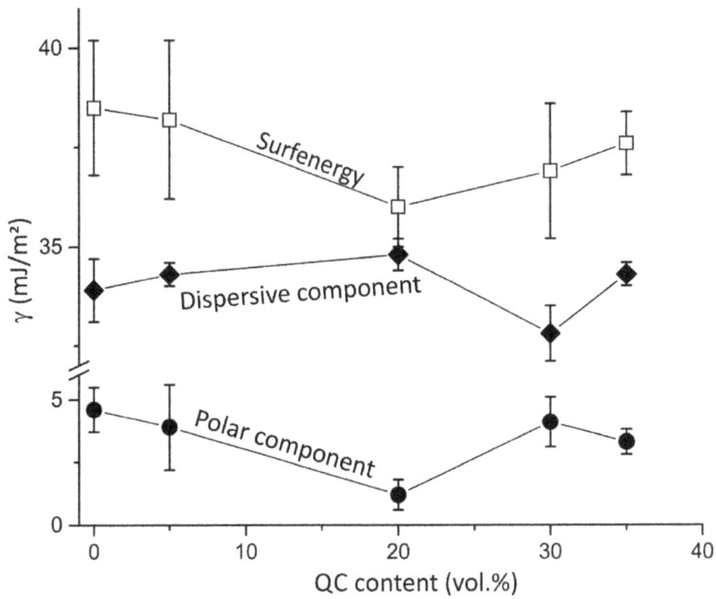

Figure 10. Evolution of the *surfenergy* (open squares), dispersive (solid diamonds) and polar (full circles) components of the PPAx composite materials. Symbols are connected by solid lines only as a guide for the eye.

5. Discussion

SEM studies were performed on a fracture surface to understand better the composite materials' microstructure and the bonding between the QC particles and the polymer matrix. Concerning practical use, QC particles may exhibit clumping during storage or extrusion mixing, resulting in a composite product with limited usability. Therefore, concerning the formation and handling of small particles of micro and nano sizes, agglomeration [29] is a challenge [30]. We observed that the QC distribution of the particles within the polymer matrix was homogenous, although some agglomerates embedded in the matrix form in composites PPA30 and PPA35. The same effect of particle distribution and the formation of agglomerates was noticeable when the quasicrystals were embedded in epoxy [17]. They emerged due to a small particle size, which facilitated agglomeration. Another factor contributing to the presence of agglomerates was an enhanced viscosity during extrusion [31] due to an increased volumetric fraction of QC particles, making mixing between the QC and the matrix more difficult. As the extrusion temperature is lower than the necessary sintering temperature, the potential sintering of individual particles into polycrystalline bodies has been eliminated.

A further observation with the SEM was the wetting of the resin in the QC particles. According to Anderson [32] wettability is defined as the fluid's tendency to spread on a surface or to adhere to a solid surface. Despite mixing materials with different properties, which form a weak or even a non-existent interface, the QC particles are essentially completely enveloped by the polymer matrix, as shown in Figure 3c,d. A favourable wetting between the matrix and the reinforcement phase is the initial requirement for the formation of a good interface and consequently for strong forces between the particles and the polymer matrix.

Another phenomenon typical for PPA–QC (polyphthalamide–quasicrystal) composites is a gradual increase in the hardness with the filler content. The observed increase in the hardness of such composite materials is governed by the very high QC hardness, which can reach values of 8–10 GPa in bulk samples [2] and up to 14 GPa in thin films at room temperature, as compared with hardened steel, which does not exceed 8 GPa.

To check whether the extrusion process leads to any degradation of the polymer or to a change in the composition of the initial QC powder, the PXRD was performed on the pure polymer and PPAx composites. Indeed, in the PXRD diffractogram of the composite material, there is evidence for a minor phase in addition to the major i-phase in the initial QC powder (Figure 4), which calls for further experiments. On the other hand, the crystallinity of polymers serves as a fingerprint for a polymers' identification, and it offers a qualitative measure of the degree of crystallinity, which defines the optical, mechanical, thermal and chemical properties. After adding the QC particles to the polymer matrix, the nature of the polymer was evidently maintained, regardless of the volumetric amount of QC filling.

We examined wear traces formed during the pin-on-disc experiment, as shown in Figure 9. Pronounced wear tracks were observed on the sintered QC sample, which was accompanied by the formation of a significant amount of debris. The debris formation was attributed to the inherent brittleness of the quasicrystals. Conversely, a polymer characterised by a notable degree of elasticity manifests only visible wear traces, with no debris formation, due to its propensity for plastic deformation. Contrary to expectations assuming the brittleness of quasicrystals, the PPA–QC composite material displays a marked plastic behaviour. This unexpected behaviour underscores the superior mechanical properties of the composite when compared to the individual constituents—pure PPA and QC—and emphasises the advantageous properties achieved through the combination of dissimilar materials in a composite, showcasing the potential for tailoring materials to achieve a balance of mechanical characteristics that surpass the individual components.

The observed increase in Young's modulus with higher QC contents can be attributed to the enhanced fracture toughness, which, on the other hand, is accompanied by a decrease in the plasticity of the PPA–QC composites due to a reduction in the elongation values at the break. The observed trend must be driven by the QC's mechanical properties and the interactions at the PPA–QC interface. However, the dependence of the impact strength, which measures the brittleness, on the QC content is slightly different. While the material PPA5 has a higher impact strength than the pure polymer, this strength decreases with additional QC filler, reaching 42.4 kJ/m^2 for PPA35. This decrease suggests a possible percolation effect.

In the context of the measured coefficient of friction, a distinct pattern emerges with PPAx materials. Initially, PPA5 shows a 20% reduction in friction compared to PPA0. This decrease is attributed to the hardness of the QC particles reinforcing the polyphthalamide matrix, which minimises plastic deformation during wear and friction tests. However, as more QC particles are added to the matrix, there is a significant increase in friction. This increase reaches up to 100% for the PPA35 sample, which contains the maximum amount of QC fillings.

As shown in Table 3, QCs, particularly Al-Cu-Fe, exhibit significantly lower friction coefficients compared to alloys of similar hardness (about 7–8 GPa, according to ref. [33]) and Young's modulus (around 100 GPa, as reported in ref. [34]). Notably, in complex crystals, friction correlates with electronic conductivity, offering insights into the unique anisotropic friction seen in decagonal quasicrystals, as discussed in refs. [2,4]. The low friction coefficient observed suggests that using these materials as binders in polymers could lead to enhanced technological applications compared to traditional hard metals. Furthermore, wear, which is closely related to friction, especially in conventional materials, shows improvement in PPA–QC composites. This improvement in wear performance is mainly due to the QC's hardness and its low friction coefficient. Additionally, literature comparisons (Table 3) suggest that among various composite materials, those combining quasicrystals with the PPA polymer still have the lowest friction coefficients.

Table 3. Friction coefficient of different materials relevant to the present study compared to the literature data.

Material	Against Material	Static Coefficient of Friction (Dry Contact)	Reference
Aluminium	Steel *	0.61	Ref. [35]
Aluminium-Bronze alloy	Steel *	0.46	Ref. [35]
Steel	Steel *	0.8	Ref. [35]
Epoxy	Steel *	0.2–1	Ref. [36]
Nylon (PA)	Steel *	0.15–0.25	Ref. [35]
Quasicrystal with excellent lattice in vacuum	Steel *	0.07	Ref. [2]
Quasicrystal with excellent lattice in the air	Steel *	0.6–0.8	Ref. [2]
Composite (epoxy resin + Al_2O_3)	Aluminium	0.4–0.6	Ref. [37]
Composite Polyamide + QCs	Steel *	0.21	Ref. [21]
Quasicrystal annealed in air 1 h 500 °C	100C6 steel	0.2	this work
Polyphthalamide PPA0	100C6 steel	0.15	this work
PPA5	100C6 steel	0.12	this work
PPA20	100C6 steel	0.13	this work
PPA30	100C6 steel	0.19	this work
PPA35	100C6 steel	0.27	this work

* For the tribology tests 100Cr6 or 100C6 steel is typically used. The given literature does not detail the steel type used in the respective study.

The observed relationship between the impact strength and the coefficient of friction based on the QC content in the polymer suggests a potential percolation effect, as shown in Figure 11.

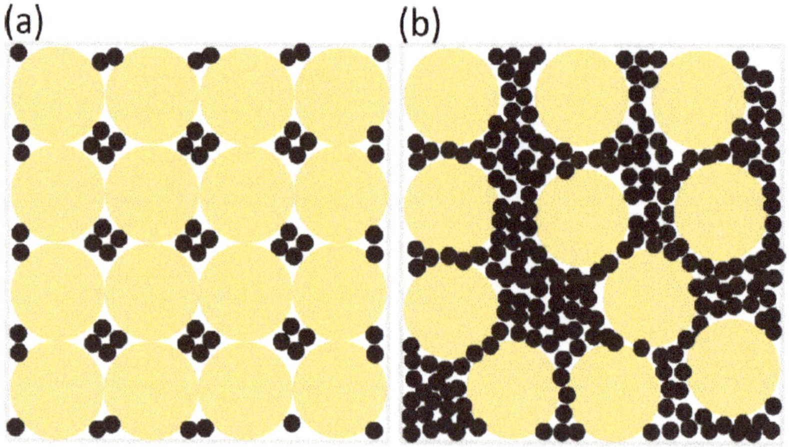

Figure 11. Idealized presentation of the percolation mechanism for (**a**) $x < x_p$ and (**b**) $x > x_p$. To simplify the figure, the QC particles are represented by black circles, whereas their actual shape is not spherical.

The friction increases with the QC content, but the value for a pure polymer is higher than that of a low QC content. A mirrored phenomenon is observed in the case of impact strength. It might be that only non-connected QC particles act as a lubricant, hence lowering the coefficient of friction, whereas the strengthening of the bonds between QC particles during the growth of a network above the percolation threshold has the opposite effect. To quantitatively support the hypothesis, we phenomenologically approximate the coefficient of friction $\mu(x)$ dependence on the QC content x using a third-order polynomial as:

$$(x) = a + bx + cx^2 + dx^3 \tag{1}$$

where a, b, c and d are unknown coefficients that are obtained from the optimum least-squares fit to the experimental data presented in Figure 9, resulting in (a = 146.944, b = −5.545, c = 0.167, and d = 0.003) × 10^{-3}, with x expressed in volume %.

We observe the minimum in the coefficient of friction, defined by the content x derivative of Equation (1) $d\mu/dx$, equals zero. It is seen as the predicted percolation threshold, the content at which a QC continuous network is formed throughout the whole sample (x_p). The derivative:

$$\frac{d}{dx} = x_p = 3dx^2 + 2cx + b \tag{2}$$

set to zero $d\mu/dx = 0$, yields x_p = 13 vol.%. We assign this minimum in the friction to an optimum filling of the PPAx composite with 13 vol.% of QC particles that undergoes a percolation of the powder grains within the polymer granules in the loading chamber of the twin-screw extruder. When the PPAx ($x > 0$) blend is introduced to this chamber and just before the pressure starts to increase, or equivalently when the QC particles are no longer free to move over long distances, the composite state is different, depending on whether the polymer granules form a continuous network ($x < x_p$) or when the polymer granules no longer form a continuous network ($x > x_p$). These two different situations are pictured in Figure 11. Let us assume that the polymer granules are close enough to a spherical shape and are all the same size. When the pressure starts to increase, these spheres form a dense packing, with a packing fraction equal to:

$$\eta = \frac{\pi}{\left(3\sqrt{2}\right)} = 0.74 \tag{3}$$

The remaining volume fraction $1-\eta$ can accommodate the QC grains, which pack randomly. Due to their irregular shape, their own packing fraction cannot exceed $\nu = 0.5$ [25]. Consequently, the maximum volume fraction that can be incorporated into the blend of polymer granules before the polymer granules no longer form a continuous network is $(1-\eta)\nu = 0.13$. Therefore, we attribute the observed minimum in the friction and the plateau in other mechanical properties of our PPAx composites to a percolation threshold that occurs around $x_p = 13$ vol.% during the preparation of the composite blend. Below this threshold, the polymer forms a continuous network, whereas, above this threshold, the powder grains have accumulated in the areas separating the granules. Consequently, the friction increases due to the brittle fracture of the QC-rich areas and debris formation, while the elongation at fracture and the impact strength decrease. Pinning effects become significant in the observed wetting behaviour.

6. Conclusions

In this study, we demonstrated the basic underlying optimisation of composites. The most valuable additional contribution is a demonstration of the link between the *surfenergy*, friction and a percolation threshold of the dual powder filling, resulting in the achievement of the optimum mechanical properties. The investigated composites exhibited a good distribution of QC particles within the polymer matrix up to the fillings of 20 vol.%. The presence of agglomerates was noticeable only for the PPA30 and PPA35. It was shown that the introduction of QC particles in the polymer matrix led to higher values of Young's modulus for the composite materials, as measured by both, the tensile and the three-point flexure tests, accompanied by an increase in the Brinell hardness. The impact strength reached the maximum for the PPA5. A decrease in plasticity was observed for the QC fillings of more than 20 vol.%, at which the material became more brittle. A good wettability of the polyphthalamide matrix around the quasicrystalline particles was achieved and confirmed by means of the SEM imaging of the PPA–QC powder interface. The friction coefficient decreased for PPA5, followed by the continuous increase for filling volumes above 20 vol.% of QC. By determining the *surfenergy*, we estimated the values of the composite's polar and dispersive components. We concluded that the only significant change in the polar component occurred for the composite PPA20. We hypothesise that the percolation threshold at about 13 vol.% QC is reflected in the observed behaviour of the impact strength and the friction coefficient. The presence of the QC powder in the polymer matrix caused a low abrasion of the steel counterpart materials and, consequently, a high wear resistance. A combination of the desired QC and polymer properties, as demonstrated in this study, indicates the investigated materials as technologically promising in applications where it is essential to simultaneously minimise the wear-induced mechanical debris and to preserve excellent mechanical properties such as strength, hardness and toughness of the PPA–QC composite. The PPA–QC composites we developed are highly promising for use in automotive and whiteware appliances, where they intersect mechanical integrity with high strength, low wear, and minimal adhesion. This is particularly beneficial in components like bearings and cogs. They are also promising in mitigating fretting, a tribological issue arising from small oscillatory movements in parts such as space antennas during launch and operation. These findings pave the way for future production of these vital components to avoid system failures.

Author Contributions: Conceptualization, M.K. (Monika Kušter); Methodology, M.K. (Monika Kušter); Validation, M.K. (Monika Kušter), Z.S., M.K. (Matej Komelj), M.H., M.B., R.K.-N., J.-M.D. and S.Š.; Formal analysis, M.K. (Monika Kušter), Z.S., M.K. (Matej Komelj), M.B. and G.P.; Investigation, M.K. (Monika Kušter); Resources, M.K. (Monika Kušter); Data curation, M.K. (Monika Kušter), Z.S., M.K. (Matej Komelj), M.H., M.B., G.P., R.K.-N., J.-M.D. and S.Š.; Writing—original draft, M.K. (Monika Kušter); Writing—review & editing, M.K. (Monika Kušter), Z.S., M.K. (Matej Komelj), M.H., M.B., R.K.-N., J.-M.D. and S.Š.; Visualization, M.K. (Monika Kušter); Supervision, J.-M.D. and S.Š.; Project administration, S.Š.; Funding acquisition, S.Š. All authors have read and agreed to the published version of the manuscript.

Funding: This research received no external funding.

Data Availability Statement: The original contributions presented in the study are included in the article, further inquiries can be directed to the corresponding author.

Acknowledgments: This work was performed within the frame of the IRP PACS2 project. The research was financially supported by Slovenian Research Agency (P2-0084), CNRS and Université de Lorraine, France, and the European Union Horizon 2020 research and innovation programme under grant agreement No. 823717—ESTEEM3. We are grateful to Borut Žužek for generously providing us with the Brinell Hardness tests, and we would like to express our gratitude to Elzbieta Krol for her invaluable assistance during the extrusion process.

Conflicts of Interest: The authors declare no conflict of interest.

References

1. Shechtman, D.; Blech, I.; Gratias, D.; Cahn, J.W. Metallic phase with long-range orientational order and no translational symmetry. *Phys. Rev. Lett.* **1984**, *53*, 1951–1953. [CrossRef]
2. Dubois, J.-M. *Useful Quasicrystals*; World Scientific Publishing Co Pte Ltd.: Singapore, 2005; p. 504.
3. Belin-Ferre, E.; Berger, C.; Quiquandon, M.; Sadoc, A. *Quasicrystals Current Topics*, 1st ed.; World Scientific Publishing Company: Singapore, 2000; p. 544.
4. Maciá-Barber, E. *Quasicrystals Fundamentals and Applications*, 1st ed.; CRC Press, Taylor & Francis Group, LLC: Boca Raton, FL, USA, 2021; p. 388.
5. Rabson, D.A. Toward theories of friction and adhesion on quasicrystals. *Prog. Surf. Sci.* **2012**, *87*, 253–271. [CrossRef]
6. Ledieu, J.; Fournée, V. Surfaces of quasicrystals. *Comptes Rendus Phys.* **2014**, *15*, 48–57. [CrossRef]
7. Jenks, C.J.; Thiel, P.A. Quasicrystals: A Short Review from a Surface Science Perspective. *Langmuir* **2002**, *14*, 1392–1397. [CrossRef]
8. Yadav, T.; Mukhopadhyay, N. Quasicrystal: A low-frictional novel material. *Curr. Opin. Chem. Eng.* **2018**, *19*, 163–169. [CrossRef]
9. Dubois, J.-M. Potential and marketed applications of quasicrystalline alloys at room temperature or above. *Rend. Lincei. Sci. Fis. e Nat.* **2023**, *34*, 689–702. [CrossRef]
10. Nanotechnology Products Database (NPD). *Sandvik Nanoflex Strip Steel Datasheet*; Sandvik Materials Technology: Sandviken, Sweeden, 2018.
11. Mittal, V. (Ed.) *High Performance Polymers and Engineering Plastics*; John Wiley & Sons: Hoboken, NJ, USA, 2011; p. 452.
12. Lin, A.-D.; Kuang, J.-H. Dynamic interaction between contact loads and tooth wear of engaged plastic gear pairs. *Int. J. Mech. Sci.* **2008**, *50*, 205–213. [CrossRef]
13. El-Sayed, A.; El-Sherbiny, M.; Abo-El-Ezz, A.; Aggag, G. Friction and wear properties of polymeric composite materials for bearing applications. *Wear* **1995**, *184*, 45–53. [CrossRef]
14. Friedrich, K.; Reinicke, P. Friction and wear of polymer-based composites. *Mech. Compos. Mater.* **1998**, *34*, 503–514. [CrossRef]
15. Khun, N.W.; Zhang, H.; Lim, L.H.; Yue, C.Y.; Hu, X.; Yang, J. Tribological properties of short carbon fibers reinforced epoxy composites. *Friction* **2014**, *2*, 226–239. [CrossRef]
16. Ajayan, P.M.; Schadler, L.S.; Giannaris, C.; Rubio, A. Single-walled carbon nanotube-polymer composites: Strength and weakness. *Adv. Mater.* **2000**, *12*, 750–753. [CrossRef]
17. Dos Santos Barros, T.P.; De Lima Cavalcante, D.G.; De Oliveira, D.F.; Caluête, R.E.; De Lima, S.J.G. Study of the surface properties of the epoxy/quasicrystal composite. *J. Mater. Res. Technol.* **2019**, *8*, 590–598. [CrossRef]
18. Bloom, P.D.; Baikerikar, K.G.; Anderegg, J.W.; Sheares, V.V. Fabrication and wear resistance of Al–Cu–Fe quasicrystal-epoxy composite materials. *Mater. Sci. Eng. A* **2003**, *360*, 46–57. [CrossRef]
19. Sakly, A.; Kenzari, S.; Bonina, D.; Corbel, S.; Fournée, V. A novel quasicrystal-resin composite for stereolithography. *Mater. Des.* **2014**, *56*, 280–285. [CrossRef]
20. Chukov, D.I.; Stepashkin, A.A.; Tcherdyntsev, V.V.; Olifirov, L.K.; Kaloshkin, S.D. Structure and properties of composites based on polyphenylene sulfide reinforced with Al-Cu-Fe quasicrystalline particles. *J. Thermoplast. Compos. Mater.* **2018**, *31*, 882–895. [CrossRef]
21. Kenzari, S.; Bonina, D.; Dubois, J.M.; Fournée, V. Quasicrystal–polymer composites for selective laser sintering technology. *Mater. Des.* **2012**, *35*, 691–695. [CrossRef]
22. Solvay Amodel®; PPA Design Guide Amodel®. *Amodel® PPA Design Guide*; Specialty Polymers: Bollate, Italy, 2019; pp. 1–86.
23. Kušter, M.; Kovač, J.; Samardžija, Z.; Komelj, M.; Semsari Parapari, S.; Podlogar, M.; Dubois, J.-M.; Šturm, S. Impact of Tuned Oxidation on the Surface Energy of Sintered Samples Produced from Atomised B-Doped Al-Cu-Fe Quasicrystalline Powders. *Crystals* **2023**, *13*, 859. [CrossRef]
24. Vitos, L.; Ruban, A.V.; Skriver, H.L.; Kollár, J. The surface energy of metals. *Surf. Sci.* **1998**, *411*, 186–202. [CrossRef]
25. Strelniker, Y.M.; Havlin, S.; Bunde, A. Fractals and Percolation. In *Encyclopedia of Complexity and Systems Science*; Meyers, R., Ed.; Springer: New York, NY, USA, 2009; p. 10398. [CrossRef]
26. Sordelet, D.J.; Bloomer, T.A.; Kramer, M.J.; Unal, O. Effects of boron on the solidification structure of an AI-Cu-Fe alloy. *J. Mater. Sci. Lett.* **1996**, *15*, 935–939. [CrossRef]

27. *ASTM D256-10;* Standard Test Methods for Determining the Izod Pendulum Impact Resistance of Plastics. ASTM International: West Conshohocken, PA, USA, 2018.
28. Adur, A.M. Upgrading performance of recycled polyamide with ZeMac®. In *Upgrading Performance of Recycled Nylon;* Vertellus Specialties Inc.: Parsippany, NJ, USA, 2011.
29. Hutchings, I.M. *Tribology, Friction and Wear of Engineering Materials,* 1st ed.; Elsevier Limited: Oxford, UK, 1992.
30. Vollath, D. Agglomerates of nanoparticles. *Beilstein J. Nanotechnol.* **2020**, *11*, 854–857. [CrossRef]
31. Iveson, S.M.; Litster, J.D.; Ennis, B.J. Fundamental studies of granule consolidation Part 1: Effects of binder content and binder viscosity. *Powder Technol.* **1996**, *88*, 15–20. [CrossRef]
32. Anderson, W. Wettability Literature Survey—Part 2: Wettability Measurement. *J. Pet. Technol.* **1986**, *38*, 1246–1262. [CrossRef]
33. Wittmann, R.; Urban, K.; Schandl, M.; Hornbogen, E. Mechanical properties of single-quasicrystalline AlCuCoSi. *J. Mater. Res.* **1991**, *6*, 1165–1168. [CrossRef]
34. Kang, S.S.; Dubois, J.M. Compression testing of quasicrystalline materials. *Philos. Mag. A* **1992**, *66*, 151–163. [CrossRef]
35. Available online: https://www.engineersedge.com/coeffients_of_friction.htm (accessed on 1 January 2024).
36. Mihu, G.; Mihalache, I.; Graur, I.; Ungureanu, C.; Bria, V. Comparative study regarding friction coefficient for three epoxy resins. *IOP Conf. Series Mater. Sci. Eng.* **2017**, *174*, 012024. [CrossRef]
37. Durand, J.M.; Vardavoulias, M.; Jeandin, M. Role of reinforcing ceramic particles in the wear behaviour of polymer-based model composites. *Wear* **1995**, *181–183*, 833–839. [CrossRef]

crystals

MDPI

Article

Heat Treatment Post-Processing for the Improved Mechanical Properties of Scalmalloy® Processed via Directed Energy Deposition

Rachel Boillat-Newport, Sriram Praneeth Isanaka and Frank Liou *

Department of Mechanical Engineering, Missouri University of Science and Technology, Rolla, MO 65409, USA
* Correspondence: liou@mst.edu

Abstract: As high-strength aluminum alloys present several processability issues with additive manufacturing (AM), Scalmalloy®, an Al-Mg-Sc-Zr-based alloy, has been developed. This alloy is age-hardenable, allowing it to precipitate out a strengthening precipitate phase, $Al_3(Sc,Zr)$. The manufacturer recommends a single-stage aging treatment at 325 °C for 4 h; however, the majority of the literature studies utilize a powder bed processing known as selective laser melting (SLM) over powder-fed processing directed energy deposition (DED). This study addresses the lack of information on heat treatments for DED fabrication by exploring the application of artificial aging temperatures of 300–400 °C for 2, 4, and 6 h to: 1. determine the impact on the microstructural evolution and mechanical performance and 2. determine whether the recommended treatment for Scalmalloy® is appropriate for DED fabrication. Tensile testing determined that low-temperature treatments exhibited no visible dependence on time (2–6 h); however, time becomes influential at higher temperatures starting at 350 °C. The temperature plays a considerable role in the mechanical and microstructural behaviors of DED Scalmalloy®. The highest tensile strength was noted at 300 °C (384 MPa, 21.6% increase), but all heat-treated cases resulted in an improvement over the as-built case. This investigation established that increasing the treatment temperature resulted in a decreasing trend for the tensile strength that held over time. Elongation at 2 h displayed a near parabolic trend that peaks at 350 °C (20%) and falls with higher temperatures. At the 4 h treatment, a slight decreasing trend was noticed for elongation. No visible change was observed for elongation at 6 h, with elongation values remaining fairly consistent. The microstructural evolution, including micron-sized and nano-sized $Al_3(Sc,Zr)$ and grain size, was examined, and coarsening effects were noted with the increase in the temperature. It is recommended that treatment be conducted at 300 °C to achieve the precipitation of the strengthening $Al_3(Sc,Zr)$ phase while minimizing coarsening.

Keywords: directed energy deposition (DED); additive manufacturing (AM); aluminum; scandium; Scalmalloy®; heat treatment; artificial aging; tensile properties; precipitation strengthening

1. Introduction

Aluminum and its alloys have received significant attention due to their application in fields that highly prize their high strength-to-weight ratio [1]. Despite this promising feature, making these alloys highly desired to use with AM, aluminum alloys exhibit high thermal conductivity [1–3], poor laser absorptivity [1–4], and a high tendency for oxide formation [1–4]. Due to these characteristics, only a select number of aluminum alloys are compatible with AM. The desired high-strength aluminum alloys do not fall into this category due to their tendency to exhibit defects, such as porosity, lack of fusion, elemental loss due to the vaporization of low boiling point elements, and cracking [1–3]. These defects have been shown to have a negative impact on mechanical behavior, producing components with considerably lower mechanical properties when compared to AM-compatible alloys and high-strength alloys fabricated using conventional processing methods [3,5].

To combat the processability issues of high-strength aluminum alloys with AM, novel alloy compositions are being developed. Scalmalloy®, a new Al-Mg-Sc-Zr-based alloy designed for AM, has garnered considerable attention for its improved processability, unique microstructural features, and ability to demonstrate mechanical properties on par or superior to its conventionally processed counterparts. The microstructure of powder bed AM-fabricated Scalmalloy® components has been shown on several occasions to be a bimodal distribution in which fine-grained and coarse-grained regions exist [6–10]. Spierings et al. hypothesized that the fine-grained region was the direct result of a large number of seed crystals acting as nucleation sites around the boundary of the melt pool [7]. Comparatively, the coarse-grained region was connected to the smaller number of seed crystals coupled with a high-temperature gradient, resulting in larger, coarse columnar grains [7]. Kuo et al. determined that the lack of Al_3Sc precipitates in the as-built microstructure led to zones in which no precipitation strengthening occurred [11].

For Scalmalloy®, the manufacturer recommends a single-step aging treatment for powder bed-fabricated components at 325 °C for 4 h [12]. Kuo et al. determined that the non-homogenous microstructure led to lower yield strengths for the as-built condition; however, the subsequent heat treatment improved the mechanical behavior through the precipitation of a strengthening Al_3Sc phase [11]. Some investigations have explored a range of times and temperatures to determine the optimal heat treatment characteristics of Scalmalloy® and other Al-Mg-Sc-based alloys [8,13,14]. Spierings et al. found the best tensile strength by aging at temperatures in the range of 325–350 °C for 4–10 h [8]. In another study, Spierings et al. noted that temperatures in excess of 400 °C for a hold time of at least four hours resulted in a decrease in hardness [7]. The lower hardness values obtained were caused by Ostwald ripening from too large $Al_3(Sc,Zr)$ precipitates where it became more favorable to introduce a dislocation rather than increase the strain around the particle [7]. Baig et al. explored heat treatments at 325, 350, and 375 °C for various times up to 24 h and noted that the highest hardness overall, 160 ± 8 HV, occurred at 325 °C at 4 h [13]. While single-stage aging for the post-processing of Scalmalloy® has been well explored for powder bed processing, little to no literature exists that explores the heat treatment of powder-fed Scalmalloy® components.

Powder bed and powder-fed processes fall into the metal additive manufacturing category but have distinct differences. Table 1 details some of the characteristics of the powder bed process, selective laser melting (SLM), and the powder-fed process, directed energy deposition (DED). Of note is the difference between the cooling rates, which can dramatically impact the final microstructure [15]. Furthermore, studies have noted the greater tendency towards defect formation using DED fabrication methods when processing Scalmalloy®, making the process difficult to use effectively [16].

Table 1. SLM vs. DED processing features.

Parameters	Selective Laser Melting (SLM)	Directed Energy Deposition (DED)
Geometric Complexity	High/Complex parts possible [17–19]	Monolithic, high/complex, multi-material structures [19,20]
Build Size	Limited by size of build plate [17,18]	Allows for large-scale parts [20,21]
Surface Roughness	Minimal/Minor post-processing may be required [17,18]	Extensive/Often requires extensive post-processing [22]
Material Requirement	Uses considerably more powder; Offers opportunities for powder reuse [18]	Minimal materials used as powder required is injected into the melt pool
Cooling Rates	10^4–10^6 K/s [18]	10^3–10^5 K/s [23]
Repair Capabilities	Limited capabilities [24]	Yes [21,25–27]

Despite the advantages of SLM, DED cannot be written off as an inferior process. DED has features that are not possible with SLM and presents several avenues for advancement and investigation. For example, DED has the capability for repair applications, in situ composition blending using multiple powder feeders, and functionally graded materials [19,21,22,25]. Furthermore, the current state of the literature has a distinct lack of heat treatment studies for DED. Most of the literature has thoroughly explored the heat treatment of SLM compositions; however, as noted above, SLM and DED are distinctly different processes and produce markedly different starting microstructures. As such, it can be stated that the microstructures, when processed using the same heat treatment, would result in different ending microstructures and diverging mechanical properties. These possibilities highlight the need for more investigation into DED processing and the post-processing of Scalmalloy® to determine avenues that can be pursued to enhance mechanical performance while maintaining the unique capabilities inherent to DED.

This investigation focuses on fabricating Scalmalloy® components via the powder-fed process DED and subsequent post-processing to enhance their mechanical behavior relative to the as-built condition and the conventionally processed high-strength 7XXX series Al-Zn alloys. As-built and heat-treated Scalmalloy® components were examined to determine their microstructural and mechanical characteristics and provide a recommendation for the heat treatment of DED Scalmalloy®. Furthermore, many DED studies use Al-Mg-Sc-Zr alloy compositions not characterized as Scalmalloy®. This study can be applied to similar compositions and act as a starting point for optimization studies for these other alloys.

2. Materials and Methods

2.1. AM Processing

Scalmalloy® was obtained from Toyal America, Inc. (Lockport, IL, USA). Figure 1a shows the morphology of the sourced Scalmalloy®. The powder exhibits high sphericity with little to no agglomerates and satellites. The powder distribution is shown in Figure 1b. Scalmalloy® components were fabricated using an in-house, custom-designed directed energy deposition (DED) system in a shielded argon environment flowing at 1 L/min, as depicted in Figure 2a. The deposition parameters for this experiment are 1625 W power, 15 mm/s scan speed, and 2.25 g/min powder feed rate. The deposited coupons were 25 mm × 25 mm × 6 mm, as presented in Figure 2b. The parameter optimization to yield the chosen deposition parameters is detailed in Boillat-Newport et al. [28]. The resulting deposits from the optimized parameters exhibited minimal defects, such as porosity and cracking. Good adhesion was noted between the deposit and substrate. The Archimedes density performed on the deposits determined that the optimized properties produced near fully dense components.

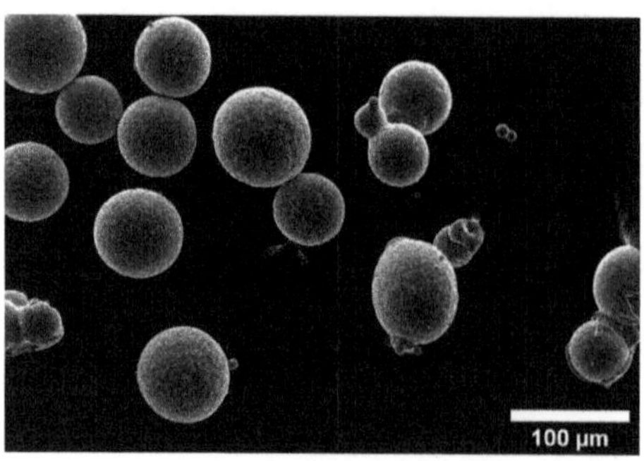

(a)

Size Metric	Particle Size (μm)
D10	58.7
D50	79.0
D90	115

(b)

Figure 1. (**a**) Scalmalloy® powder sourced from Toyal America, Inc. Reprinted with permission from Ref. [28], 2024, MDPI. (**b**) Particle size distribution of the Scalmalloy® feedstock material.

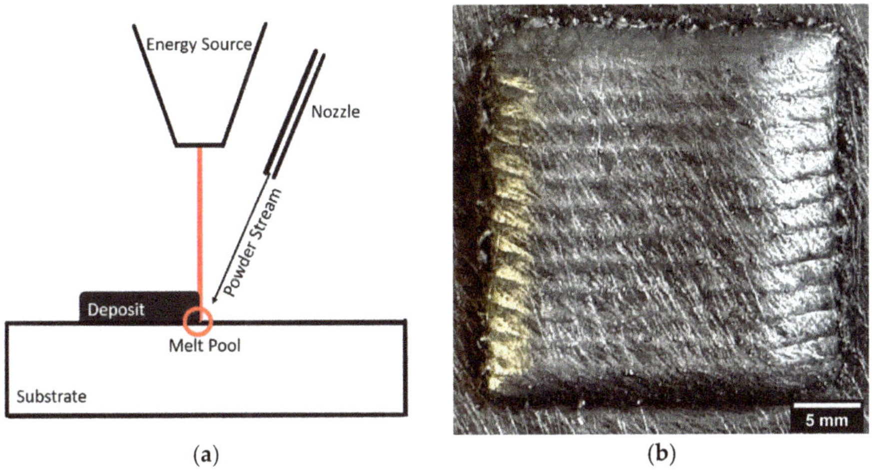

Figure 2. (**a**) DED process schematic. Reprinted with permission from Ref. [29], 2021, MDPI. (**b**) DED-fabricated Scalmalloy® coupon. Reprinted with permission from Ref. [28], 2024, MDPI.

2.2. Heat Treatment

Scalmalloy® deposits were heat-treated across a range of times and temperatures to identify a direct aging treatment time and temperature that would work for DED components. Deposits were heat-treated at 300, 325, 350, 375, and 400 °C for 2, 4, and 6 h. For each treatment, a ramp from 25 °C (room temperature) to the target temperature occurred at 10 °C per minute. After the specified hold for each case, the deposits were air-cooled to room temperature.

2.3. Microstructural and Mechanical Characterization

Microscopy specimens were polished to 0.05 microns using a colloidal alumina suspension. The microstructure was characterized using a Thermo Fisher Scientific, Inc. (TFS) (Waltham, MA, USA) PRISMA scanning electron microscope (SEM) and TFS HYDRA for electron microscopy imaging and TFS, previously FEI, Helios NanoLab 600 with an Oxford Aztec detector for energy dispersive spectroscopy (EDS). For each scope, the manufacturer's recommended working distance was utilized (10 mm for TFS PRISMA, 4 mm for TFS HYDRA, and 5 mm for TFS Helios NanoLab 600). Electron backscatter diffraction (EBSD) was performed by JH Analytical using a Thermo Fisher Scientific (Brno, Czech Republic) Apreo S SEM with an Oxford Instruments (High Wycombe, Buckinghamshire, UK) Symmetry S2 EBSD with an accelerating voltage of 30 kV.

Miniature tensile specimens, as shown in Figure 3, were cut from the center of each deposit in the XY plane using a Sodick Inc. (Schaumburg, IL, USA) VL600QH electrical discharge machine (EDM). Miniature tensile specimens were chosen for experimentation over the larger, more traditional tensile specimens due to the ability to strategically locate the specimens spatially and the fact that the interaction volumes are more sensitive to any defects present in AM-fabricated samples, providing a unique understanding of AM-induced defect mechanism behavior. Furthermore, the smaller size of the tensile samples eliminates the need for very large, cumbersome, and time-consuming deposits. Tensile testing was conducted using a Tinius Olsen Universal Testing Machine Model 25ST operating at a 0.25 mm/min positional rate.

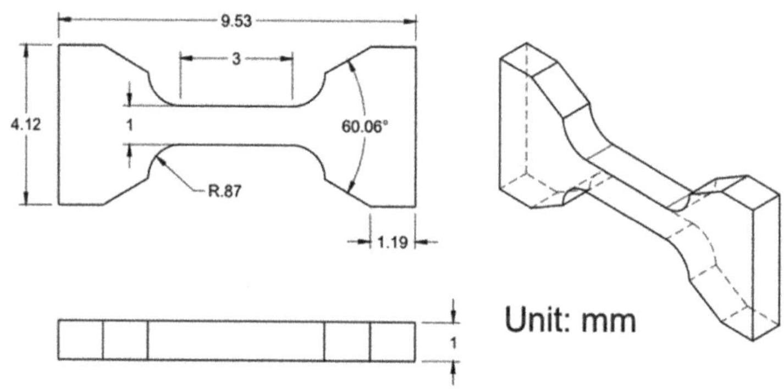

Figure 3. Miniature tensile specimen part drawing [30].

It is worth noting that miniature specimens can impact the results as they tend to pick up on any inhomogeneities and anisotropies in the deposits. This is crucial for AM, especially as this allows for tensile specimens that are more sensitive to defects present, providing an in-depth understanding of AM critical flaws. Studies by Karnati et al. [31] and Michopoulos et al. [32] have explored the use of miniature tensile specimens with additive manufacturing. Karnati et al. [31] note that miniature tensile specimens display a wider spread in the data, but the overall strength values are comparable to ASTM tensile samples. The spread in the data is attributed to the greater sensitivity inherent in the sample design [31].

The data in this study utilized two significant figures in correlation with ASTM E29-22 Section 7.6 [33] for average and standard deviation values. Tensile strength is reported to the nearest 1 MPa in accordance with ASTM E8 Section 7.13.1 [34].

3. Results

3.1. Tensile Properties

The ultimate tensile strength (UTS) and elongation for each temperature at 2, 4, and 6 h are presented in Figure 4. A decreasing trend is noted for UTS across all times, but of note is that a minimal change is seen across different hold times for low temperatures. For example, at 300 °C, the median UTS values are 380, 371, and 384 MPa for 2, 4, and 6 h hold times, respectively. At 2 h, the elongation displays a parabolic trend, increasing until 350 °C, where the trend decreases with higher temperatures. This trend is not seen at 4 and 6 h. Instead, a slight decreasing trend is observed at 4 h, while no visible trend is observed at 6 h.

3.2. Microstructural Characterization

Five samples were selected for microstructural analysis based on the results presented in Section 3.1. Of these five samples, the as-built condition and 325 °C—4 h were chosen for analysis to act as standard cases. The as-built condition served as a basis for the comparison of all heat-treated cases to determine the success of the treatment, while the 325 °C—4 h case was selected as it is the manufacturer's recommended treatment. The three remaining cases of the five chosen are 300 °C—2 h, 325 °C—2 h, and 350 °C—2 h. These three cases were selected for further analysis as the highest strengths were seen at 300 and 325 °C, with time playing a minimal role in the effect against tensile strength. From the data at 350 °C, time durations longer than 2 h were shown to affect the measured mechanical performance; so, the 350 °C—2 h case was selected as the pivotal point. Furthermore, for the cases at 300 °C and 350 °C, the chosen time duration was also 2 h to minimize the effects of time and allow for a clearer comparison and analysis of the temperature effects.

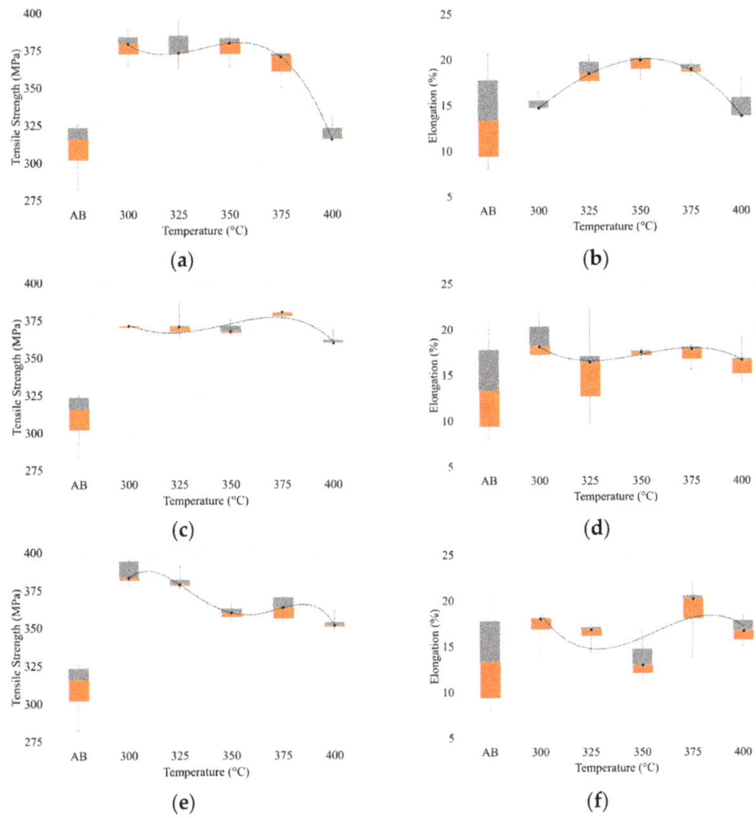

Figure 4. UTS and elongation of Scalmalloy® treated at temperatures in the range of 300–400 °C at (**a**,**b**) 2 h, (**c**,**d**) 4 h, and (**e**,**f**) 6 h. Note: Orange block corresponds to the 25th percentile, the gray block the 75th percentile, and the interface between the orange and gray blocks is the median value. A trendline is plotted across the heat treatments to highlight the impact of changing temperature at constant time.

3.2.1. As-Built Microstructural Phenomenon of DED Scalmalloy®

Before the examination of the microstructures of interest in this study, it is important to briefly discuss the as-built condition and the impact that DED processing has on the $Al_3(Sc,Zr)$ phase during cooling. Figure 5 shows the microstructure of the as-built DED Scalmalloy®. Boillat-Newport et al. [28] examined the microstructure of DED Scalmalloy® and found that the as-built case had a very prevalent secondary phase, determined to be $Al_3(Sc,Zr)$. The $Al_3(Sc,Zr)$ phase formed during solidification rather than being induced by an age-hardening heat treatment [28]. This was also noted in other studies and linked to the cooling rates of DED, which are not fast enough to maintain the supersaturated solid solution, resulting in the precipitation of micron-sized $Al_3(Sc,Zr)$ [35]. Figure 5a highlights the presence of zones with larger and fewer precipitates in contrast to smaller and many precipitates found in other zones. This type of bimodal distribution is noted in SLM studies and is exaggerated, leading to the formation of coarse- and fine-grained regions. A higher magnification micrograph is presented in Figure 5b, which shows a representation of the morphology of the $Al_3(Sc,Zr)$ secondary phase that is heavily present in DED Scalmalloy®. The regions noted in Figure 5a are further detailed in the melt pool schematic in Figure 5c. This schematic highlights that these regions with many small particles of $Al_3(Sc,Zr)$ and a few large particles of $Al_3(Sc,Zr)$ are located primarily at the interface between the remelted and non-remelted sections of the deposits. The large particles were seen in the remelted regions as the result of growth during thermal cycling, while the small particles were located in the parts of the deposit that were not remelted.

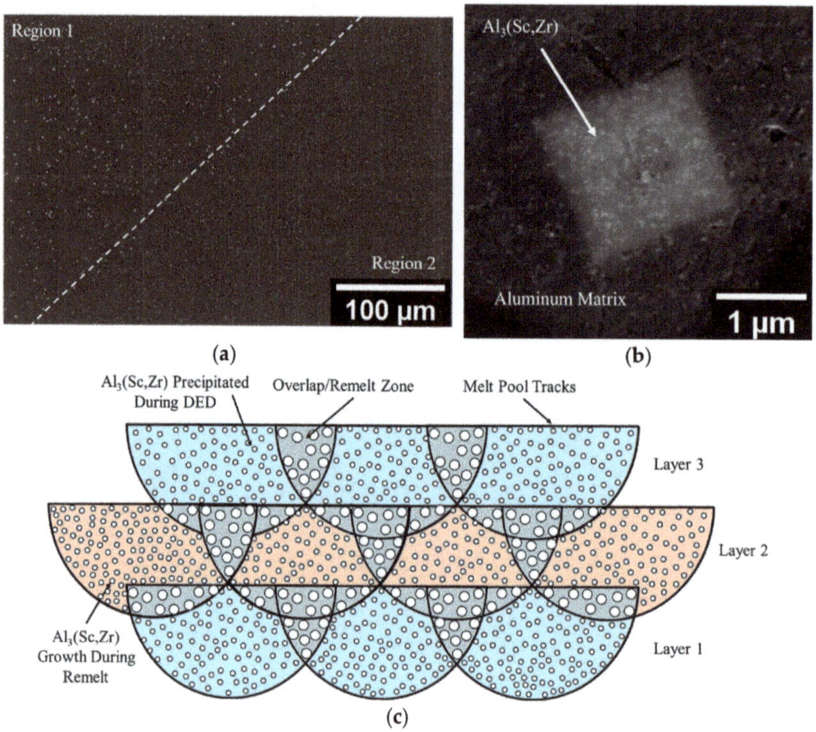

Figure 5. As-built microstructure of DED Scalmalloy® showcasing the presence of an $Al_3(Sc,Zr)$ secondary phase: (**a**) micrograph showing two different regions with differing $Al_3(Sc,Zr)$ precipitates. Adapted with permission from Ref. [28], 2024, MDPI. (**b**) High magnification image of the $Al_3(Sc,Zr)$ precipitate. (**c**) Melt pool schematic showing the relationship between the regions shown in (**a**) and the melt track morphology, where the circles represent $Al_3(Sc,Zr)$ particles in two sizes, smaller in the non-remelted region and larger in the remelted. The dark line represents melt-pool boundaries across three layers.

From the analysis of the as-built condition for DED and the literature examination of the microstructure for as-built SLM, it is apparent that there are significant differences between the two processes. Baker et al. [36], in their study on AM titanium, noted that heat treatments for conventionally processed titanium may not provide optimal properties as expected with other manufacturing processes due to the variation in the microstructure of the starting material. Currently, the manufacturer recommends that Scalmalloy® be treated at 325 °C for 4 h to produce high tensile strengths above 500 MPa. However, as noted by Baker et al. [36], the likelihood of obtaining the same result through a single heat treatment of two differently fabricated samples is low. To date, SLM Scalmalloy® has been shown to produce the desired high strengths, while DED does not. Considering the differences between the processes and starting microstructure, it is clear that optimized properties cannot be achieved without process-specific treatments.

3.2.2. $Al_3(Sc,Zr)$ Phase Particles

Figure 6a provides a representational micrograph showing the presence of a secondary micron and sub-micron-sized phase. This phase has been noted in all heat-treated cases, as well as the as-built case, with the growth of the phase found with heat treatment due to coarsening. ImageJ v1.53k software was utilized in conjunction with SEM micrographs sourced from the TFS PRISMA to investigate the evolution of the phase particles from the as-built to heat-treated case (300 °C—2 h, 325 °C—2 h, 325 °C—4 h, and 350 °C—2 h). ImageJ was used to threshold the micrographs so that all unwanted features were eliminated and only the particles under investigation were visible. At least 1000 particles were analyzed for all cases, and the Feret diameter was determined.

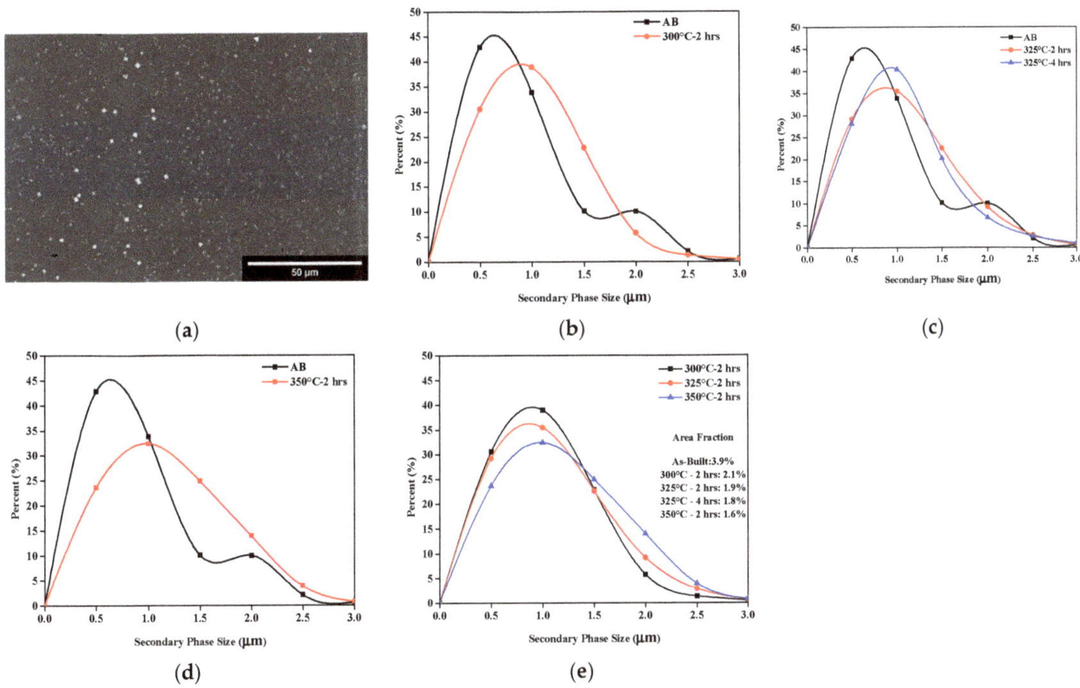

Figure 6. Microstructural evolution of secondary phase particles: (**a**) representational micrograph at 300 °C—2 h depicting the micron- and sub-micron-sized particles; (**b**) AB vs. 300 °C; (**c**) AB vs. 325 °C; (**d**) AB vs. 350 °C; and (**e**) heat-treated case comparison.

Figure 6 also shows the size distribution for each case plotted against the as-built case. All heat-treated cases show an overall increase in the particle size compared to the as-built ones, as shown in Figure 6b–d. Figure 6e details the size evolution of each heat treatment relative to each other. The 300 °C—2 h case exhibited the narrowest distribution of the heat-treated cases. All other cases produced a broader distribution, with the 350 °C—2 h case exhibiting the widest distribution, highlighting the growth influenced by heat treatment. Figure 6e also shows the area fraction of the particles. Overall, it can be seen that, with the heat treatment, the particle size increased, while the number of particles decreased.

EDS analysis, shown in Figure 7, was utilized to identify the particle phase in Figure 6. The phase was primarily composed of aluminum, scandium, and zirconium. The standardized EDS composition of the phase is presented in Figure 7a. A small amount of magnesium is detected, but this is due to signals from the surrounding matrix material rather than the phase itself. Considering the EDS results and carefully examining the literature [37] leads to the conclusion that these particles are $Al_3(Sc,Zr)$. $Al_3(Sc,Zr)$ is a common phase in

Scalmalloy® and provides a major strengthening mechanism that works by pinning grain boundaries. In addition to EDS, X-ray diffraction (XRD) was performed to further confirm the designation as $Al_3(Sc,Zr)$; however, due to the small content of the phase, it was not detected by XRD. This phenomenon has been noted by other researchers and is concluded to be related to the detection limit of the equipment utilized [28,38,39]. The EDS maps in Figure 7 show the elemental segregation of scandium and zirconium at the secondary $Al_3(Sc,Zr)$ phase. The matrix material around $Al_3(Sc,Zr)$ is primarily an aluminum solid solution, as noted by XRD.

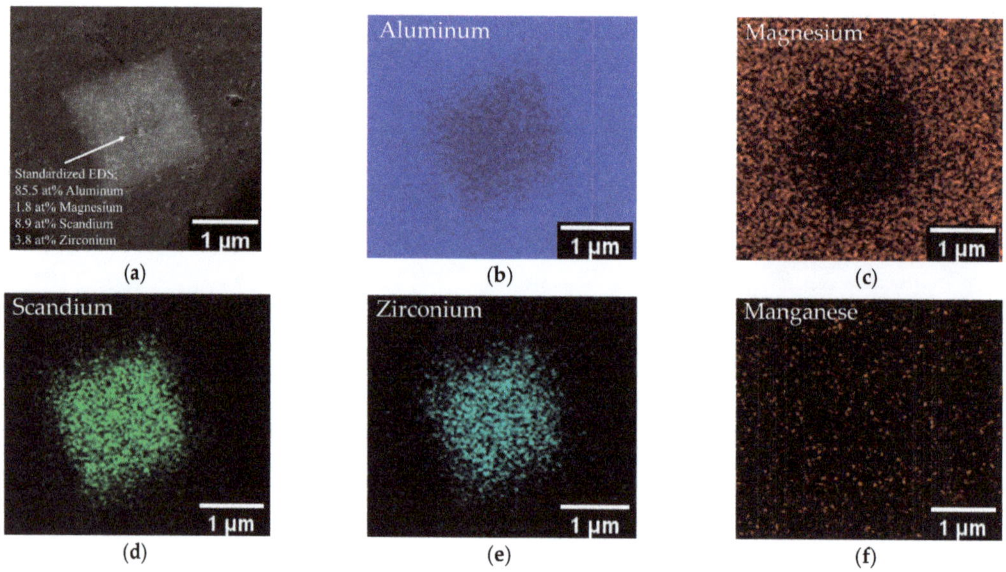

Figure 7. Elemental EDS mapping of secondary phase particle in DED Scalmalloy®: (**a**) SEM image of secondary phase particle with quantitative EDS; (**b**) aluminum map; (**c**) magnesium map; (**d**) scandium map; (**e**) zirconium map; and (**f**) manganese map.

3.2.3. $Al_3(Sc,Zr)$ Nanoprecipitates

Figure 8a shows a representation of the nanoprecipitates analyzed using ImageJ to produce the size distributions presented in Figure 8b–e. When compared to the as-built condition, all heat treatments result in the growth of the nanoparticle phase. Furthermore, as noted at the micron scale, size evolves with the increase in the treatment temperature. Of note is that a difference can be seen between the distribution at 325 °C—2 h and 325 °C—4 h. This indicates that, while the effects of time at the lower temperatures are fairly negligible for tensile strength and micron-sized $Al_3(Sc,Zr)$, time considerably impacts the growth of the nano-sized precipitates.

3.2.4. Grain Evolution

Figure 9 shows representative images of EBSD grain maps for the as-built, 300 °C—2 h, and 350 °C—2 h cases. Compared to the as-built case, all heat-treated cases show an increased grain size with a greater coarsening exhibited by increased temperatures. Table 2 details the average and maximum grain size for each case, indicating that, as the temperature increased, the grain size also increased. A similar trend was noted with the size evolution of the secondary $Al_3(Sc,Zr)$ micron phase. Cases that exhibited an increase in $Al_3(Sc,Zr)$ size also exhibited a greater grain growth, indicating that, as the strengthening phase grew, it was not able to pin the grain boundaries effectively.

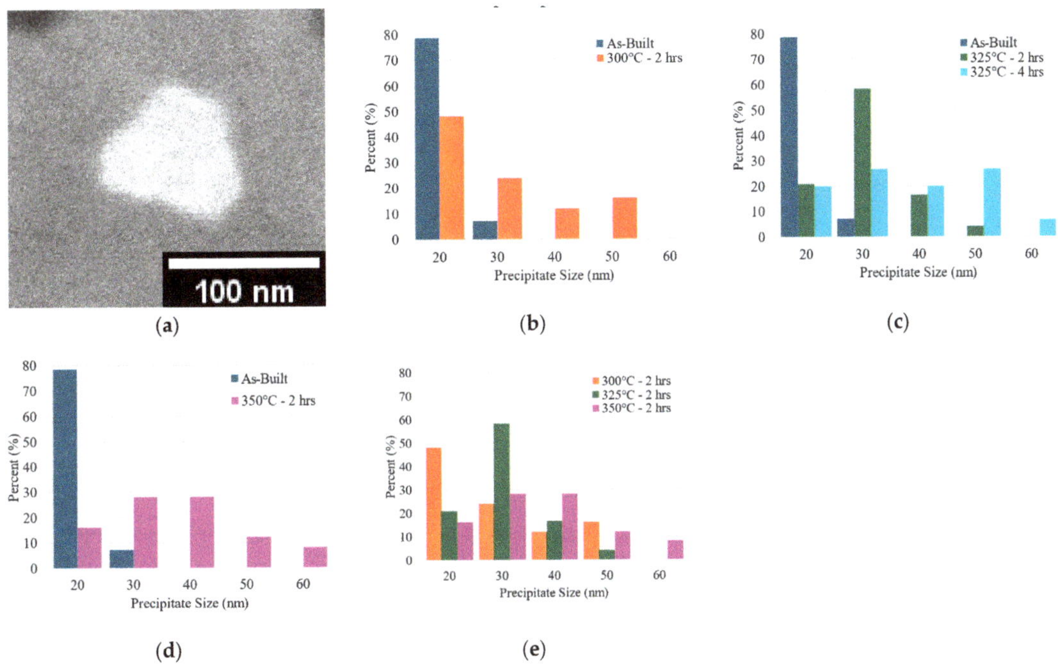

Figure 8. Evolution of Al_3(Sc,Zr) nanoprecipitates: (**a**) representational image of nanoprecipitates analyzed. Note: Micron-scale Al_3(Sc,Zr) has a cut-off of 500 nm. Under 500 nm, it falls into the category of nanoprecipitates. (**b**) As-built vs. 300 °C; (**c**) as-built vs. 325 °C; (**d**) as-built vs. 350 °C; and (**e**) 2 h heat-treated case comparison.

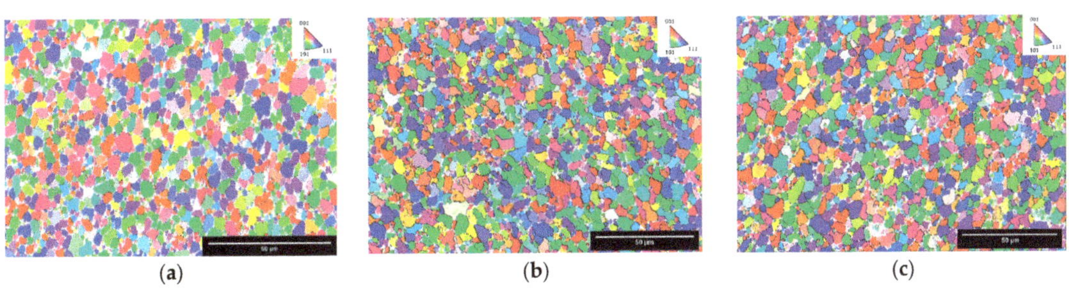

Figure 9. EBSD maps depicting the grain size evolution and texture orientation: (**a**) as-built. Reprinted with permission from Ref. [28], MDPI, 2024; (**b**) 300 °C—2 h; and (**c**) 350 °C—2 h showing an increase in the grain size with higher aging temperatures.

Table 2. Equivalent circle diameter grain size determined through EBSD analysis.

Sample	Area Weight Mean (μm)	Max. Grain Size (μm)
As-Built	4.2 ± 1.5	10
300 °C—2 h	5.0 ± 1.8	11
325 °C—4 h	5.3 ± 1.9	12
350 °C—2 h	5.7 ± 2.1	12

4. Discussion

4.1. Impact of Heat Treatment on Tensile Behavior

Several studies have noted the peak aging condition for Scalmalloy® fabricated via SLM to be around 325–350 °C, and treatment at higher temperatures deteriorated the properties [8,40]. Spierings et al. noted that aging at high temperatures and long hold times can result in overaging and a lower mechanical strength [8]. The use of excessive heat treatment temperatures negatively impacts the strengthening of the alloy, which can be linked to a diminished tensile strength [8,40]. Ma et al. found that increased temperatures increase the precipitation and coarsening rate for secondary $Al_3(Sc,Zr)$ particles [40]. Furthermore, when aging is performed at high temperatures (above 350 °C) and for long times, the alloy strength starts to decrease due to the coarsening of precipitated $Al_3(Sc,Zr)$ particles and loss of coherency between the particles and the matrix [40]. This weakens the strengthening mechanism, as dislocation motion and grain growth cannot be effectively hindered [40]. The trends noted in the literature correlate with the trends shown here and are best seen with the 2 h treatment condition shown in Figure 4a, where treatment at 375 °C and 400 °C results in property deterioration compared to that at lower temperatures at 2 h. This trend is still present at 4 and 6 h, but the temperature at which the properties begin to deteriorate is lower due to the addition of the time factor at longer aging hold times.

While similar trends are shown in this study and the literature regarding diminished strength when aged above the peak aging temperatures, the strength values of SLM Scalmalloy® are distinctly higher than those of DED Scalmalloy®. It is worth noting that there is a distinct lack of data on heat-treated DED Scalmalloy® as most studies only present the as-built case. This is in sharp contrast to SLM Scalmalloy® studies, in which extensive analyses of heat treatments have been conducted. Table 3 details the average tensile properties from the literature studies that fabricated Scalmalloy® and similar Al-Mg-Sc-Zr-based alloys via AM processing.

Table 3. Tensile properties of AM-fabricated Scalmalloy® and similar Al-Mg-Sc-Zr-based compositions.

Name	Composition	Process	Treatment	UTS (MPa)	Elongation (%)	Ref.
Scalmalloy®	(Mg: 4.5–4.9 wt%, Sc: 0.68–0.78 wt%, Zr: 0.2–0.4 wt%)	SLM	325 °C—4 h	512 496	14 16	[41]
			AB	349 ± 4 350 ± 7	20 ± 3 19 ± 4	[42]
			325 °C—4 h	512 ± 6 521 ± 1	12 ± 1 13 ± 4	
Other Al-Mg-Sc-Zr Alloys	(Mg: 6.52 wt%, Sc: 0.48 wt%, Zr: 0.27 wt%)		AB	400	9.4	[15]
	(Mg: 4.66 wt%, Sc: 0.72 wt%, Zr: 0.33 wt%)		AB	401 ± 5		[43]
	(Mg: 4.66 wt%, Sc: 0.48 wt%, Zr: 0.72 wt%)		AB	403 ± 4	27.5 ± 1.7	[44]
			325 °C—4 h	541 ± 5	15.9 ± 0.7	
Scalmalloy®	(Mg: 4.5–4.9 wt%, Sc: 0.68–0.78 wt%, Zr: 0.2–0.4 wt%)	DED	AB	<300	~3.5	[16]
Other Al-Mg-Sc-Zr Alloys	(Mg: 6.52 wt%, Sc: 0.48 wt%, Zr: 0.27 wt%)		AB	280	5.9	[15]
	(Mg: 4.66 wt%, Sc: 0.72 wt%, Zr: 0.33 wt%)		AB	278 ± 2		[43]
	(Mg: 4.66 wt%, Sc: 0.48 wt%, Zr: 0.72 wt%)		AB	328 ± 4	27.1 ± 1.7	[44]
			325 °C—4 h	395 ± 2	22.7 ± 0.5	

As is clear from examining the table, the values of DED-fabricated Scalmalloy® across the board, this study included, are surpassed by the properties made possible with SLM fabrication. As discussed above, SLM and DED are two different processes with considerable differences that can cause property value differences. SLM has a considerably faster cooling rate that can more effectively form the supersaturated solid solution upon solidification, which is the precursor to precipitation during aging. DED, being a slower cooling process, cannot maintain the supersaturated solid solution, and an $Al_3(Sc,Zr)$ phase can form before precipitation during aging [35,38]; refer to Figure 5. This limits the potential for nanoprecipitates, which provide much of the strengthening for Scalmalloy® and pin the grain boundaries to promote Hall–Petch strengthening [35]. This is detailed further in the next section. The tensile properties in this study are on the higher end of the properties reported for DED-fabricated Scalmalloy®. The research by Awd et al. explored both the SLM and DED fabrication of Scalmalloy® [16]. The tensile properties reported by Awd et al. highlight the detrimental effect of defects on mechanical behavior. Compared to this study, the tensile behavior seen by Awd et al. is surpassed due to the minimization of defects and application of post-processing heat treatment, demonstrating the potential for the DED processing of Scalmalloy®. Furthermore, DED researchers, understanding the potential of compositions with scandium and zirconium, have explored other Al-Mg-Sc-Zr alloys, as seen in Table 3. Of note is that some studies show similar results to those presented in this investigation. Thus, this study can act as the first step toward optimizing the heat treatment for each unique alloy.

4.2. Heat-Treated Microstructural Evolution

Figures 6 and 9 show that heat treatment plays a crucial role in the microstructural evolution of a Scalmalloy® deposit. The micron- and sub-micron-sized $Al_3(Sc,Zr)$ phase noted in Figure 6 showed an increase in size and decrease in number with heat treatment. In general, higher temperatures and longer times result in larger $Al_3(Sc,Zr)$ phase particles, with fewer of them being present. This indicates that the application of either higher temperatures or longer times than needed to facilitate the nanoprecipitation of $Al_3(Sc,Zr)$ from solid solution impacts coarsening and promotes the growth of large particles at the expense of smaller ones. Similarly, the nanoprecipitates, seen in Figure 8, show a strong temperature dependence, with the nanoprecipitate size increasing as the heat treatment temperature increases. Furthermore, unlike the micron-sized $Al_3(Sc,Zr)$, time impacts the nano-scale precipitates' size evolution. This is seen well when considering the nanoprecipitate evolution at 325 °C—2 h and 325 °C—4 h. After 4 h at 325 °C, the nanoprecipitates showed larger sizes and a wider size distribution than found at 2 h, indicating that small time differences play an effect. However, this effect is not significant enough to be shown on a larger scale. It is theorized that this would be much more drastic and found in both the micron-scale $Al_3(Sc,Zr)$ and tensile behaviors at considerably longer times.

The grain size of the heat-treated cases, as seen in Figure 9 and Table 2, surpasses the grain size of the as-built condition, indicative of coarsening occurring with the application of heat during the artificial aging treatment to promote nanoprecipitation. The grain size is strongly tied to the $Al_3(Sc,Zr)$ phase particles (micron and sub-micron) and nanoprecipitates. $Al_3(Sc,Zr)$ has been found in many studies to be a very effective strengthening mechanism, stabilizing the microstructure by pinning grain boundaries and promoting strengthening via the Hall–Petch effect. As $Al_3(Sc,Zr)$ evolves with the application of heat treatments, the effectiveness of grain boundary pinning and strengthening is impacted. Cases with high temperatures and/or long processing times provide coarsening during and after precipitation, leading to larger sizes for $Al_3(Sc,Zr)$, as seen in Figure 6, while lower temperatures and shorter times lead to less coarsening.

4.3. Microstructure and Tensile Behavior

This study highlights that the largest grain size, nanoprecipitates, and micron/sub-micron $Al_3(Sc,Zr)$ phase occurs at higher temperatures, as seen at 350 °C, relative to lower temperatures. Furthermore, as the temperature increases, the tensile strength decreases. This can be linked to microstructural evolution and the weakening of the strengthening potential. Much of Scalmalloy®'s strength stems from the ability of the nanoprecipitates to pin the grain boundaries, preventing dislocation motion. As the size of the precipitates increases, the number of them decreases, and the ability to hinder dislocation motion becomes stunted.

From this study, the data indicate that there is a threshold where treatment yields an overaged condition and provides properties below the peak aged properties possible. This occurs at temperatures above 350 °C for shorter treatment times (such as 2 h) or at 350 °C for longer times. Heat treatment promotes the growth of nanoprecipitates from the supersaturated solid solution and provides a coarsening effect that grows the precipitates and, subsequently, the grains. This study shows that the tensile strength is maximized at 300 °C and 325 °C; however, the growth of the grains and Al_3(Sc,Zr) is greater at 325 °C.

Marquis and Seidman explored the impact of the temperature on the coarsening rate of an Al-0.3 wt%Sc alloy and determined that the coarsening rate increases dramatically with the temperature [45]. The coarsening rate dependence on the temperature from Marquis and Seidman shows that the coarsening rate relative to the temperature is not linear [45]. Each temperature increase, even 50 °C, produces a substantial increase in coarsening [45]. Thus, it is desired to utilize a treatment temperature that can still produce the nanoprecipitates that provide strengthening while minimizing the coarsening rate. From this study, it is recommended for DED Scalmalloy® that heat treatment be performed at 300 °C to allow for nanoprecipitation to occur, providing strengthening and improving tensile behavior, while minimizing the coarsening rate. This recommendation for the treatment at 300 °C differs from the manufacturer's 325 °C at 4 h. It is theorized that the need for lower temperatures is not just linked to the minimization of coarsening but also influenced by the fact that the cooling rates for DED are slower than with SLM, resulting in Al_3(Sc,Zr) falling out of solution and forming before artificial aging. Thus, the amount of solute available for the precipitation of nanoparticles is lowered and could be a strong driving force for the application of a lower temperature treatment to promote precipitation and control coarsening. This highlights the distinct differences between DED and SLM and the need for process-specific treatments.

5. Conclusions

This study investigated the application of post-processing heat treatments to improve the tensile strength of DED-processed Scalmalloy®. The examination of the results of this study allows the following conclusions to be drawn:

1. The post-processing heat treatment of DED Scalmalloy® components improved UTS relative to the as-built case. After treatment at 300 °C, the UTS improved by 20.4%.
2. Tensile strength decreased with the increase in the treatment temperatures, with the effect being significantly noticed at temperatures above 325 °C. At 300 and 325 °C, the time was noted not to impact tensile behavior significantly; however, at temperatures of 350 °C and above, time becomes more important and influences behavior.
3. Micron-sized and nano-sized Al_3(Sc,Zr) exhibited a strong temperature dependence, with the size increasing as the treatment temperature increased. The examination of the micron-sized phase at 325 °C—2 h and 325 °C—4 h shows minimal time dependence at low temperatures and shorter periods. However, this was not the case when examining the nanoprecipitates treated at 325 °C over 2 and 4 h. The nanoprecipitate distribution at 2 h relative to the 4 h case was narrower. The 4 h case exhibited a wider distribution and larger sizes, indicating that, while no impact is seen for low temperatures and short hold times on the micron scale, there is, in fact, an effect that is visible at the nano-scale.
4. The examination of the microstructure indicates that the grain size increases with higher treatment temperatures. Compared to the as-built case, the treatments at 300 °C—2 h, 325 °C—4 h, and 350 °C—2 h yielded 19.1, 26.2, and 35.7% increases, respectively. The grain size evolution correlates with the increase in micron- and nano-sized Al_3(Sc,Zr), exhibiting the weakening ability for grain pinning and the

subsequent diminishment of the Hall–Petch strengthening made possible by smaller grains. This can be linked to the decreasing trend noted for the tensile strength.

From the results found in this study, it is recommended that the heat treatment for DED Scalmalloy® occur at short times and as low temperatures as possible. It was shown that the tensile strength when treated at 300° was on par with the treatment at 325 °C. The examination of the microstructure shows the growth of Al_3(Sc,Zr) and the grains with a small increase in the temperature from 300 to 325 °C. It is suggested that the treatment be performed at 300 °C rather than 325 °C to minimize coarsening. While this study did not yield the same high properties noted with SLM, this is the beginning step towards the optimization of the DED process to improve the resulting process. Furthermore, this study specifically targeted Scalmalloy®; however, the results shown in this paper can act as a guideline and promote the optimization of similar Al-Mg-Sc-Zr-based compositions.

Author Contributions: Conceptualization, R.B.-N. and S.P.I.; methodology, R.B.-N. and S.P.I.; formal analysis, R.B.-N.; writing—original draft preparation, R.B.-N.; writing—review and editing, R.B.-N. and S.P.I.; supervision, F.L.; project administration, F.L.; funding acquisition, S.P.I. and F.L. All authors have read and agreed to the published version of the manuscript.

Funding: This study was supported by DoEdu GAANN Grant # P200A210100, NSF Grants CMMI 1625736, and NSF EEC 1937128, Navair Contract # N6833520C0029 by way of Product Innovation and Engineering, LLC., the Intelligent Systems Center (ISC), and Material Research Center (MRC) at Missouri S&T. Their financial support is greatly appreciated.

Data Availability Statement: Data are contained within the article.

Acknowledgments: The authors would like to acknowledge the support from Nam Pham, project manager for Navair Contract # N6833520C0029. Additional thanks to the Intelligent Systems Center (ISC) and Materials Research Center (MRC) for their help in sample preparation and testing. Lastly, the authors also want to acknowledge the contributions of Ranjit Joy and Braden Mclain for their assistance in sample fabrication and tensile testing, Clarissa Wisner from the Advanced Materials Characterization Lab (AMCL) at Missouri S&T for her assistance with SEM analysis, and JH Technologies for EBSD microscopy.

Conflicts of Interest: The authors declare no conflicts of interest.

References

1. Altıparmak, S.C.; Yardley, V.A.; Shi, Z.; Lin, J. Challenges in additive manufacturing of high-strength aluminium alloys and current developments in hybrid additive manufacturing. *Int. J. Light. Mater. Manuf.* **2021**, *4*, 246–261. [CrossRef]
2. Yang, H.; Sha, J.; Zhao, D.; He, F.; Ma, Z.; He, C.; Shi, C.; Zhao, N. Defects control of aluminum alloys and their composites fabricated via laser powder bed fusion: A review. *J. Mater. Process. Technol.* **2023**, *319*, 118064. [CrossRef]
3. Dixit, S.; Liu, S. Laser Additive Manufacturing of High-Strength Aluminum Alloys: Challenges and Strategies. *J. Manuf. Mater. Process.* **2022**, *6*, 156. [CrossRef]
4. Zafar, F.; Reis, A.; Vieira, M.; Emadinia, O. Additively Manufactured High-Strength Aluminum Alloys: A Review. In *Recent Advancements in Aluminum Alloys*; Rajendrachari, D.S., Ed.; IntechOpen: Rijeka, Yugoslavia, 2023; ISBN 978-1-83768-510-3.
5. Martin, J.H.; Yahata, B.D.; Hundley, J.M.; Mayer, J.A.; Schaedler, T.A.; Pollock, T.M. 3D printing of high-strength aluminium alloys. *Nature* **2017**, *549*, 365. [CrossRef] [PubMed]
6. Ekubaru, Y.; Gokcekaya, O.; Ishimoto, T.; Sato, K.; Manabe, K.; Wang, P.; Nakano, T. Excellent strength–ductility balance of Sc-Zr-modified Al–Mg alloy by tuning bimodal microstructure via hatch spacing in laser powder bed fusion. *Mater. Des.* **2022**, *221*, 110976. [CrossRef]
7. Spierings, A.B.; Dawson, K.; Heeling, T.; Uggowitzer, P.J.; Schäublin, R.; Palm, F.; Wegener, K. Microstructural features of Sc- and Zr-modified Al-Mg alloys processed by selective laser melting. *Mater. Des.* **2017**, *115*, 52–63. [CrossRef]
8. Spierings, A.B.; Dawson, K.; Kern, K.; Palm, F.; Wegener, K. SLM-processed Sc- and Zr-modified Al-Mg alloy: Mechanical properties and microstructural effects of heat treatment. *Mater. Sci. Eng. A* **2017**, *701*, 264–273. [CrossRef]
9. Cabrera-Correa, L.; González-Rovira, L.; de Dios López-Castro, J.; Castillo-Rodríguez, M.; Botana, F.J. Effect of the heat treatment on the mechanical properties and microstructure of Scalmalloy® manufactured by Selective Laser Melting (SLM) under certified conditions. *Mater. Charact.* **2023**, *196*, 112549. [CrossRef]
10. Schimbäck, D.; Kaserer, L.; Mair, P.; Palm, F.; Leichtfried, G.; Pogatscher, S.; Hohenwarter, A. Deformation and fatigue behaviour of additively manufactured Scalmalloy® with bimodal microstructure. *Int. J. Fatigue* **2023**, *172*, 107592. [CrossRef]

11. Kuo, C.N.; Peng, P.C.; Liu, D.H.; Chao, C.Y. Microstructure evolution and mechanical property response of 3d-printed scalmalloy with different heat-treatment times at 325 °C. *Metals* **2021**, *11*, 555. [CrossRef]
12. Carpenter Additive Scalmalloy Datasheet. Available online: https://www.carpenteradditive.com/hubfs/Resources/Data%20 Sheets/Scalmalloy_Datasheet.pdf (accessed on 18 June 2024).
13. Baig, S.; Ghiaasiaan, S.R.; Shamsaei, N. Effect of Heat Treatment on the Microstructure and Mechanical Properties of LB-PBF AlSi10Mg and Scalmalloy. In Proceedings of the Light Metals 2021; Perander, L., Ed.; Springer International Publishing: Cham, Switzerland, 2021; pp. 119–125.
14. Li, R.; Chen, H.; Zhu, H.; Wang, M.; Chen, C.; Yuan, T. Effect of aging treatment on the microstructure and mechanical properties of Al-3.02Mg-0.2Sc-0.1Zr alloy printed by selective laser melting. *Mater. Des.* **2019**, *168*, 107668. [CrossRef]
15. Deng, C.; Li, R.; Yuan, T.; Niu, P.; Wang, Y. Microstructure and Mechanical Properties of a Combination Interface between Direct Energy Deposition and Selective Laser Melted Al-Mg-Sc-Zr Alloy. *Metals* **2021**, *11*, 801. [CrossRef]
16. Awd, M.; Tenkamp, J.; Hirtler, M.; Siddique, S.; Bambach, M.; Walther, F. Comparison of Microstructure and Mechanical Properties of Scalmalloy® Produced by Selective Laser Melting and Laser Metal Deposition. *Mater* **2017**, *11*, 17. [CrossRef] [PubMed]
17. Gao, B.; Zhao, H.; Peng, L.; Sun, Z. A Review of Research Progress in Selective Laser Melting (SLM). *Micromachines* **2022**, *14*, 57. [CrossRef] [PubMed]
18. Gokuldoss, P.K.; Kolla, S.; Eckert, J. Additive Manufacturing Processes: Selective Laser Melting, Electron Beam Melting and Binder Jetting-Selection Guidelines. *Materials* **2017**, *10*, 672. [CrossRef] [PubMed]
19. Kim, J.; Wakai, A.; Moridi, A. Materials and manufacturing renaissance: Additive manufacturing of high-entropy alloys. *J. Mater. Res.* **2020**, *35*, 1963–1983. [CrossRef]
20. Gibson, B.T.; Mhatre, P.; Borish, M.C.; Atkins, C.E.; Potter, J.T.; Vaughan, J.E.; Love, L.J. Controls and process planning strategies for 5-axis laser directed energy deposition of Ti-6Al-4V using an 8-axis industrial robot and rotary motion. *Addit. Manuf.* **2022**, *58*, 103048. [CrossRef]
21. Gouveia, J.R.; Pinto, S.M.; Campos, S.; Matos, J.R.; Sobral, J.; Esteves, S.; Oliveira, L. Life Cycle Assessment and Cost Analysis of Additive Manufacturing Repair Processes in the Mold Industry. *Sustainability* **2022**, *14*, 2105. [CrossRef]
22. Svetlizky, D.; Das, M.; Zheng, B.; Vyatskikh, A.L.; Bose, S.; Bandyopadhyay, A.; Schoenung, J.M.; Lavernia, E.J.; Eliaz, N. Directed energy deposition (DED) additive manufacturing: Physical characteristics, defects, challenges and applications. *Mater. Today* **2021**, *49*, 271–295. [CrossRef]
23. Dass, A.; Moridi, A. State of the Art in Directed Energy Deposition: From Additive Manufacturing to Materials Design. *Coatings* **2019**, *9*, 418. [CrossRef]
24. Yap, C.Y.; Chua, K.; Dong, Z.; Liu, Z.; Zhang, D.; Loh, L.E.; Sing, S.L. Review of selective laser melting: Materials and applications. *Appl. Phys. Rev.* **2015**, *2*, 41101. [CrossRef]
25. Ahn, D.-G. Directed Energy Deposition (DED) Process: State of the Art. *Int. J. Precis. Eng. Manuf. Technol.* **2021**, *8*, 703–742. [CrossRef]
26. Herzog, D.; Seyda, V.; Wycisk, E.; Emmelmann, C. Additive manufacturing of metals. *Acta Mater.* **2016**, *117*, 371–392. [CrossRef]
27. Najmon, J.C.; Raeisi, S.; Tovar, A. Review of Additive Manufacturing Technologies and Applications in the Aerospace Industry. In *Additive Manufacturing for the Aerospace Industry*; Froes, F., Boyer, R., Eds.; Elsevier: Amsterdam, The Netherlands, 2019; ISBN 978-0-1281-4062-8.
28. Boillat-Newport, R.; Isanaka, S.P.; Kelley, J.; Liou, F. Heat Treatments for Minimization of Residual Stresses and Maximization of Tensile Strengths of Scalmalloy® Processed via Directed Energy Deposition. *Materials* **2024**, *17*, 1333. [CrossRef] [PubMed]
29. Boillat, R.; Isanaka, S.P.; Liou, F. The Effect of Nanostructures in Aluminum Alloys Processed Using Additive Manufacturing on Microstructural Evolution and Mechanical Performance Behavior. *Crystals* **2021**, *11*, 524. [CrossRef]
30. Kelley, J. Influence of Alloy Composition on the Process Robustness of Steels Consolidated via Laser-Directed Energy Deposition, Missouri University of Science and Technology. 2023. Available online: https://repositories.lib.utexas.edu/items/63015677-378b-4f1e-adbf-76d56a5ab816 (accessed on 18 June 2024).
31. Karnati, S.; Isanaka, S.P.; Zhang, Y.; Liou, F.F.; Schulthess, J.L. A Comparative Study on Representativeness and Stochastic Efficacy of Miniature Tensile Specimen Testing. *Mater. Perform. Charact.* **2022**, *11*, 424–439. [CrossRef]
32. Michopoulos, J.G.; Iliopoulos, A.P.; Steuben, J.C.; Kittur, M.; Phan, N.; A Salem, A.; Satko, D.P.; Karnati, S.; Isanaka, S.P.; Liou, F. Multiscale Data Driven Methodology for Accelerating Qualification and Certification of Additively Manufactured Parts. In *Encyclopedia of Materials: Metals and Alloys Volume 3*; Caballero, F.G., Ed.; Elsevier: Oxford, UK, 2022; pp. 223–244. ISBN 978-0-12-819733-2.
33. *ASTM E29-22 Standard*; Practice for Using Significant Digits in Test Data to Determine Conformance with Specifications. ASTM International: West Conshohocken, PA, USA, 2022.
34. *ASTM E8-24 Standard*; Test Methods for Tension Testing of Metallic Materials. ASTM International: West Conshohocken, PA, USA, 2024.
35. Michi, R.A.; Plotkowski, A.; Shyam, A.; Dehoff, R.R.; Babu, S.S. Towards high-temperature applications of aluminium alloys enabled by additive manufacturing. *Int. Mater. Rev.* **2022**, *67*, 298–345. [CrossRef]
36. Baker, A.H.; Collins, P.C.; Williams, J.C. New Nomenclatures for Heat Treatments of Additively Manufactured Titanium Alloys. *JOM* **2017**, *69*, 1221–1227. [CrossRef]

37. Pan, D.; Zhou, S.; Zhang, Z.; Li, M.; Wu, Y. Effects of Sc(Zr) on the microstructure and mechanical properties of as-cast Al–Mg alloys. *Mater. Sci. Technol.* **2017**, *33*, 751–757. [CrossRef]
38. Martucci, A.; Aversa, A.; Manfredi, D.; Bondioli, F.; Biamino, S.; Ugues, D.; Lombardi, M.; Fino, P. Low-Power Laser Powder Bed Fusion Processing of Scalmalloy®. *Materials* **2022**, *15*, 3123. [CrossRef] [PubMed]
39. Jeyaprakash, N.; Yang, C.-H.; Kumar, M.S. Influence of coherent intermetallic nano-precipitates on the nano-level mechanical and tribological properties of the Laser-Powder bed fused Scalmalloy. *Mater. Charact.* **2022**, *193*, 112269. [CrossRef]
40. Ma, R.; Peng, C.; Cai, Z.; Wang, R.; Zhou, Z.; Li, X.; Cao, X. Manipulating the microstructure and tensile properties of selective laser melted Al–Mg-Sc-Zr alloy through heat treatment. *J. Alloys Compd.* **2020**, *831*, 154773. [CrossRef]
41. Schimbäck, D.; Mair, P.; Kaserer, L.; Perfler, L.; Palm, F.; Leichtfried, G.; Pogatscher, S. An improved process scan strategy to obtain high-performance fatigue properties for Scalmalloy®. *Mater. Des.* **2022**, *224*, 111410. [CrossRef]
42. Shakil, S.I.; González-Rovira, L.; Cabrera-Correa, L.; de Dios López-Castro, J.; Castillo-Rodríguez, M.; Botana, F.J.; Haghshenas, M. Insights into laser powder bed fused Scalmalloy®: Investigating the correlation between micromechanical and macroscale properties. *J. Mater. Res. Technol.* **2023**, *25*, 4409–4424. [CrossRef]
43. Wang, Z.; Lin, X.; Kang, N.; Hu, Y.; Chen, J.; Huang, W. Strength-ductility synergy of selective laser melted Al-Mg-Sc-Zr alloy with a heterogeneous grain structure. *Addit. Manuf.* **2020**, *34*, 101260. [CrossRef]
44. Wang, Z.; Zeng, L.; Lin, X.; Wang, J.; Feng, Z.; Dang, C.; Li, H.; Wang, Y.; Huang, W. Microstructural evolution and strengthening mechanisms of additively-manufactured Al-Mg-Sc-Zr alloys: Laser directed energy deposition versus laser powder bed fusion. *J. Alloys Compd.* **2024**, *985*, 173946. [CrossRef]
45. Marquis, E.A.; Seidman, D.N. Nanoscale structural evolution of Al_3Sc precipitates in Al(Sc) alloys. *Acta Mater.* **2001**, *49*, 1909–1919.

crystals

MDPI

Article

Simulation Analysis of Three-Point Bending Fracture Process of Yellow River Ice

Yu Deng [1,2], Juan Wang [3,*], Yuhan Meng [1,2] and Yong Zhu [1,2]

1 Yellow River Institute of Hydraulic Research, Zhengzhou 450003, China; lanson0201@aliyun.com (Y.D.)
2 Research Center on Levee Safety Disaster Prevention, Ministry of Water Resources, Zhengzhou 450003, China
3 School of Water Conservancy and Transportation, Zhengzhou University, Zhengzhou 450001, China
* Correspondence: wangjuan@zzu.edu.cn

Abstract: During the ice flood period of the Yellow River, the fracture and destruction of river ice can easily lead to the formation of ice jams and ice dams in the curved and narrow reaches. However, the occurrence and development mechanism of river ice fracture remain incompletely understood in the Yellow River. Therefore, based on the three-point bending physical test of the Yellow River ice, a three-point bending fracture numerical model of the Yellow River ice was constructed. The fracture failure process of the Yellow River ice under three-point bending was simulated, and the effects of the crack-to-height ratio and ice grain size on the fracture properties of the river ice were analyzed. By comparing the results with those of physical tests on river ice, it is evident that the fracture model can effectively simulate the cracking process of river ice. Within the confines of the simulated sample size spectrum, as the crack-to-height ratio varies from 0.2 to 0.8, the fracture toughness value of the Yellow River ice spans a range from 115.01 to 143.37 KPa·m$^{1/2}$. Correspondingly, within the simulated calculation values ranging from 5.38 mm to 24.07 mm for ice crystal size, the fracture toughness value of the Yellow River ice exhibits a range from 116.89 to 143.37 KPa·m$^{1/2}$. The findings reveal that an increase in the crack-to-depth ratio leads to a decrement in the fracture toughness of river ice. Within the scale range encompassed by the model calculations, as the average size of the ice crystal grains augments, the fracture toughness of the river ice exhibits a gradual ascending trend. The research results provide a parameter basis for studying the fracture performance of the Yellow River ice using a numerical simulation method and lays a foundation for investigating the cracking process of river ice from macro and micro multi-scales.

Keywords: river ice; fracture; numerical simulation; Yellow River

1. Introduction

The ice disaster in the Yellow River is a complex physical process, affected by many factors such as hydraulics, thermodynamics, and river topography. Most ice disasters occur during the ice break-up period, which is a phenomenon that the ice cover melts and breaks in the frozen river, resulting in ice flow [1]. The ice blocks produced by the fracture and destruction of river ice flow downstream with the water and are prone to form ice jams and ice dams in curved and narrow river sections. This result further threatens the safety of the embankment, and even leads to the occurrence of ice flood disasters. Currently, the mechanism of ice break-up in the Yellow River during spring is not yet fully understood, and the physical and mechanical properties of the river ice fracture and failure process still need to be further explored.

Yellow River ice, characterized as a quasi-brittle and crystalline material, presents one of the most formidable challenges within the domain of river ice research. Investigations into matters pertaining to Yellow River ice can be retrospectively traced to the 1950s. Among these early contributions, Chen Zanting [2] provided a seminal overview of the observational theory concerning Yellow River ice. Subsequently, Sun Zhaochu, Sui Jueyi, and their colleagues [3–5], through meticulous observations of river ice during the ice-jam period in the Hequ section of the Yellow River, formulated the inaugural mathematical model that correlates ice jam thickness with ice flow volume and hydraulic conditions. Furthermore, Wang Jun and his team [6–9], based on rigorous flume experiments, deduced the intricate relationship between initial ice jam thickness and various factors, including flow conditions and ice discharge. In another noteworthy study, Gu Lihua [10], utilizing a multiple exponential regression model, discovered that the refined cumulative negative air temperature method can predict ice cover thickness with remarkable accuracy. Additionally, Gao Guoming et al. [11] devised calculations for sub-ice flow and conducted a thorough analysis of the variation patterns of sub-ice velocity.

With the relentless progression of technology, significant milestones have been achieved in the realms of Yellow River ice monitoring [12–14], ice forecasting [15–17], and ice disaster prevention and mitigation [18–20]. Moreover, a steady stream of outcomes pertaining to river ice reviews has emerged [21,22]. However, research delving into fundamental issues, such as the physical and mechanical properties of the fracture and failure processes of Yellow River ice, remains in its nascent stages, primarily hindered by the challenges associated with ice sample collection and experimental conditions. Given that alterations in the internal microstructure of river ice are pivotal factors that lead to its intricate macroscopic manifestations, an in-depth exploration of the internal microstructure and material properties of river ice is imperative to unravel the macroscopic fracture mechanical behavior of river ice break-up from its fundamental essence [23].

The fracture process of Yellow River ice is a multi-scale problem that combines macroscopic and microscopic perspectives. The study of the mechanism underlying the initiation and propagation of microscopic cracks within river ice will contribute to explaining its macroscopic fracture mechanical behavior [24]. However, due to multiple factors such as ice sample collection and experimental conditions, studying the fracture mechanical properties of Yellow River ice through physical experiments still faces objective limitations. On the other hand, numerical simulation methods combined with the mesoscopic mechanics theory have continuously produced research results in simulating and analyzing the fracture process of river ice [25–27].

The finite element method is one of the commonly employed calculation methods. For instance, Kendall put forward a novel crushing failure criterion grounded in fracture energy balance theory [28]. Evans et al. revealed the changing trends of crack location, crack propagation load, and spallation that occur during the fracture process [29]. Xiao et al. provided a damage calculation model specific to ice [30]. Han Lei et al. conducted a systematic analysis of the entire process of destruction in ice–cone interaction using Ansys and obtained the variation law of the ice force–deflection relationship curve with respect to failure modes [31].

Another frequently used calculation method is the discrete element method. Cundall previously introduced a discrete element model for damage and fracture [32]. Jirasek et al. simulated and analyzed the collision process between floating ice moving at varying speeds and rigid obstacles [33,34]. Korlie applied the principles of particle mechanics to analyze the generation, propagation, and fracture process of three-dimensional ice plate cracks under pressure [27].

Furthermore, Li Zhijun et al. performed numerical simulation analysis on the issue of sea ice fracture [35], Wang Jun et al. investigated the formation and evolution mechanism of ice jams [36], Yang Kailin and Fu Hui et al. delved into channel ice problems [37,38], and Mao Zeyu et al. studied the variation of water flow movement under ice cover [39]. These researchers have achieved notable outcomes in their respective fields. However, despite utilizing numerical simulation methods to explore the macroscopic cracking of river ice from a mesoscopic scale based on mesoscopic ice mechanics theory, the existing research results still require further enrichment.

For the basis of this study, the experimental data from the existing research results presented in reference [40] were utilized. Based on the parametric programming language of Ansys software 12.0 (ANSYS Parametric Design Language, APDL), a meso-scale numerical model of river ice was developed to simulate the three-point bending fracture of Yellow River ice. The model considers two influencing factors: crack to height ratio and grain size and simulates the three-point bending fracture process of river ice. The model construction mainly includes the establishment of the meso-structure of Yellow River ice, the determination of constitutive relation and failure criterion, and the selection of model parameters. The numerical model and calculation analysis results simulate and demonstrate the cracking process of river ice and provide a theoretical basis for grasping the fracture properties of Yellow River ice and revealing the fracture failure mechanism of Yellow River ice from a macro and meso multi-scale perspective.

2. Establishment of a Numerical Model for River Ice Fracture

With the aim of delving deeper into the influence of mesoscale crystal alterations on the macro-mechanical attributes of river ice, and surmounting the constraints imposed by physical testing environments, a mesoscale numerical simulation method was adopted to analyze the fracture process of river ice. At the mesoscale, Yellow River ice can be considered as consisting of ice crystals, ice crystal boundaries, and initial defects. The development of the mesoscale model for Yellow River ice mainly involves the generation of ice crystals and the formation of grain boundaries and initial defects, constitutive relationships, and failure criteria.

2.1. Generation of Ice Crystals

Taking into account the irregular nature of river ice crystals, the Voronoi polygon method is adopted to simulate the river ice grains. Based on the observational results obtained from the mesostructure experiments conducted on Yellow River ice [40] and by incorporating the characteristics of the Voronoi polygon method, the ice grains of the Yellow River were simulated. Ice crystal specimens were generated randomly, and their structure is depicted in Figure 1. Within this figure, Figure 1a is a typical sample image of the crystal structure of Yellow River ice, and Figure 1b is the numerical model diagram of the crystal structure of Yellow River ice.

(**a**) Typical sample image of the crystal structure of Yellow River ice (**b**) Numerical model diagram of the crystal structure of Yellow River ice

Figure 1. Structure distribution figure of Yellow River ice grains.

2.2. Generation of Ice Grain Boundaries and Initial Defects

The destruction of river ice structures is usually caused by the evolution of dislocation and slip at the ice crystal interface. Therefore, the influence of ice grain boundaries cannot be ignored in the numerical model. In the model, a thin solid element is used to simulate the ice crystal boundary. The initial defects inside the Yellow River ice include bubbles, impurities, micro-cracks, and sediment particles, among which bubbles are the most common and account for the largest proportion. In this paper, various initial defects in Yellow River ice are assumed to be a unified substance and randomly distributed in river ice grains and grain boundaries. In the numerical model, equivalent initial damage elements are used for simulation. Based on the measured initial defect content from the physics experiment results [40–42], the number of equivalent crack elements is determined. The performance of equivalent initial damage elements is determined according to the size of initial defects and grid size. The methods and steps employed in this study are outlined as follows: (a) It is assumed that the initial defect takes a circular shape, and its diameter is considered equivalent to the diameter of the bubble, denoted as d. (b) A further assumption is made that each equivalent initial damage unit contains only a single initial defect. (c) The number of initial defects, denoted as N, is estimated based on the initial defect content and the defect size d. (d) Subsequently, N units are randomly selected from the grain and boundary units, and these are designated as equivalent initial damage units. (e) The unit properties of the equivalent initial damage units are then modified accordingly. The specific performance parameters of these units are set as described in Section 2.4. The finite element model of river ice with initial defects is shown in Figure 2. Within this figure, Figure 2a is the crystal mesh division diagram of the Yellow River ice model, Figure 2b is the initial defects diagram of the Yellow River ice model, and Figure 2c is the locally enlarged crystal boundary diagram of the Yellow River ice model.

(**a**) Crystal mesh division diagram of Yellow River ice model (**b**) Initial defects diagram of Yellow River ice model (**c**) Locally enlarged crystal boundary diagram of Yellow River ice model

Figure 2. Schematic figure of the finite element model of Yellow River ice.

2.3. Constitutive Relation and Failure Criterion

River ice exhibits strong brittleness during failure. The mesoscopic components of river ice are regarded as homogeneous linear elastic materials, where both grains and grain boundaries are elastic-brittle materials and isotropic within the same cross-section. Through an analysis of the four commonly used failure criteria for fracture problems–maximum tensile stress theory, maximum linear strain theory, maximum shear stress theory, and shape change ratio energy theory, it can be seen that the maximum tensile stress theory is more suitable for analyzing the river ice fracture process [30,43]. Therefore, the maximum tensile stress criterion is adopted as the failure criterion for river ice [30,43]. For elements with an initial defect, the corresponding failure criteria for grains and grain boundaries are adopted according to the location of the element. It is assumed that the element loses its bearing capacity after failure, and its material properties are modified [44]. The elastic modulus of the failed element is taken as a very small value (10^{-5} to 10^{-4} of the original value), ignoring its contribution to the stiffness matrix [44].

2.4. Determination of Mesoscopic Parameters

There are two types of parameters that need to be determined in the river ice calculation model: mesoscopic structural parameters and mesoscopic material parameters. The selection of river ice parameters has been discussed in detail in the literature [44]. Among them, the mesoscopic structural parameters mainly include the size and distribution of ice crystal grains, the composition of boundaries, and the determination of initial defects. The mesoscopic material parameters mainly include the grain anisotropy, elastic modulus of ice grains, fracture strength, mechanical properties of ice grain boundaries, etc. The main calculation parameter values for the analysis of the three-point bending fracture process of river ice are shown in Table 1.

Table 1. Values of main parameters for mesoscopic calculation of river ice.

Temperature (°C)	Strain Rate (s^{-1})	Grain Elastic Modulus (GPa)	Grain Strength (MPa)	Grain Poisson's Ratio	Grain Boundary Elastic Modulus (GPa)	Grain Boundary Strength (MPa)	Initial Defect Content (%)	Grain Size (mm)
−10	10^{-4}~10^{-5}	9	1	0.3	5	0.5	6	2^{-10}

3. Simulation Analysis of Three-Point Bending Fracture Process of River Ice

3.1. Analysis of Simulation Results

Based on the physical experiments of the three-point bending of Yellow River ice, a numerical model for the three-point bending fracture of Yellow River ice was constructed. Figure 3 shows the typical failure mode during the model calculation process. As can be seen from Figure 3a, during the initial cracking stage of the ice sample, the crack at the central notch commences its slow upward progression. Simultaneously, displacement begins to manifest at the support positions on the left and right sides of the lower interface of the beam, as well as at the loading position on the upper interface of the beam. Figure 3b illustrates that, as the load increases, the crack at the central notch of the beam gradually propagates upwards, ultimately forming a macroscopically visible crack. As can be further seen from Figure 3b, with a continued increase in load, the crack at the central notch of the beam stabilizes, accompanied by the emergence of macro cracks at the loading position on the upper interface and the support positions on the lower interface of the beam. As can be seen from Figure 3c, the final fracture form of the river ice is characterized by a main crack formed in the center of the specimen, along with typical wing-shaped cracks developing in the connection areas between the two supports and the loading point. Lastly, Figure 3d presents a representative image of a failed sample from the physical experiment of the three-point bending of the river ice, with its failure mode resembling the one obtained from the simulation calculation.

3.2. Analysis of the Impact of Crack-to-Height Ratio on the Three-Point Bending Fracture

Due to objective factors, the fracture toughness calculation results derived from physical experiments on Yellow River ice exhibit considerable scattering. Among these factors, the prefabricated crack length of the incision is one of the more sensitive factors when measuring the fracture toughness of river ice through fracture toughness tests. The crack length, the remaining propagation zone length, and the distance from the crack tip to the loading point must all conform to the precise description of linear elastic fracture theory. If the crack is too short or too long, the incision's sensitivity will decrease. Consequently, this paper puts forward a constructed numerical model of river ice fracture to simulate ice samples with varying incision lengths. It then analyzes the river ice fracture toughness calculated for different crack-to-height ratios (i.e., crack incision length/beam height) based on the simulation results. As shown in Figure 4, the sample height is 70 mm, the thickness is 70 mm, the span-to-depth ratio S/W is 4, and the crack-to-height ratios are 0.2, 0.3, 0.4, 0.5, 0.6, 0.7, and 0.8. Among Figure 4, the green line represents the trend line of fracture toughness change for simulated values, while the yellow line represents the trend line of fracture toughness change for experimental values. According to the analysis of Figure 4, the fracture toughness value of Yellow River ice is the highest when the crack-to-height ratio is 0.2. As the crack-to-height ratio increases, the fracture toughness value of Yellow River ice decreases slightly, but the overall decrease is not substantial. Within the simulated sample size range, as the crack-to-height ratio varies from 0.2 to 0.8, the fracture toughness value of Yellow River ice ranges from 115.01 to 143.37 $\text{KPa}\cdot\text{m}^{1/2}$. At the same time, to validate the model calculation results, similar experimental data from Dempsey's literature [45] are introduced in Figure 4 for comparison. The dimensions ($W \times h \times L$) of the three-point bending beam ice sample in the literature [45] are 102 mm × 45 mm × 306 mm. Through comparative analysis, it can be seen that, although there are differences in fracture toughness results due to different experimental influencing factors, the trend of change in the fracture toughness of ice is consistent as the crack-to-height ratio increases.

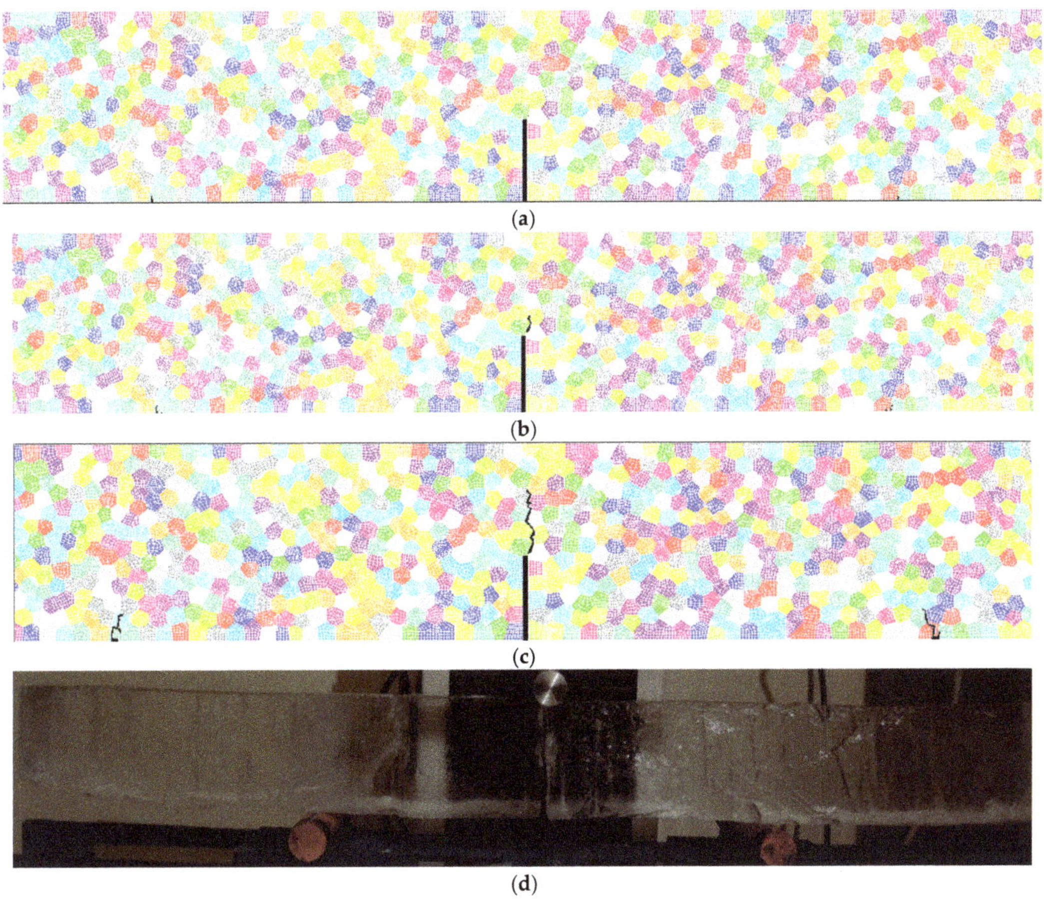

(a)

(b)

(c)

(d)

Figure 3. Diagram of the cracking process and physical experiment failure modes of river ice under three-point bending load: (**a**) initial crack initiation; (**b**) development of the macroscopic main crack; (**c**) final failure mode; (**d**) physical experimental failure mode.

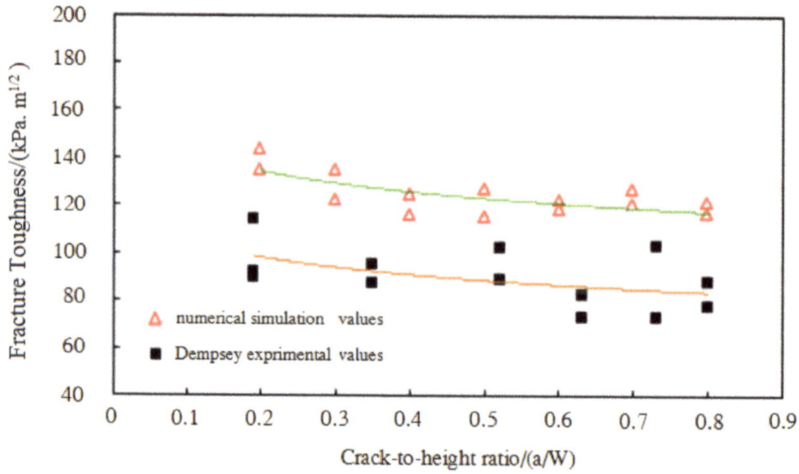

Figure 4. Impact of initial defects on the failure mode of river ice.

3.3. Analysis of the Ice Crystal Size on the Three-Point Bending Fracture

Based on the established numerical model of Yellow River ice fracture, a simulation analysis was conducted to examine the three-point bending fracture process of Yellow River ice under varying ice crystal sizes. Figure 5 illustrates a typical grain size distribution diagram, with average ice crystal sizes (d_{av}) of 4.8 mm, 7.2 mm, 10.6 mm, 13 mm, and 24 mm. Specifically, Figure 5a,b depict the mechanical properties of granular ice, Figure 5d,e portray the mechanical properties of columnar ice, and Figure 5c represents the mechanical properties of the transitional layer between these two ice types. The outcomes of computational analyses conducted on models with different ice crystal size distributions are organized and presented in Figure 6. In Figure 6, the ice crystal sizes for physical tests are derived from the average particle sizes obtained through statistical analyses of granular and columnar ice, while the simulated calculation values range from 5.38 mm to 24.07 mm for ice crystal size. The fracture toughness value of Yellow River ice exhibits a range from 116.89 to 185.54 KPa·m$^{1/2}$. The simulated fracture toughness values of river ice under different ice crystal sizes align well with the experimental values, with an average error of approximately 5% between the two datasets. As the average size of ice crystals increases, the simulated calculation values exhibit a gradual upward trend. This trend can be attributed to the following factors: granular ice from the Yellow River contains more internal bubbles and impurities, which objectively reduces its hardness and strength, ultimately leading to a decrease in its fracture toughness value. This objective reality is accounted for in the calculation model by incorporating increased initial defects. On the other hand, columnar ice from the Yellow River is primarily formed through thermodynamic growth, with relatively fewer internal bubbles and impurities, resulting in a slight enhancement of its strength and brittleness. Overall, the calculated fracture toughness of columnar ice in the Yellow River is slightly larger than that of granular ice. When the average size of ice crystals further increases (d_{av} = 24.07 m), the simulated fracture toughness value rises to 185.54 kPa.m$^{1/2}$. This specific working condition surpasses the size range of the naturally occurring columnar ice samples collected from the Yellow River.

Figure 5. Three-point bending beam model figure of river ice with different ice crystal grain distributions: (a) average grain size d_{av} = 4.8 mm; (b) average grain size d_{av} = 7.2 mm; (c) average grain size d_{av} = 10.6 mm; (d) average grain size d_{av} = 13 mm; (e) average grain size d_{av} = 24 mm.

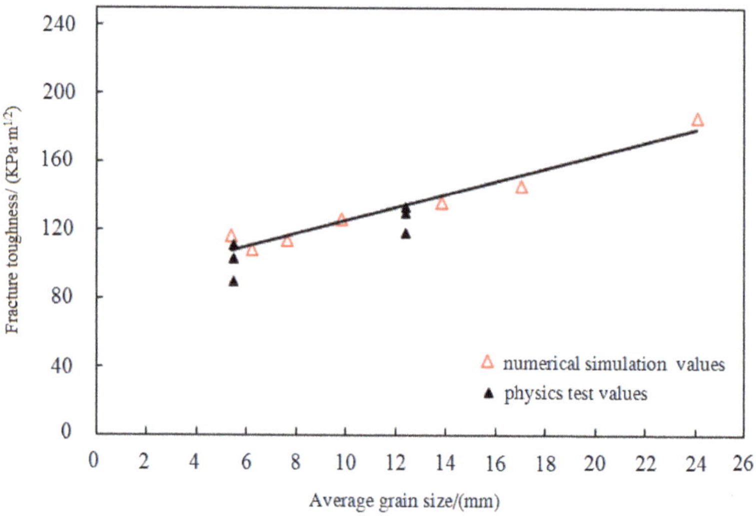

Figure 6. Comparison of simulated and experimental values of three-point bending fracture toughness of river ice with different ice crystal sizes.

4. Conclusions

This paper develops a numerical model for simulating the fracture of river ice, incorporating considerations of ice crystal grains, ice crystal boundaries, and initial defects. It simulates the cracking process of river ice under three-point bending conditions and analyzes the impact of factors such as the crack-to-depth ratio and ice crystal size on the fracture behavior of river ice subject to various loads. The primary conclusions are summarized as follows:

(1) Voronoi polygons are utilized to represent ice crystal grains, and a numerical model for simulating the fracture of river ice, accounting for initial defects, is established through secondary development using ANSYS software.

(2) The cracking process of river ice under diverse loads is simulated. The three-point bending failure of river ice exhibits a typical wing-shaped crack, with an average error between simulated and experimentally obtained values of approximately 5%.

(3) Based on the outcomes derived from the numerical simulations, an analysis is conducted to elucidate the impact of two pivotal factors—the crack-to-depth ratio and the size of ice crystals—on the fracture behavior of river ice. Within the confines of the simulated sample size spectrum, as the crack-to-height ratio varies from 0.2 to 0.8, the fracture toughness value of Yellow River ice spans a range from 115.01 to 143.37 KPa·$m^{1/2}$. Correspondingly, within the simulated calculation values ranging from 5.38 mm to 24.07 mm for ice crystal size, the fracture toughness value of Yellow River ice exhibits a range from 116.89 to 185.54 KPa·$m^{1/2}$.

(4) The findings reveal that an increase in the crack-to-depth ratio leads to a decrement in the fracture toughness of river ice. Within the scale range encompassed by the model calculations, as the average size of ice crystal grains augments, the fracture toughness of river ice exhibits a gradual ascending trend.

The results presented in this paper provide numerical calculations for the fracture behavior of Yellow River ice under specific conditions. Further research is necessary to investigate the fracture process under more complex loading conditions.

Author Contributions: Conceptualization, Y.D. and J.W.; methodology, Y.D.; software, J.W.; data curation, Y.M. and Y.Z.; writing—original draft preparation, Y.D. and J.W. All authors have read and agreed to the published version of the manuscript.

Funding: This research was supported by the National Natural Science Foundation of China (Grant Nos. U23A2012) and the Special Funds for Basic Scientific Research of the Yellow River Institute of Hydraulic Research (HKY-JBYW-2022-08).

Data Availability Statement: The original contributions presented in the study are included in the article.

Conflicts of Interest: The authors declare no conflicts of interest.

References

1. Chen, Z.T. Research on ice conditions of rivers, lakes, and seas in China. *J. Glaciol. Geocryol.* **1988**, *10*, 279–283.
2. Chen, Z. Overview of the development of research on Yellow River ice observation. *Yellow River Sci. Technol. Inf.* **1989**, *1*, 15.
3. Sun, Z.; Sui, J. Research on river ice jam and its significance. *Adv. Earth Sci.* **1990**, *3*, 51–54.
4. Sui, J. Riverbed evolution during the freeze-up period in the Longkou-Tianqiao section of the middle Yellow River. *Acta Geogr. Sin.* **1992**, *5*, 209–219.
5. Sui, J.; Fang, D.; Zhou, Y. Analysis and calculation of ice jam water level in the Hequ section of the Yellow River. *Hydrology* **1994**, *63*, 18–24.
6. Wang, J. Experimental study on the incipient velocity of granular sediment under ice cover. *J. Water Resour. Water Transp. Res.* **1998**, *2*, 164–169.
7. Wang, J. Experimental study on ice transport in balanced ice jam. *J. Hydroelectr. Eng.* **2002**, *1*, 61–67.
8. Wang, J. Experimental study on the relationship between the thickness of balanced ice jam and flow conditions and ice discharge. *J. Lanzhou Univ. (Nat. Sci.)* **2002**, *38*, 117–121.
9. Wang, J.; Fu, H.; Yi, M.; Yin, Y.; Gao, Y. Analysis of ice jam water level. *Adv. Water Sci.* **2007**, *18*, 102–107.
10. Gu, L. Analysis of Ice Thickness during the Stable Freeze-Up Period and the Water Level during Ice Break-Up. Master's Thesis, Hefei University of Technology, Hefei, China, 2007.
11. Gao, G.; Ma, Z.; Li, S.; Deng, Y. Study on the influence of ice cover on the vertical velocity distribution of laminar flow. *Yellow River* **2018**, *40*, 15–17.
12. Qin, J.; Shen, B. A new type of ice thickness sensor based on the electrical conductivity of ice and water. *Sens. Technol.* **2004**, *23*, 55–56.
13. Qin, J. Theoretical and applied research on ice thickness detection based on the conductivity of air, ice, and water. Master's Thesis, Xi'an University of Technology, Xi'an, China, 2005.
14. Qin, J.; Cheng, P.; Qin, M. Ice thickness sensor and its detection method. *Adv. Water Sci.* **2008**, *3*, 418–421.
15. Ji, H. Analysis of the Causes of Ice Jam Flooding in the Inner Mongolia section of the Yellow River and Research on the Forecast Model of Freeze-Up and Break-Up Dates. Ph.D. Thesis, Inner Mongolia Agricultural University, Hohhot, China, 2002.
16. Wang, L.; Liu, J.; Xie, W. Research progress on ice jam forecast models in the Ningxia-Inner Mongolia section of the Yellow River. *Yellow River* **2014**, *36*, 8–10.
17. Liu, J.; Huo, S.; Wang, C. Research on ice jam forecast and demand for ice prevention on the Yellow River. *China Flood Drought Manag.* **2017**, *6*, 10–13.
18. Meng, W.; Liu, X.; Hu, J. Numerical simulation and experimental study on ice jam blasting based on cumulative penetration technology. *J. North China Inst. Water Conserv. Hydroelectr. Power* **2013**, *3*, 44–47.
19. Yang, X.; Tong, Z.; Song, C. Research on blasting ice-breaking technology for disaster reduction during the ice jam period of the Yellow River. In Proceedings of the 125th Forum on Engineering Sciences and Technologies in China-Explosion Synthesis of New Materials and Key Science and Engineering Technologies for Efficient and Safe Blasting, Nanjing, China, 19 September 2011; pp. 297–303.
20. Song, C.; Yang, X.; Yan, J.; Liang, Q.; Tong, Z. Research on ice-breaking technology using vehicle-mounted rocket explosion belt. *Eng. Blasting* **2013**, *8*, 50–53.
21. Gao, G.; Deng, Y.; Tian, Z.; Li, S.; Zhang, B. Brief review and prospects of recent research on Yellow River ice jam. *Yellow River* **2019**, *41*, 77–81.
22. Wang, E.; Xu, C.; Yu, J. Research progress on mechanical properties of freshwater ice in China. *Water Resour. Sci. Cold Reg. Eng.* **2019**, *2*, 37–43.
23. Schulson, E.M. An analysis of the brittle to ductile transition in polycrystalline ice under tension. *Cold Reg. Sci. Technol.* **1979**, *1*, 87–91. [CrossRef]
24. Nemat, S.; Horii, H. Rock failure in compression. *Int. J. Eng. Sci.* **1984**, *22*, 999–1011. [CrossRef]
25. Wu, M.S.; Juan, N. Micromechanical prediction of the compressive failure of ice: Model development. *Mech. Mater.* **1995**, *20*, 9–32. [CrossRef]
26. Wu, M.S.; Niu, J. Micromechanical prediction of the compressive failure of ice: Numerical simulations. *Mech. Mater.* **1995**, *20*, 33–58. [CrossRef]
27. Korlie, M.S. 3D simulation of cracks and fractures in a molecular solid under stress and compression. *Comput. Math. Appl.* **2007**, *54*, 638–650. [CrossRef]
28. Kendall, K. Complexities of compression failure. *Proc. R. Soc. A Math. Phys. Eng. Sci. Lond.* **1978**, *361*, 245–263.

29. Evans, A.G.; Palmer, A.C.; Goodman, D.J.; Ashby, M.F.; Williams, G.J. Indentation spalling of edge loaded ice sheets. In Proceedings of the 7th IAHR Ice Symposium, Hamburg, Germany, 27–31 August 1984; Volume 1, pp. 113–121.
30. Xiao, J.; Jordaan, I.J. Application of damage mechanics to ice failure in compression. *Cold Reg. Sci. Technol.* **1996**, *24*, 305–322. [CrossRef]
31. Han, L.; Li, F.; Yue, Q.J. Finite element simulation of the entire process of ice cone interaction damage. *China Offshore Platf.* **2007**, *22*, 22–27.
32. Cundall, P.A. A computer model for simulating progressive large-scale movements in block rock systems. In Proceedings of the International Symposium on Rock Fracture, Nancy, France, 4–6 October 1971; Volume 1, pp. 8–11.
33. Jirasek, M.; Bazant, Z.P. Macroscopic fracture characteristics of random particle systems. *J. Eng. Mech.* **1995**, *69*, 201–228. [CrossRef]
34. Jirasek, M.; Bazant, Z.P. Particle model for quasibrittle fracture and application to sea ice. *J. Eng. Mech.* **1995**, *121*, 1016–1025. [CrossRef]
35. Lu, P.; Li, Z.J.; Han, H.W. Introduction of parameterized sea ice drag coefficients into ice free-drift modeling. *J. Oceanol.* **2016**, *35*, 53–59. [CrossRef]
36. Wang, J.; He, L.; Chen, P.; Sui, J. Numerical simulation of mechanical breakup of river ice-cover. *J. Hydrodyn. Ser. B* **2013**, *25*, 415–421. [CrossRef]
37. Yang, K. Research progress on river and canal ice hydraulics, ice observation, and forecasting. *J. Hydraul. Eng.* **2018**, *49*, 81–91.
38. Fu, H.; Yang, K.; Guo, X.L.; Guo, Y.X.; Wang, T. Safe operation of inverted siphon during ice period. *J. Hydrodyn. Ser. B* **2015**, *27*, 204–209. [CrossRef]
39. Mao, Z.; Ma, J.; She, Y.; Wu, J.J. Research on resistance of frozen rivers. *J. Hydraul. Eng.* **2002**, *5*, 59–64.
40. Wang, J.; Huang, Y.; Deng, Y.; Zhijun, L.; Yaodan, Z. Research on ice fracture of the Yellow River performance based on digital image correlation method. *J. Hydrol.* **2021**, *52*, 1036–1045.
41. Yaodan, Z.; Zhijun, L.; Yuanren, X.; Chunjiang, L.; Baosen, Z.; Yu, D. Microstructural characteristics of frazil particles and the physical properties of frazil ice in the Yellow River, China. *Crystals* **2021**, *11*, 617. [CrossRef]
42. Yu, D.; Zongkun, L.; Juan, W.; Lukai, X. The microstructure of Yellow River ice in the freezing period. *Crystals* **2019**, *9*, 484. [CrossRef]
43. Zou, B.; Xiao, J.; Jordaan, I.J. Ice fracture and spalling in ice structure interaction. *Cold Reg. Sci. Technol.* **1996**, *24*, 213–220. [CrossRef]
44. Deng, Y.; Wang, J.; Xiao, Z. Selection of meso-parameters in the construction of a meso-model for river ice. *Yellow River* **2017**, *10*, 27–31.
45. Dempsey, J.P.; Wei, Y.; Defranco, S.J. Notch sensitivity and brittleness in fracture testing of S2 columnar freshwater ice. *Int. J. Fract.* **1992**, *53*, 101–120. [CrossRef]

crystals

MDPI

Article

Multiscale Simulation Study on the Spallation Characteristics of Ductile Metal Ta under High Strain Rate Impact

Yuntian Wang [1,2,3], Yuzhu Guo [4,*], Sheng Li [5,*], Yushu Li [1], Min Yang [1] and Weimei Shi [1,2]

1 School of Urban Construction, Chengdu Polytechnic, Chengdu 610041, China; wangyuntian@cdp.edu.cn (Y.W.); scatcliyushu@gmail.com (Y.L.); yangmin1975@foxmail.com (M.Y.); shiweimei@cdp.edu.cn (W.S.)
2 Sichuan Thermoelectric Materials and Devices Application Engineering Research Center, Chengdu 610041, China
3 Luzhou Key Laboratory of Intelligent Construction and Low-Carbon Technology, Luzhou 646000, China
4 School of Civil Engineering, Changsha University of Science and Technology, Changsha 410114, China
5 College of Civil Engineering, Lanzhou Jiaotong University, Lanzhou 730070, China
* Correspondence: guoyuzhu5185@163.com (Y.G.); ligwin@126.com (S.L.)

Abstract: This work employs a multiscale simulation framework to systematically explore the spallation behavior of ductile tantalum (Ta) subjected to high strain rate impacts. The approach integrates macroscopic simulations, utilizing both the Lagrangian mesh and Smoothed Particle Hydrodynamics (SPH) methods, with microscopic molecular dynamics (MD) simulations to dissect the dynamic failure processes of tantalum. The macroscopic simulations, validated against experimental data, demonstrate the effectiveness of the SPH method in accurately capturing the spallation process. An exponential correlation between spallation strength and tensile strain rate has been established. An in-depth analysis of the free surface velocity profile indicates that the pullback signal is associated with microvoid nucleation, where the velocity drop signifies the initiation conditions for microvoid development. Additionally, the rebound rate following the pullback signal reflects the progression of damage within the spallation region. By integrating results across macro- and microscales, this work offers comprehensive insights into the complex spallation behavior of ductile tantalum under extreme conditions, advancing the understanding of its failure mechanisms at high strain rates.

Keywords: ductile metal Ta; multiscale simulation; high strain rate impact; spallation characteristics; molecular dynamics

1. Introduction

The dynamic failure of ductile metals under high strain rate impact is a critical issue in impact dynamics. With ongoing advancements in fields such as national defense, automotive engineering, and aerospace, gaining a deeper understanding of the dynamic damage mechanisms of ductile metals, particularly under high strain rate conditions, has become increasingly important [1]. Spallation is a prevalent form of dynamic damage in these metals, and the free surface velocity profiles obtained from spallation experiments are crucial for understanding the underlying failure processes [2,3].

Among the various methods used to study spallation in ductile metals, the Planar Plate Impact (PPI) experiment is particularly significant. The propagation of wave systems during such experiments is illustrated in Figure 1a. At time t_0, the flyer impacts the specimen at a certain velocity, generating shock waves that propagate through both the flyer and the specimen, as illustrated in Figure 1b. When the shock waves reach the free surface of the specimen and the rear surface of the flyer, they reflect back as rarefaction waves. Since the flyer is thinner than the specimen, the two rarefaction waves meet within the specimen, creating a tensile region. If the tensile stress exceeds the dynamic strength limit of the material, spallation phenomena occur. Previous studies have established that spallation in ductile metals is a complex dynamic process driven by the nucleation, growth, and coalescence of microdamages, culminating in the formation of a spallation plane [3].

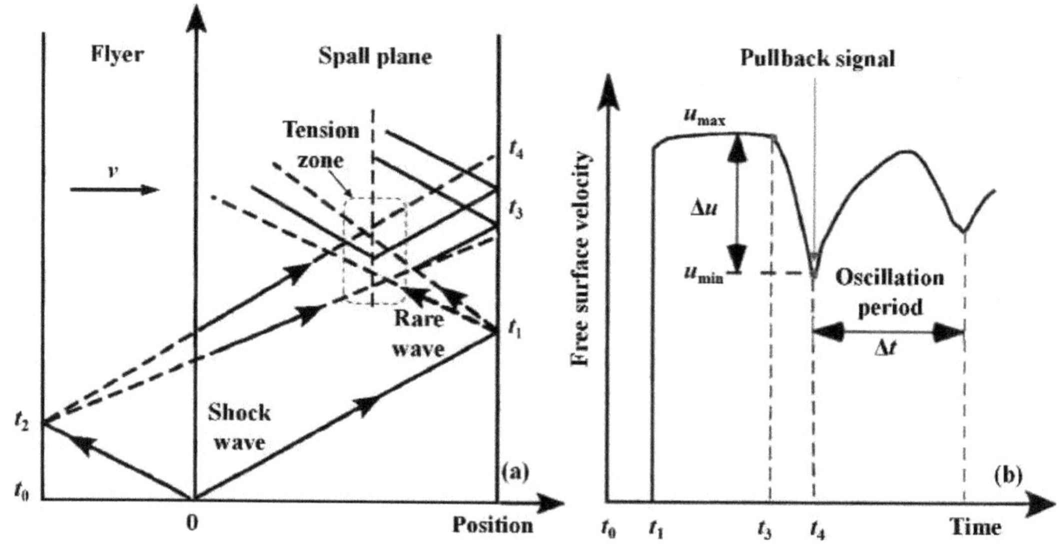

Figure 1. Depiction of the fundamental concept of the flat plate impact experiment, along with a schematic representation of the free surface velocity profile. (**a**) The principle of the spallation failure experiment, (**b**) Typical free surface velocity profile.

Despite the valuable insights provided by experimental studies, they inherently lack the ability to directly measure the temporal evolution of internal physical quantities in ductile metals [4–6]. As a result, researchers often rely on indirect data, such as free surface velocity profiles, to infer the internal damage evolution in materials.

With the development and application of computational techniques, numerical simulations have become a powerful complement to experimental research. Currently, two primary numerical methods are employed in impact mechanics: mesh-based and meshless methods. Among the meshless methods, Smoothed Particle Hydrodynamics (SPH) has gained widespread acceptance [7]. These numerical simulations have shown strong agreement with experimental results, offering greater flexibility in loading conditions, faster computations, and intuitive insights into the macroscopic dynamic failure of materials [8,9]. However, their accuracy depends on factors such as the equation of state, strength models, and failure criteria used. Moreover, investigating spallation mechanisms at the microscopic scale remains limited due to the influence of element or particle scales.

At the microscopic scale, molecular dynamics (MD) simulations have become increasingly prominent as a method for studying material properties [10–12]. MD simulations offer a detailed perspective on the microscopic mechanisms underlying dynamic damage, shedding light on the stress and damage progression within materials. However, due to computational limitations, it is not yet feasible to fully replicate spallation phenomena at larger scales.

The typical free surface velocity profile shown in Figure 1b provides a wealth of critical information about the macroscopic response of ductile metals to dynamic damage, including loading stress amplitude, tensile strain rate, spallation strength, and spallation plane thickness [2,13]. However, the spallation process in ductile metals is highly complex, involving the dynamic evolution of internal damage over both time and space. Understanding the relationship between macroscopic response and microscopic damage development remains challenging, leading to debates over the interpretation of specific features in free surface velocity profiles.

Most current research focuses on either simulating spallation phenomena or analyzing microscopic damage evolution at a single scale [12,14,15], with few studies addressing the characteristics of free surface velocity profiles from a multiscale perspective. This paper aims to bridge this gap by investigating the spallation characteristics of tantalum, a ductile metal, under high strain rate impacts through multiscale simulation. By integrating simulation results from both macroscopic and microscopic scales, this study provides a comprehensive analysis of the spallation characteristics of ductile tantalum, revealing the damage evolution laws and physical implications embedded in the typical features of the free surface velocity profile.

2. Compositional Method and Models

This study leverages both macroscale numerical simulations and microscale molecular dynamics (MD) simulations, enabling a comprehensive multiscale analysis. To facilitate comparison and validation with experimental data, the full-scale macroscopic model was established based on the planar plate impact experiment [16]. The free surface velocity profile was obtained by extracting data from the central point of the sample's free surface, as illustrated in Figure 2a. At the microscopic scale, the MD model concentrated on the tensile area in the sample's center, as shown in Figure 2b.

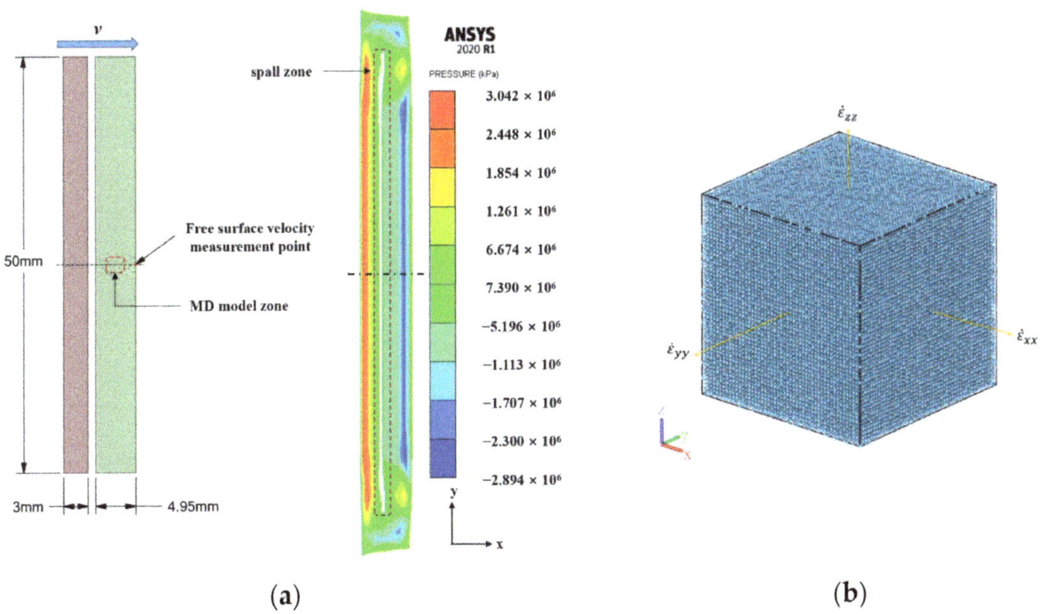

Figure 2. Schematic of the macro− and microscale models. (**a**) Macro−level Model, (**b**) Micro−level model.

2.1. Macroscale Numerical Simulation

2.1.1. Equation of State and Material Constitutive Model

The Mie–Grüneisen equation of state is a prevalent model for characterizing the dynamic response of metals under shock conditions, describing the interplay between pressure, density, and internal energy. [17].

The Mie–Grüneisen equation of state is expressed as follows:

$$p = \frac{\rho_0 c_0^2 \mu (1+\mu)}{[1-(S_1-1)\mu]} + \gamma \rho_0 e \tag{1}$$

$$\mu = \frac{\rho}{\rho_0 - 1} \tag{2}$$

where ρ is the material density after shock, ρ_0 is the initial material density, S_1 is a material constant, γ is the Grüneisen coefficient, and e is the specific internal energy.

The parameters used for the Mie–Grüneisen equation of state in this study are listed in Table 1 [18].

Table 1. Parameters of the Mie–Grüneisen equation of state applied in the simulation.

Material	ρ_0 (kg/m^3)	c_0 (m/s)	S_1	γ
Ta	16,690	3340	1.20	1.67

To model the dynamic failure behavior of ductile metals under intense dynamic loading, three established constitutive models were employed in this study: the Johnson–Cook (JC) [19], Steinberg–Cochran–Guinan (SCG) [20], and Zerilli–Armstrong (ZA) models [21].

(1) Johnson–Cook (JC) Model

The Johnson–Cook (JC) model is extensively employed due to its capacity to represent strain rate hardening, strain hardening, and thermal softening phenomena. It is formulated as follows:

$$\sigma = (A + B\varepsilon^n)\left(1 + C\ln\frac{\dot{\varepsilon}}{\dot{\varepsilon}_0}\right)[1 - (T^*)^m] \tag{3}$$

$$T^* = \frac{T - T_r}{T_m - T_r} \tag{4}$$

Here, A, B, C, n, and m represent material constants, T_r denotes the reference temperature, T_m is the melting temperature, and $\dot{\varepsilon}_0$ is the reference strain rate. The parameters for the Johnson–Cook model are listed in Table 2 [22].

Table 2. JC model parameters used in simulation.

Material	A/MPa	B/MPa	n	C	m
Ta	142	164	0.31	0.057	0.88

(2) Zerilli–Armstrong (ZA) Model

The Zerilli–Armstrong (ZA) model, grounded in thermal activation theory and dislocation dynamics, considers the influence of temperature, strain rate, and grain size on material strength. It is defined as follows:

$$\sigma = C_0 + k_1\lambda^{-1/2} + C_2\exp(-C_3T + C_4T\ln\dot{\varepsilon}) + C_5\varepsilon^n \tag{5}$$

where C_0, k_1, C_2, C_3, C_4, C_5 are material constants, T is the absolute temperature, ε is the plastic strain, and λ is the grain size. The parameters for the ZA model are listed in Table 3 [21].

Table 3. ZA model parameters used in simulation.

Material	C_0/MPa	k_1/MPa·m$^{-3/2}$	C_2/MPa	C_3/10^{-3} K^{-1}	C_4/10^{-3} K^{-1}	C_5/MPa	n
Ta	1125	10	178	5.35	0.327	310	0.44

(3) Steinberg–Cochran–Guinan (SCG) Model

The SCG model incorporates the influence of hydrostatic pressure on shear modulus and yield strength, as well as thermal softening effects [20]. The yield strength Y and shear modulus G are defined as follows:

$$G = G_0\left[1 + \left(\frac{G'_p}{G_0}\right)\frac{p}{\eta^{1/3}} + \left(\frac{G'_T}{G_0}\right)(T - 300)\right] \quad (6)$$

$$Y = Y_0(1 + \beta\varepsilon_p)^n \frac{G}{G_0} \quad (7)$$

where η is the compression ratio, ε_p is the equivalent plastic strain, T is the material temperature, p is the pressure, G_0 and Y_0 are the initial shear modulus and yield strength, respectively (at $T = 300$ K, $p = 0$, $\epsilon = 0$), and G'_p, G'_T, β, n are material constants. The yield strength Y is also subject to the following condition:

$$Y_0(1 + \beta\varepsilon_p)^n \leq Y_{max} \quad (8)$$

where Y_{max} is the maximum yield strength. The SCG model parameters are presented in Table 4 [20].

Table 4. SCG Model parameters applied in the simulation.

Material	G_0/GPa	Y_0/GPa	Y_{max}/GPa	β	n	G'_p	G'_T/(MPa K^{-1})	T_{m0}/K
Ta	69	0.77	1.10	10	0.1	1.005	−8.97	4340

2.1.2. Spallation Model

Grady [23], based on fracture mechanics theory, proposed an energy balance fragmentation model for spallation, relating the spallation strength p_s to the average size of spall fragments and the strain rate. For ductile metals, the spallation strength varies with the impact loading process and is expressed as follows:

$$p_s = \left(2\rho c_0^2 Y \varepsilon_c\right)^{1/2} \quad (9)$$

where ρ is the material density, Y is the yield strength, c_0 is the material sound speed, and ε_c is the critical failure strain, generally taken as 0.15 for metals [23].

2.2. Spallation Characters Analysis

The spallation strength is a critical indicator of a material's performance under intense dynamic loading. Novikov's acoustic approximation method was employed to calculate the spallation strength using the following formula [24]:

$$\sigma_{spall} = \frac{1}{2}\rho_0 c_b \Delta u_s \quad (10)$$

Here, ρ_0 denotes the initial material density, c_b represents the initial bulk sound speed, and Δu_s is defined as the difference between the peak free surface velocity and the velocity at the first pullback.

The average tensile strain rate in the spallation zone can be determined using the following equation [25]:

$$\dot{\varepsilon}_s = \frac{\Delta u_s}{\Delta t_s}\frac{1}{2c_b} \quad (11)$$

where Δu_s and Δt_s denote the velocity and time differences between the maximum value and the first minimum value on the free surface velocity profile, respectively.

The velocity rebound rate, defined as the slope between the minimum and subsequent peak velocities, is calculated as follows:

$$\dot{\varepsilon}_r = \frac{\Delta u_r}{\Delta t_r}\frac{1}{2c_b} \quad (12)$$

where Δu_r and Δt_r represent the velocity and time differences between the initial minimum value and the subsequent peak value on the free surface.

The thickness of the spallation layer h can be computed with the following equation:

$$h = \Delta t \cdot \frac{c_l}{2} \quad (13)$$

where Δt is the duration of the first spallation oscillation cycle, and c_l is the longitudinal elastic wave speed.

2.3. Microscale MD Model

At the microscopic level, the interaction between tantalum atoms was simulated using the embedded atom method (EAM) potential function developed by Ravelo et al. [25]. Molecular dynamics (MD) simulations were conducted with the LAMMPS software(Aug 2023) [26], employing a model size of 16.53 nm × 16.53 nm × 16.53 nm, comprising approximately 250,000 atoms. Periodic boundary conditions were imposed in all directions. Prior to loading, the model was equilibrated for 30 ps using the NPT ensemble to achieve system stability. During loading, the NVE ensemble was utilized, applying uniform strain rates in all directions. The strain rate in the MD model was aligned with the tensile strain rate of the macroscale model.

3. Macroscale Model Validation and Analysis

3.1. Model Setup and Parameter Selection

The macroscale simulations were performed in ANSYS/AUTODYN. To validate the macroscale model, the spallation of tantalum was simulated using both the Lagrangian mesh method and the SPH method in conjunction with three different strength models: Johnson–Cook (JC), Zerilli–Armstrong (ZA), and Steinberg–Cochran–Guinan (SCG). The models were configured as two-dimensional axisymmetric models, with a Lagrangian mesh size of 0.05 mm and an SPH particle size of 0.1 mm. The sample had a diameter of 50 mm, and the impact velocity was set to 306 m/s. The information regarding model size parameters, impact velocity, and strength models used for validation is listed in Table 5.

Table 5. Parameter settings of various simulation models ("V" is an abbreviation of "Validation", representing the number of the Validation model).

Model No.	Flyer Thickness (mm)	Sample Thickness (mm)	Strength Model	Method
V-01	3	4.95	JC	Lagrange
V-02	3	4.95	JC	SPH
V-03	3	4.95	ZA	Lagrange
V-04	3	4.95	ZA	SPH
V-05	3	4.95	SCG	Lagrange
V-06	3	4.95	SCG	SPH

3.2. Comparison of Simulation Results with Experimental Data

The free surface velocity profiles generated by various models were compared against experimental data to assess the accuracy of the simulations. Figure 3 presents the free surface velocity profiles obtained from different models. The comparison indicates that both the Lagrangian mesh method and the SPH meshless method offer distinct advantages in numerical simulations.

(1) Lagrangian Method: The Lagrangian method's results exhibit good agreement with experimental data, particularly in the initial time range (0–2 μs), where the free surface velocity increased from 0 m/s to its maximum value. This method effectively captured the Hugoniot elastic limit signal, reflecting the elastic–plastic transition of tantalum.

(2) SPH Method: While the SPH method did not capture the Hugoniot elastic limit signal as distinctly as the Lagrangian method, it provided a more accurate description of the overall free surface velocity profile, particularly during the pullback and rebound phases. The SPH method's results were more consistent with experimental observations, indicating its feasibility for simulating the spallation behavior of tantalum.

(3) The experimental data used in this study are sourced from the work in [16]. High-purity, initially void-free tantalum samples were employed. Plate impact tests were conducted at room temperature using a single-stage gas gun. The velocity profiles were recorded using Doppler laser interferometry. The impact velocity was 306 m/s, with a flyer plate thickness of 3 mm and a target thickness of 4.95 mm.

The comparative analysis suggests that combining both methods can yield a more complete and accurate free surface velocity profile. Additionally, the SCG model was identified as the most suitable for simulating tantalum spallation among the models tested.

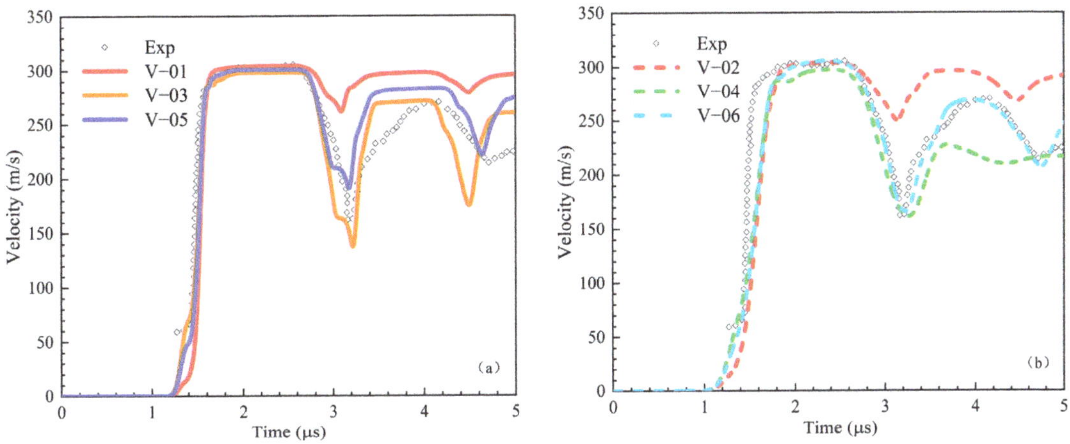

Figure 3. Comparison of simulated and experimental free surface velocity profiles [16]. (**a**) Lagrange method, (**b**) SPH method.

3.3. Analysis of Spallation Characteristics under Different Tensile Strain Rates

To analyze the spallation characteristics under different tensile strain rates, the tensile strain rate in the spallation region was varied by changing the loading conditions (the tensile strain rate ranging from 2.13×10^4–5.40×10^4 s^{-1} and the pressure ranging from 6.19 GPa to 12.25 GPa), and the simulation parameters are shown in Table 6.

Table 6. Parameters and results of PPI Simulations at different strain rates ("S" is an abbreviation of "simulation", representing the number of the simulation model).

Model No.	Flyer Thickness (mm)	Sample Thickness (mm)	Loading Velocity (m/s)	p/GPa	$\dot{\varepsilon}_s$/s^{-1}	σ_{spall}/GPa	$\dot{\varepsilon}_r$/s^{-1}
S-01	2	4.95	306	8.84	5.40×10^4	4.92	3.57×10^4
S-02	2	4.95	250	7.05	4.69×10^4	4.7	3.25×10^4
S-03	3	4.95	410	12.25	3.92×10^4	4.14	2.32×10^4
S-04	3	4.95	306	8.84	3.28×10^4	3.97	1.74×10^4
S-05	3	4.95	210	6.19	2.68×10^4	3.71	1.69×10^4
S-06	4	4.95	306	8.84	2.31×10^4	3.34	1.34×10^4

Figure 4 presents the macroscale numerical simulation results of tantalum spallation at various tensile strain rates. The findings indicate that the maximum free surface velocity closely corresponds to the impact velocity of the flyer plate. Furthermore, the width of the velocity plateau on the free surface velocity profile increases proportionally with the flyer plate thickness, with thicker plates leading to a broader plateau.

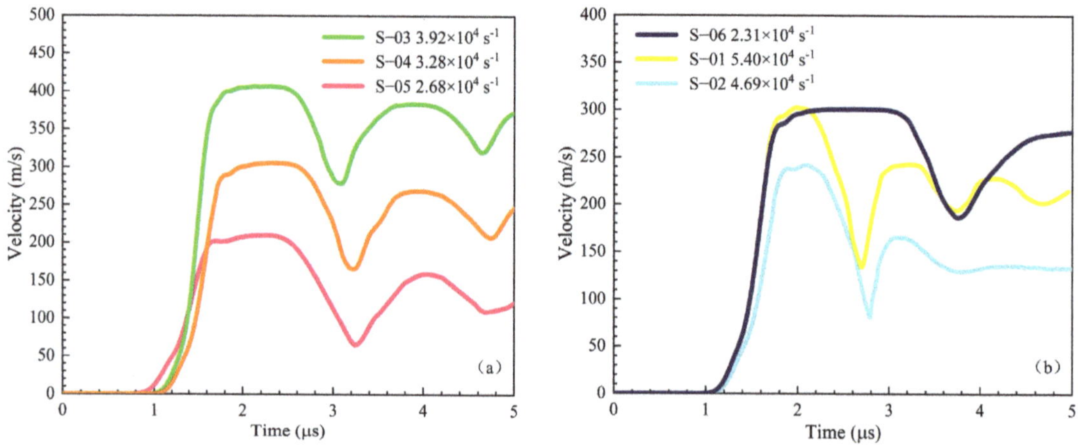

Figure 4. Free surface velocity at different tensile strain rates on the macroscale. (**a**) Free surface velocity of S−03 to S−05, (**b**) Free surface velocity of S−01, S−02 and S−06.

3.4. Spallation Strength Analysis

Understanding spallation strength is essential for evaluating the dynamic damage characteristics of metallic materials. By comparing simulation outcomes and the established spallation strength–strain rate relationship for copper [3], a corresponding relationship for tantalum was formulated:

$$\sigma_{\text{spall}} = 0.042 \times \dot{\varepsilon}^{0.44} \tag{14}$$

Figure 5 illustrates the relationship between tantalum's spallation strength and tensile strain rate. The analysis revealed that spallation strength is not solely influenced by the loading velocity. At identical loading velocities, spallation strength increased with tensile strain rate, indicating a strain rate-dependent characteristic.

Under identical loading velocities, the spallation strengths of S-06, S-04, and S-01 increase with the rise in tensile strain rate. Similarly, although the loading velocities differ significantly for S-05, S-03, and S-02, a proportional relationship between spallation strength and tensile strain rate is still observed. This indicates that spallation strength exhibits a strain-rate-dependent characteristic, increasing as the tensile strain rate rises.

Figure 6 presents the relationship between spallation strength σ_{spall} and tensile strain rate $\dot{\varepsilon}_s$ in logarithmic coordinates. It is apparent from the figure that spallation strength displays a clear strain rate dependency in logarithmic coordinates. Spallation strength's logarithmic dependence on strain rate can be attributed to the strain-rate sensitivity of microvoid nucleation and growth processes. Higher strain rates increase the material's resistance to dynamic failure due to limited time for void formation, similar to the logarithmic relations seen in work hardening and strain hardening, where increased strain rates lead to greater dislocation interactions and material strength. This suggests that both phenomena share underlying mechanisms related to the strain-rate sensitivity of the material's microstructural evolution. We compared our results with experimental data at higher tensile strain rates and found that spallation strength also increases with strain rate under these conditions, further validating the accuracy of the numerical simulation results.

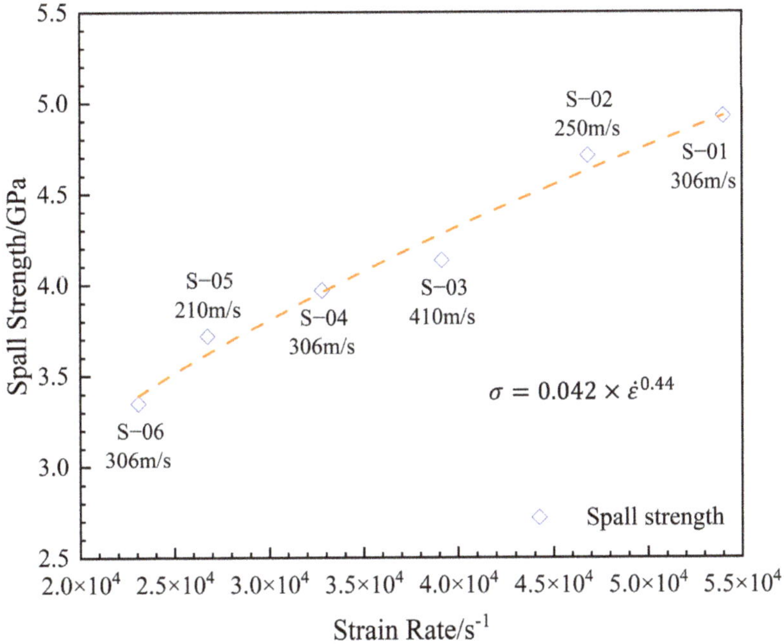

Figure 5. Correlation between spallation strength and tensile strain rate.

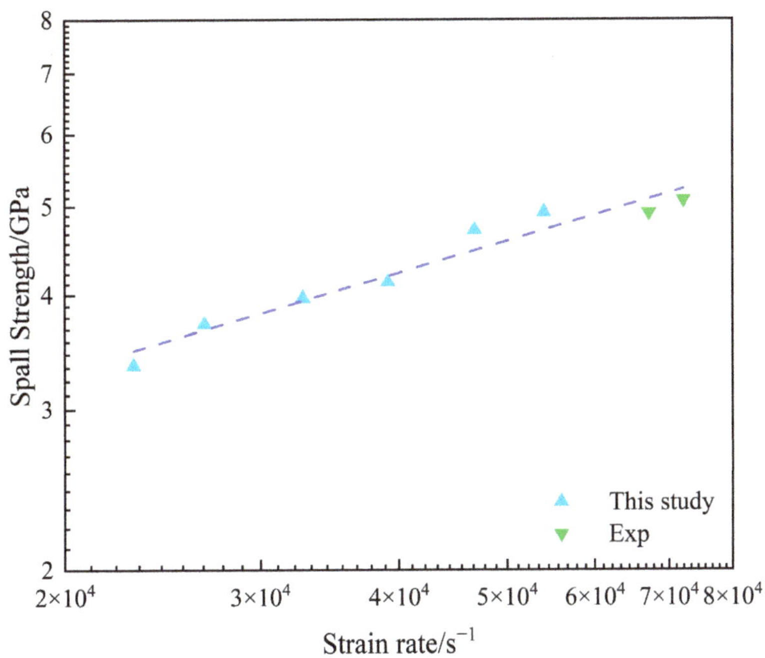

Figure 6. Correlation between spallation strength and tensile strain rate on logarithmic scale [27].

The spallation strength analyzed earlier was calculated using equation (11), which is a widely used acoustic approximation method. However, it is not the only calculation method available. Stepanov [28] pointed out that in ductile metals, during a planar impact, the shock wave propagates at the elastic longitudinal wave speed c_l. The preceding incident rarefaction plastic wave propagates at the bulk sound speed c_b. When the elastic wave's influence is primarily considered, the effective sound speed c_e can be expressed as

$$c_e = 2c_l c_b/(c_l + c_b) \tag{15}$$

Based on this, a modified form of the spallation strength can be derived:

$$\sigma_{spall}^{(1)} = \rho_0 c_b \Delta u \left(\frac{1}{1 + c_b/c_l} \right) \tag{16}$$

Here, ρ_0 and c_b are the material density, and Δu denotes the velocity difference between the maximum and first minimum values on the free surface velocity profile. Additionally, Kanel [29] proposed that the thickness of the spallation layer should be included when calculating spallation strength. Incorporating the spallation layer thickness, the formula is modified as follows:

$$\sigma_{spall}^{(2)} = \frac{1}{2}\rho_0 c_b \Delta u + \delta_2 \tag{17}$$

$$\delta_2 = \left(\frac{h_{sp}}{c_b} - \frac{h_{sp}}{c_l} \right) \cdot \frac{|\dot{u}_1 \dot{u}_2|}{|\dot{u}_1| + \dot{u}_2} \cdot \frac{1}{2}\rho_0 c_b \tag{18}$$

where $\dot{u}_1$ represents the velocity change rate before the appearance of the pullback signal on the free surface velocity–time curve, $\dot{u}_2$ represents the velocity change rate of the spallation rebound signal, and h_{sp} is the thickness of the spallation layer. The spallation strength values calculated using the three formulas are compared in Table 7. The analysis shows that the results from Equations (17) and (11) differ by up to 8%, while the results from Equation (18) are quite similar to those from Equation (17). This indicates that the method chosen to calculate spallation strength from the free surface velocity profile is significant, and the differences between various calculation models should be carefully considered during analysis.

Table 7. Comparison of spallation strength data obtained from different calculation formulas.

Model No.	σ_{spall}/GPa	$\sigma_{spall}^{(1)}$/GPa	$\sigma_{spall}^{(2)}$/GPa
S-01	4.92	5.25	5.36
S-02	4.71	5.02	5.32
S-03	4.13	4.40	4.48
S-04	3.97	4.23	4.32
S-05	3.72	3.96	4.07
S-06	3.35	3.57	3.58

4. Multiscale Analysis of Spallation Characteristics

The analysis of spallation characteristics at the macroscale reveals that the free surface velocity profile reflects the interaction between the internal damage evolution within the material and the macroscopic response field during the spallation process. Understanding the free surface velocity profile comprehensively and accurately is crucial for studying spallation phenomena. However, current research methods face challenges in in situ observation of damage evolution during spallation, making it difficult to intuitively and accurately understand the relationship between the material's macroscopic dynamic response and microstructural evolution during spallation. This has led to some controversies regarding the interpretation of the free surface velocity profile, such as the significance of the pullback signal, the nature of damage evolution, and the meaning of the rebound rate. To address these issues, molecular dynamics (MD) simulations were employed to provide insights into the damage evolution process during spallation, offering a qualitative interpretation of some typical features on the free surface velocity profile from a microscopic perspective.

4.1. Pullback Signal Analysis

The pullback signal on the free surface velocity profile is a critical indicator for determining whether spallation has occurred. However, there has been debate regarding whether the material is fully separated at this point. Zurek et al. [30] suggested that the material is completely separated when the pullback signal appears, while Kanel et al. [31] observed that a pullback signal can occur even when the damage at the spallation plane is minimal. Bonora et al. [2] hypothesized that the pullback signal is associated with the nucleation of microvoids, although this has not been empirically verified.

In this study, the loading condition of model S-04 was employed to explore the significance of the pullback signal from a microscopic perspective. Molecular dynamics (MD) simulations were used to analyze the temporal evolution of the free surface velocity, internal stress, and damage progression within the spallation region.

Figures 7 and 8 present the temporal evolution of the free surface velocity after the decrease, along with the internal stress and damage evolution within the spallation region as captured by the MD model. The analysis accounts for potential errors in identifying the pullback signal by extending the pullback signal time range to 0.05 μs. As shown in Figure 7, during the time range when the pullback signal appears, the stress within the spallation region reaches its peak. After the pullback signal, the stress in the spallation region rapidly decreases, indicating stress relaxation. When comparing this with the damage evolution within the spallation region, it can be observed that the time range corresponding to the pullback signal aligns with the initial stage of damage development, specifically the nucleation stage of microvoids. In Figure 8, the internal damage evolution is visually depicted, showing that microvoids begin to gradually form during the pullback signal period, with the volume fraction of internal voids being very small at this stage. This indicates that the spallation region in tantalum has not fully separated within the time range of the pullback signal and that the pullback signal corresponds to the nucleation process of internal microvoids. This microscopic-scale evidence supports Bonora et al.'s hypothesis [2].

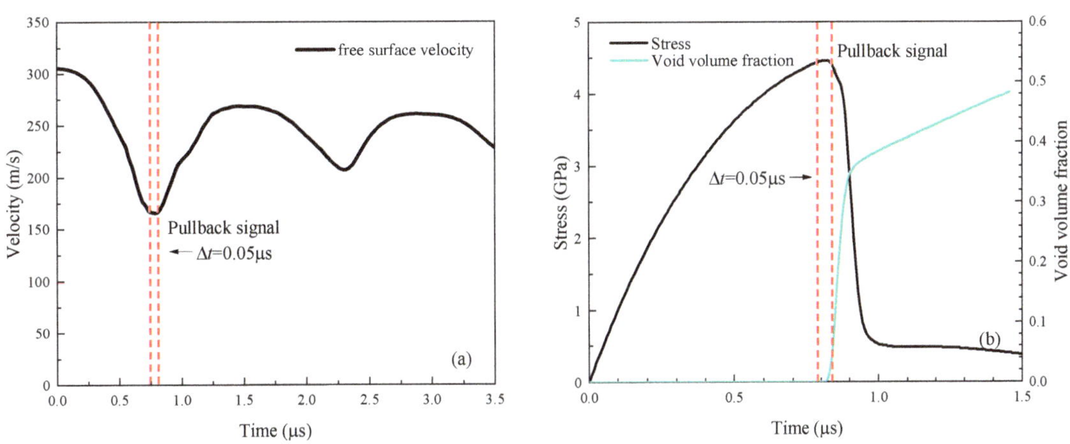

Figure 7. Time evolution of free surface velocity and stress in the spallation zone. (**a**) Free surface velocity curve, (**b**) MD model stress and damage evolution curve.

Furthermore, when calculating spallation strength from the free surface velocity profile, the velocity drop associated with the pullback signal (Δu in Figure 1b) serves as an important basis. The previous analysis indicates that in the spallation of tantalum, the pullback signal on the free surface velocity profile corresponds to the microvoid nucleation process, suggesting that Δu actually reflects the conditions for microvoid nucleation. Therefore, the spallation strength derived from Δu should more accurately be interpreted as the strength resisting the initiation of damage or as a representation of microvoid nucleation resistance.

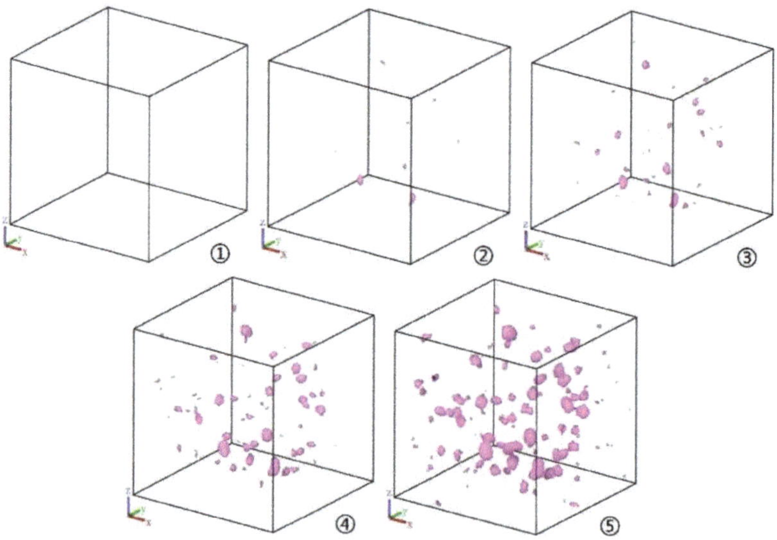

Figure 8. Visualization of damage evolution in the spallation zone during the pullback signal phase.

4.2. Damage Evolution Process

The damage evolution process within the spallation region is challenging to observe directly at the macroscale during experiments. However, by utilizing the microscale MD model, this study analyzed the entire damage evolution process during spallation, from microvoid nucleation to complete fracture.

Figures 9–11 illustrate the volume fraction of internal voids, the microvoid count, and the distribution of internal voids during the damage evolution process. This process can be divided into three distinct stages:

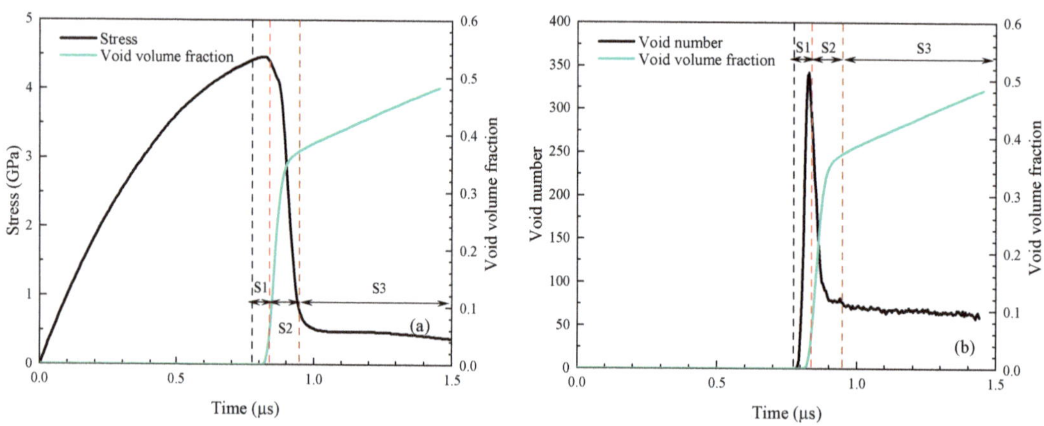

Figure 9. Correlation between void volume evolution, stress, and void count. (**a**) Stress and void volume, (**b**) Void numbers and volume.

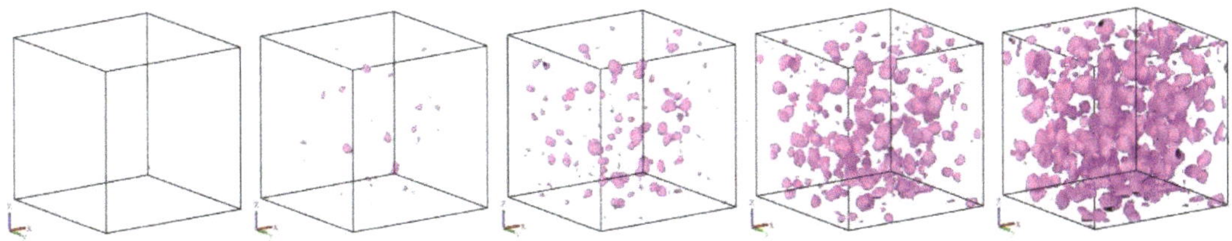

Figure 10. Evolution of void count during Stage 1.

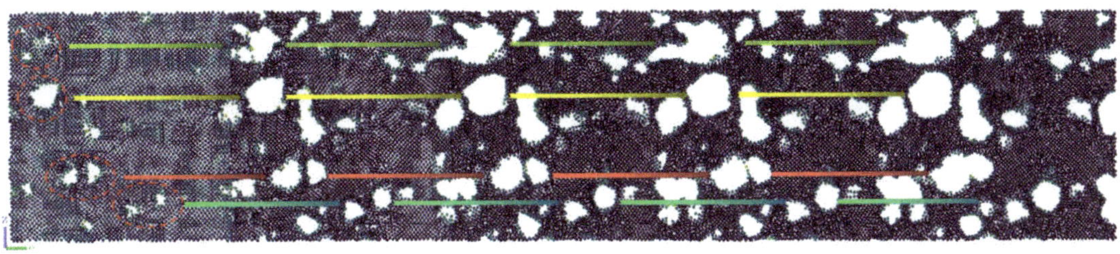

Figure 11. Void growth and coalescence during Stage 2 and Stage 3.

(1) Nucleation (S1): As internal pressure nears its peak, microvoids start to form. In this stage, the number of voids increases rapidly, while the void volume fraction remains low, as illustrated in Figure 10.

(2) Growth (S2): Following nucleation, stress rapidly decreases, leading to energy dissipation and a significant increase in void volume fraction. During this stage, the number of voids decreases, indicating that void growth, rather than new void formation, drives the increase in volume.

(3) Coalescence (S3): Voids coalesce, resulting in the complete fracture of the material. During this stage, the growth rate of the void volume fraction decreases, and the number of voids stabilizes, indicating that the final spallation is driven by the merging of larger voids, as depicted in Figure 11.

4.3. Rebound Rate Analysis

In previous analyses, the focus was often on the pullback signal and the preceding segment of the free surface velocity profile, with little attention given to the rebound curve following the pullback signal. Starting from the pullback signal, we analyzed the variations in the free surface velocity rebound curve under different strain rate conditions, as shown in Figure 12.

The results indicate that the rebound curve does not show a clear trend with changes in loading velocity or flyer plate thickness. However, as the tensile strain rate increases, the curve becomes steeper, and the slope increases, indicating a strain rate-dependent characteristic.

Figure 13 illustrates the relationship of rebound rate with spallation strength and tensile strain rate. The results show that the relationship between spallation strength and rebound rate was found to be nearly linear, suggesting that the rebound rate is also reflective of spallation behavior. Furthermore, the rebound rate increases approximately linearly with the tensile strain rate, implying that the damage evolution rate within the sample during spallation increases with the tensile strain rate.

As shown in Figure 9, during the rebound phase following the pullback signal, damage within the spallation region rapidly intensifies, accompanied by a significant increase in the volume fraction of microvoids. This highlights a correlation between the rebound of free surface velocity and the evolution of damage. Kanel et al. [31] examined the relationship between rebound rate and damage evolution rate using the characteristic line method, concluding that they are proportional. Figure 14 illustrates the stress evolution and damage progression under various strain rates.

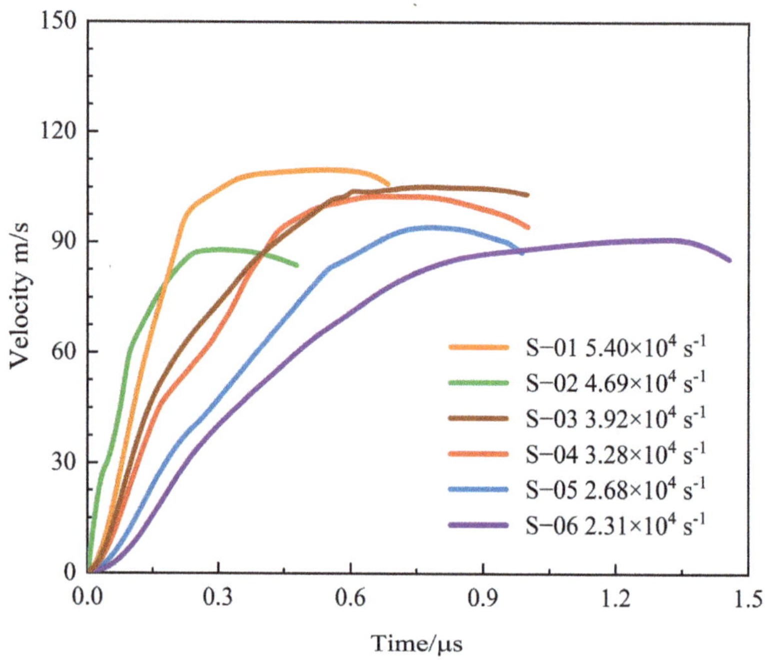

Figure 12. Free surface velocity curve derived from spallation signal.

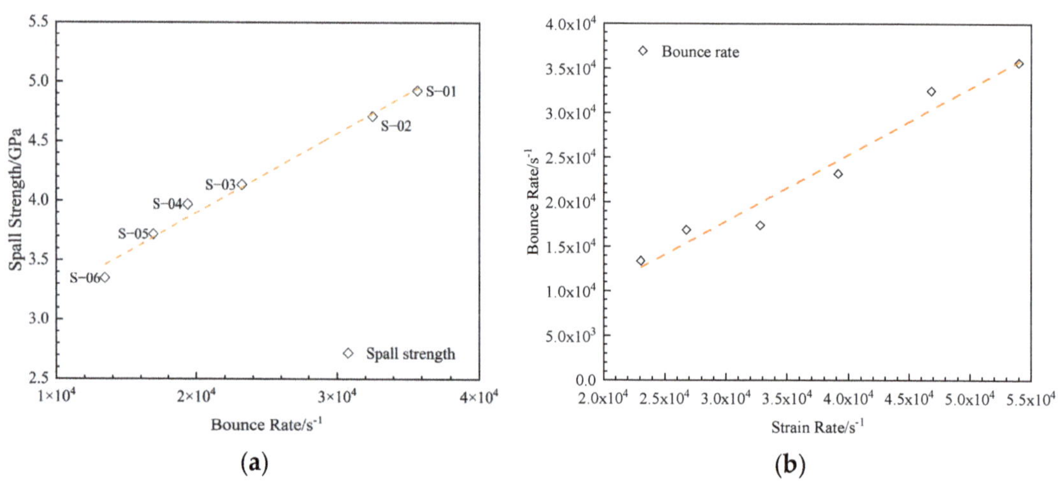

Figure 13. The relationship of rebound rate with spallation strength and tensile strain rate. (**a**) spalling strength vs. rebound rate, (**b**) rebound rate vs. tensile strain rates.

From the previous analysis, we found that the rebound rate and tensile strain rate exhibit an approximately linear relationship, with higher tensile strain rates corresponding to higher rebound rates. Figure 14a shows the stress evolution under different strain rates. A macroscopic wave analysis indicates that the free surface velocity rebound rate correlates with the stress relaxation rate. The stress relaxation rate increases with higher strain rates, suggesting a positive correlation between rebound rate and stress relaxation rate. Faster stress relaxation indicates quicker energy dissipation, which is associated with the damage evolution rate. Figure 14b shows that with increasing strain rates, the speed of microvoid growth during the growth stage accelerates. Overall, the free surface velocity rebound rate is a macroscopic reflection of the damage evolution rate within the spallation region.

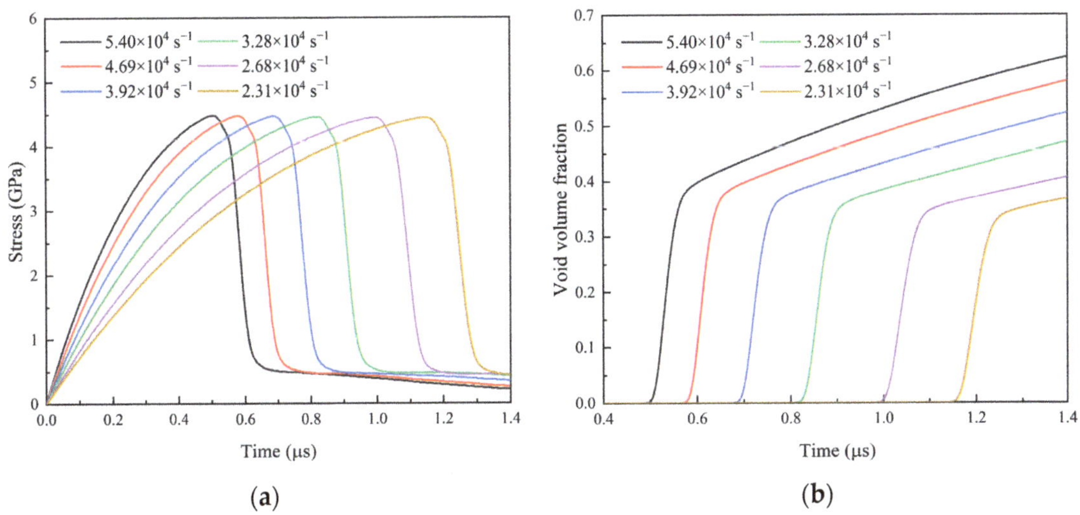

Figure 14. Stress and damage evolution under different strain rates. (**a**) Stress evolution, (**b**) Damage evolution.

5. Conclusions

This study investigated the spallation characteristics of ductile tantalum under high strain rate impacts using a multiscale simulation approach. The key findings are summarized as follows:

(1) Both the Lagrangian mesh method and the Smoothed Particle Hydrodynamics (SPH) method effectively captured the spallation behavior of tantalum. The Lagrangian method accurately simulated the initial elastic–plastic transition, evidenced by the Hugoniot elastic limit signal. The SPH method provided a comprehensive depiction of the free surface velocity profile, particularly during the pullback and rebound phases.

(2) Spallation strength showed a strain-rate-dependent characteristic, following an exponential relationship: $\sigma_{spall} = 0.042 \times \dot{\varepsilon}^{0.44}$. For instance, the spallation strength increased from 3.34 GPa to 4.92 GPa as the tensile strain rate rose from 2.31×10^4 to $5.40 \times 10^4 \cdot s^{-1}$.

(3) The microscale molecular dynamics (MD) simulations identified three distinct stages of damage evolution: nucleation, growth, and coalescence of microvoids. The pullback signal on the free surface velocity profile corresponded to the nucleation of microvoids. The velocity drop before the pullback signal indicated the initiation of microvoids, while the rebound rate reflected the rate of damage progression.

(4) A near-linear relationship between the rebound rate and both spallation strength and tensile strain rate was established. This indicates that the rebound rate can serve as an effective indicator of spallation behavior and the dynamic damage resistance of the material.

By integrating macro- and microscale results, this study bridges the gap between macroscopic observations and microscopic damage mechanisms, providing a deeper understanding of tantalum's spallation behavior under extreme conditions.

Author Contributions: Conceptualization, Y.W. and Y.G.; methodology, Y.W.; software, Y.W. and W.S.; resources, Y.L. and M.Y.; writing—original draft preparation, Y.W.; writing—review and editing, S.L. All authors have read and agreed to the published version of the manuscript.

Funding: The authors would like to acknowledge the support provided by Key Research Program of Ningxia(2022BEG02056); the Luzhou Key Laboratory of Intelligent Construction and Low-carbon Technology; the Scientific Research Platform of Chengdu Polytechnic (No. 23KYPT01, No. 23KYPT02) and the Chengdu Polytechnic Scientific Research Project (No. 23CZYG047, No: 21CZYG020).

Data Availability Statement: The data that support the findings of this study are available from the corresponding author, Yuzhu Guo (guoyuzhu5185@163.com), upon reasonable request.

Conflicts of Interest: The authors declare no conflict of interest.

References

1. Lu, K. The future of metals. *Science* **2010**, *328*, 319–320. [CrossRef]
2. Peng, H. Discussion on the physical meaning of free surface velocity curve in ductile spallation. *Acta Phys. Sin.* **2015**, *64*, 034601.
3. Curran, D.R.; Seaman, L.; Shockey, D.A. Dynamic failure of solids. *Phys. Rep.* **1987**, *147*, 253–388. [CrossRef]
4. Antoun, T.; Seaman, L.; Curran, D.R.; Kanel, G.I.; Razorenov, S.V.; Utkin, A.V. *Spall Fracture*; Springer Science & Business Media: Berlin/Heidelberg, Germany, 2006.
5. Thomas, S.A.; Hawkins, M.C.; Matthes, M.K.; Gray, G.T.; Hixson, R.S. Dynamic strength properties and alpha-phase shock Hugoniot of iron and steel. *J. Appl. Phys.* **2018**, *123*, 175902. [CrossRef]
6. Koller, D.D.; Hixson, R.S.; Gray, G.T.; Rigg, P.A.; Addessio, L.B.; Cerreta, E.K.; Maestas, J.D.; Yablinsky, C.A. Influence of shock-wave profile shape on dynamically induced damage in high-purity copper. *J. Appl. Phys.* **2005**, *98*, 103518. [CrossRef]
7. Liu, M.B.; Liu, G.R. Smoothed particle hydrodynamics (SPH): An overview and recent developments. *Arch. Comput. Methods Eng.* **2010**, *17*, 25–76. [CrossRef]
8. Glam, B.; Strauss, M.; Eliezer, S.; Moreno, D. Shock compression and spall formation in aluminum containing helium bubbles at room temperature and near the melting temperature: Experiments and simulations. *Int. J. Impact Eng.* **2014**, *65*, 1–12. [CrossRef]
9. Zhou, C.E.; Liu, G.R.; Lou, K.Y. Three-dimensional penetration simulation using smoothed particle hydrodynamics. *Int. J. Comput. Methods* **2007**, *4*, 671–691. [CrossRef]
10. Rawat, S.; Raole, P.M. Molecular dynamics investigation of void evolution dynamics in single crystal iron at extreme strain rates. *Comput. Mater. Sci.* **2018**, *154*, 393–404. [CrossRef]
11. Yang, X.; Zeng, X.; Wang, J.; Wang, J.; Wang, F.; Ding, J. Atomic-scale modeling of the void nucleation, growth, and coalescence in Al at high strain rates. *Mech. Mater.* **2019**, *135*, 98–113. [CrossRef]
12. Chen, J.; Fensin, S.J. Associating damage nucleation and distribution with grain boundary characteristics in Ta. *Scr. Mater.* **2020**, *187*, 329–334. [CrossRef]
13. Chen, D.; Fan, C.; Xie, S.; Hu, J.; Wu, S.; Wang, H.; Ta, H.; Yu, Y. Study on constitutive relations and spall models for oxygen-free high-conductivity copper under planar shock tests. *J. Appl. Phys.* **2007**, *101*, 063532.
14. Remington, T.P.; Hahn, E.N.; Zhao, S.; Flanagan, R.; Mertens, J.; Sabbaghianrad, S.; Langdon, T.; Wehrenberg, C.; Maddox, B.; Swift, D.; et al. Spall strength dependence on grain size and strain rate in tantalum. *Acta Mater.* **2018**, *158*, S1359645418305846. [CrossRef]
15. Wang, Y.; Zeng, X.; Yang, X.; Xu, T. Shock-induced spallation in single-crystalline tantalum at elevated temperatures through molecular dynamics modeling. *Comput. Mater. Sci.* **2022**, *201*, 110870. [CrossRef]
16. Czarnota, C.; Jacques, N.; Mercier, S.; Molinari, A. Modelling of dynamic ductile fracture and application to the simulation of plate impact tests on tantalum. *J. Mech. Phys. Solids* **2008**, *56*, 1624–1650. [CrossRef]
17. Heuzé, O. General form of the Mie–Grüneisen equation of state. *Comptes Rendus. Mécanique* **2012**, *340*, 679–687. [CrossRef]
18. Marsh, S.P. *LASL Shock Hugoniot Data*; Univ of California Press: Oakland, CA, USA, 1980.
19. Johnson, G.R.; Cook, W.H. Fracture characteristics of three metals subjected to various strains, strain rates, temperatures and pressures. *Eng. Fract. Mech.* **1985**, *21*, 31–48. [CrossRef]
20. Steinberg, D.J.; Cochran, S.G.; Guinan, M.W. A constitutive model for metals applicable at high-strain rate. *J. Appl. Phys.* **1980**, *51*, 1498–1504. [CrossRef]
21. Zerilli, F.J.; Armstrong, R.W. Dislocation-mechanics-based constitutive relations for material dynamics calculations. *J. Appl. Phys.* **1987**, *61*, 1816–1825. [CrossRef]
22. Wang, Y.; Zeng, X.; Chen, H.; Yang, X.; Wang, F.; Zeng, L. Modified Johnson-Cook constitutive model of metallic materials under a wide range of temperatures and strain rates. *Results Phys.* **2021**, *27*, 104498. [CrossRef]
23. Grady, D.E. The spall strength of condensed matter. *J. Mech. Phys. Solids* **1988**, *36*, 353–384. [CrossRef]
24. Forquin, P.; Lukić, B. On the processing of spalling experiments. Part I: Identification of the dynamic tensile strength of concrete. *J. Dyn. Behav. Mater.* **2018**, *4*, 34–55. [CrossRef]

25. Ravelo, R.; Germann, T.C.; Guerrero, O.; An, Q.; Holian, B.L. Shock-induced plasticity in tantalum single crystals: Interatomic potentials and large-scale molecular-dynamics simulations. *Phys. Rev. B—Condens. Matter Mater. Phys.* **2013**, *88*, 134101. [CrossRef]
26. Plimpton, S. Fast parallel algorithms for short-range molecular dynamics. *J. Comput. Phys.* **1995**, *117*, 1–19. [CrossRef]
27. Zhang, L. Impact response of ductile materials: A new model of dynamic damage and fracture, structural phase transition. Ph.D. Thesis, China Academy of Engineering Physics, Mianyang, China, 2005.
28. Stepanov, K.L.; Stanchits, L.K.; Stankevich, Y.A. Radiation of strong shock waves in air: Part I. *High Temp.* **2000**, *38*, 182–190. [CrossRef]
29. Kanel, G.I. Dynamic strength of materials. *Fatigue Fract. Eng. Mater. Struct.* **1999**, *22*, 1011–1020.
30. Zurek, A.K.; Thissell, W.R.; Johnson, J.N.; Tonks, D.; Hixson, R. Micromechanics of spall and damage in tantalum. *J. Mater. Process. Technol.* **1996**, *60*, 261–267. [CrossRef]
31. Kanel, G.I.; Razorenov, S.V.; Bogatch, A.; Utkin, A.; Grady, D.E. Simulation of spall fracture of aluminum and magnesium over a wide range of load duration and temperature. *Int. J. Impact Eng.* **1997**, *20*, 467–478. [CrossRef]

crystals

MDPI

Article

Effects of Temperature and Secondary Orientations on the Deformation Behavior of Single-Crystal Superalloys

Sujie Liu [1,2], Cui Zong [3], Guangcai Ma [4], Yafeng Zhao [1,5], Junjie Huang [1,2], Yi Guo [1,*] and Xingqiu Chen [1]

1 Shenyang National Laboratory for Materials Science, Institute of Metal Research, Chinese Academy of Sciences, Shenyang 110016, China; sjliu20s@imr.ac.cn (S.L.); yfzhao22h@imr.ac.cn (Y.Z.); jjhuang22s@imr.ac.cn (J.H.); xingqiu.chen@imr.ac.cn (X.C.)

2 School of Materials Science and Engineering, University of Science and Technology of China, Shenyang 110016, China

3 National Key Laboratory of Advanced High Temperature Structural Materials, Beijing Institute of Aeronautical Materials, Beijing 100095, China; czong19@hotmail.com

4 Analysis and Test Center, Institute of Metal Research, Chinese Academy of Sciences, Shenyang 110016, China; gcma@imr.ac.cn

5 School of Materials Science and Engineering, Northeastern University, Shenyang 110819, China

* Correspondence: yguo@imr.ac.cn

Abstract: The tensile behavior of single-crystal superalloys was investigated at room temperature (RT) and 850 °C, focusing on various secondary orientations. Transmission electron microscopy (TEM) and quasi in situ electron backscatter diffraction (EBSD) were employed to study the deformation mechanisms across length scales. Deformation at 850 °C enhanced the tensile ductility of the samples, evidenced by the more uniform coverage of dislocations across the γ and γ' phases, and the fracture mode switched from pure cleavage at room temperature to mixed mode due to accelerated void growth. The influence of secondary orientations on mechanical properties is insignificant at room temperature. However, the ductility of the different secondary orientation samples shows significant variations at 850 °C, among which the one with [001] rotated 37° demonstrated superior ductility compared to others.

Keywords: nickel-based superalloy; secondary orientation; quasi in situ EBSD; geometrically necessary dislocations (GNDs); deformation mechanisms

1. Introduction

Nickel-based single-crystal (Ni-based SC) superalloys have been widely employed in aircraft engines and gas turbines due to their structural stability, excellent mechanical properties, and corrosion resistance at high temperatures. The excellent elevated temperature performance of these alloys originates from the unique γ/γ' two-phase coherent microstructure, where the ordered cubic phase γ' ($L1_2$ structure) is coherent with the disordered γ matrix [1,2]. In terms of micromechanics, the strengthening phase γ' and the coherent γ/γ' interface to impede dislocation motions, and thus improving the overall mechanical properties of the alloy [3,4]. Tensile properties are regarded as a representative indicator of the comprehensive properties for SC superalloys, which can provide a reliable reference for fatigue life and creep properties. There have been many studies on the deformation mechanism of Ni-based SC superalloys under various experimental conditions, but the temperature-dependent deformation mechanism is not yet clearly understood. Especially at room and intermediate temperatures, different configurations of dislocations are closely related to the temperature, and these have significant impacts on the properties of the alloys [5–8].

Ni-based SC superalloys are mainly casted by directional solidification [9]. Due to the elimination of grain boundaries and the face-centered cubic (FCC) structure, the mechanical properties of the alloys show anisotropic characteristics, and it has been shown that the [001] orientation has the best overall performance [10,11]. During the past decades, a lot of studies have focused on the effects of orientations on mechanical properties, but most of them have been based on the influences of primary orientation [12,13]. However, studies in recent years have found that the secondary orientation also affects the mechanical properties [14–21]. Arakere et al. [14] have predicted that the appropriate secondary orientation can significantly increase the fatigue crack extension resistance of the component, without additional weight or cost. Sabnis et al. [15], using finite element analysis, argued that secondary orientation has a significant effect on stress distribution, and the slip mode as well as that the size and shape of the plastic zone are strongly dependent on the secondary orientation of the notch, which may have a significant impact on crack initiation. Zhou et al. [17] demonstrated that the tensile strength and fracture strain at room temperature are affected by the cooling holes and secondary orientations, with the (110) samples without holes exhibiting slightly higher tensile strength than the (100) samples. Suzuki et al. [19] tested fatigue crack propagation experiments on notched samples with secondary orientations (001) and (110) at different temperatures; the results showed the difference in crack propagation rates between (100) orientations was small at both 450 °C and 700 °C, while there was a significant difference in crack propagation rate for the (110) orientations, which consistently exceeded that of the (100) orientations at a variety of temperatures. Due to the anisotropy of the secondary orientation, Zhou et al. [20] demonstrated that samples with holes exhibited variations in tensile properties with secondary orientations, among which the (100) specimen demonstrated better tensile strength and plasticity compared to the (110) orientations. In summary, the optimal secondary orientation for different compositions of Ni-based SC superalloys may not be unique, and the activated slip systems vary for samples at different temperatures [19], thus requiring case-specific testing for different alloys.

Therefore, the main objective of this study is to test the diverged opinions on the effects of secondary orientations using a second-generation Ni-base superalloy. Additionally, we conducted the tests at elevated temperatures, hoping to uncover any related temperature effect. The influence of temperatures on the secondary orientation effect were explored by quasi in situ electron back scattering diffraction (EBSD), which was combined with TEM to investigate the dislocation mechanisms underlying secondary orientation effects at different temperatures.

2. Materials and Methods

2.1. Material Preparation

A second-generation Ni-based SC superalloy was used in the present work, and its composition is listed in Table 1. Directionally solidified plate ingots were subjected to a solution treatment at 1300 °C for 2 h, followed by three-stage aging at 1120 °C, 1080 °C, and 900 °C for 4 h sequentially. All the cooling methods involved air cooling and temperature was controlled within a ±10 °C range. In order to determine the crystal orientation of the ingot, a small sample was cut from the top of the ingot and observed by EBSD. The results show that the solidification direction of the single crystal ingot was ~6° misaligned with the [001] orientation. Based on this orientation information, secondary orientation sampling was performed on the ingot. The outer surface of the plate ingot was used as the 0° reference plane; four different secondary orientations were, respectively, rotated around the [001] main axis by 5°, 37°, 47°, and 57° (Figure 1a). These orientations were chosen such that different slip systems are activated based on Schmid factors. Tensile specimens were prepared using electrical discharge machining (EDM), with the geometry shown in Figure 1b. Specimens were ground using SiC sandpapers ranging from 150# to 3000#, followed by polishing with 2.5 μm diamond paste and fine polishing with an 80 nm colloidal silica suspension until the sample surface was scratch-free. Subsequently, electrolytic polishing was performed at −20 °C using a 10% perchloric acid +90% alcohol solution for 50 s at −20 °C and 15 kV. The microstructures of the samples were shown in

Figure 2a–c. As illustrated in Figure 2a, the scanning electron microscope (SEM) images show that a number of carbides are dispersed in the matrix in a needle or platelet-like morphology. The area fraction of carbides on the surfaces of samples in various secondary orientations was statistically analyzed, as depicted in Figure 2b. Variations in carbide proportions were observed among these orientations.

Table 1. Chemical composition of the alloy.

	C	Cr	Co	W	Al	Ta	Mo	Hf	B	Re	Ni
wt.%	0.05	7.05	7.5	5.4	6.24	6.58	1.51	0.16	0.0047	2.96	Bal

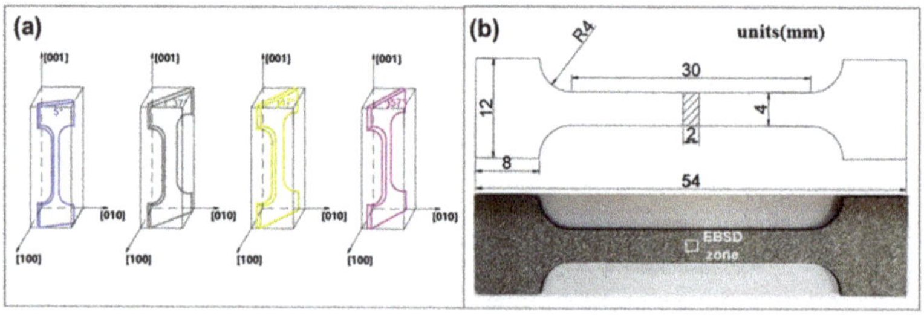

Figure 1. (**a**) Schematic diagram of the four secondary orientations of 5°, 37°, 47°, and 57°, sampled on the ingot; (**b**) geometry of the tension sample with the location for the EBSD scans indicated by the box.

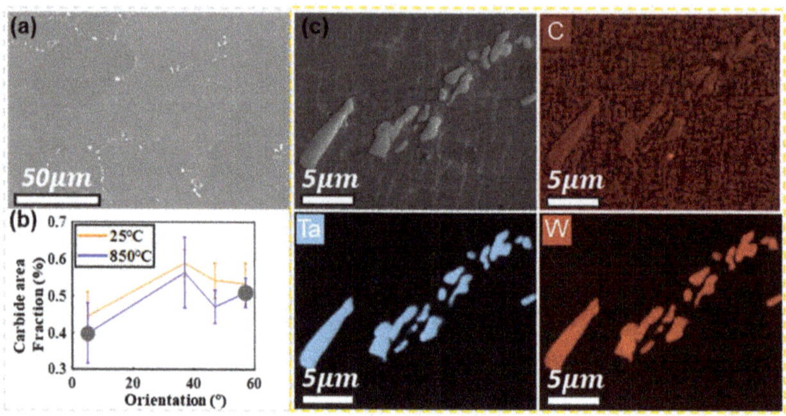

Figure 2. Microstructures of the alloy: (**a**) SEM image; (**b**) the carbide area fraction varies with different secondary orientations; (**c**) EDS images indicating the carbide compositions.

2.2. Mechanical Testing

The uniaxial tensile tests were performed using a MTS E45.105 electronic universal testing machine (MTS SANS, Shenzhen, China) at room temperature (RT) and a high temperature (HT, 850 °C) along the [001] orientation. Firstly, four specimens, each corresponding to one of the stated secondary orientations, were tested under a strain rate of $1.1 \times 10^{-3} s^{-1}$ until rupture to generate the stress–strain curve. Then, based on the stress–strain data, interrupted tensile tests were carried out under the same loading conditions, with the RT tests interrupted at 4%, 8%, and 12% of the accumulate strains, and the meso-scale GND density distribution was observed by EBSD at each of the strain levels. Tensile tests at 850 °C were interrupted at 4%, 8%, and 18% of the accumulate strains, and due to oxidation at the high temperature, the samples were ground and electro-polished before EBSD. The tensile rig compliance was corrected using procedures described in reference [22].

2.3. Microstructure Analysis

After the tensile test, the temperature dependence of the fracture surface morphology was observed using a Gmini-SEM460 scanning electron microscope (SEM) (Zeiss, Oberkochen, Germany). For TEM study, thin foils were cut parallel to the (001) orientation at 6 mm away from the fracture surface on the 37° orientation at RT and 850 °C. The foils were ground to a thickness of approximately 50 μm, and polished through a Leibow TJ100-SE-TMS (LEBO Science, Wuxi, China) standard temperature controlled electrolytic double sprayer, then thinned in a Gatan 695 ion thinning apparatus, and the TEM experiments were performed in a FEI Talos F200X (FEI, Hillsboro, OR, USA).

2.4. Quasi In Situ EBSD

The samples were ground, chemical-mechanical polished, and electrolytic polished before the tensile testing. EBSD was performed using a FEI Versa5 scanning electron microscope (FEI, Hillsboro, OR, USA) and an EDAX velocity probe (EDAX, Mahwah, NJ, USA) under 20 kV, 13 nA, with a step size of 0.7 μm. For those tensile tests conducted at 850 °C, electro-polishing was conducted after the pre-determined strain before EBSD in order to remove the oxidation. All EBSD scans were maintained at the same region of interest during the interrupted testing. Data processing was performed using Matlab and GND density distributions were obtained using Mtex-5.7.0.

3. Results

3.1. Comparison of Tensile Properties

The true stress–strain curves are demonstrated in Figure 3. Obviously, the alloy exhibits completely different tensile behaviors at different temperatures, where the flow stress and work hardening at room temperature and 850 °C exhibit different features after yielding. The RT specimens all demonstrate a similar yield strength of ~867 MPa, an initial slow work hardening spanning ~1% of the plastic strain, followed by a work hardening rate of ~3910 MPa until fracture. The HT specimen demonstrated similar yield strength at ~865 MPa compared to the RT specimens, but the 37° sample demonstrated a higher loss of yield strength by 2.5% compared to the average of the other three orientations. In general, for the HT samples, the work hardening rate increased from ~267 MPa to ~731 MPa after yielding, and then gradually decreased after reaching 9% strain. Compared to the RT tests, the HT tests show strain softening before fracture, possibly because of necking due to the growth of voids. Significant improvement of ductility is seen from the samples tested at the HT, with an average fracture strain of 9.6% at RT compared to an average of 17% at the HT. A feature to notice is that among the secondary orientation samples tested at the HT, the fracture strain shows significantly higher dispersion (2.32 standard deviations) compared to those tested at RT (0.32 standard deviation). It is also interesting to note that the 37° sample has higher fracture strain at both the HT and RT.

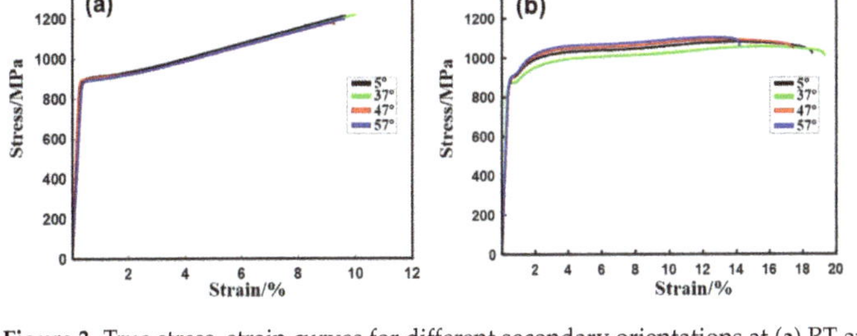

Figure 3. True stress–strain curves for different secondary orientations at (**a**) RT and (**b**) 850 °C.

3.2. Fracture Patterns at Different Temperatures

The fracture surfaces at different temperatures are displayed in Figure 4. On the fracture surface at RT (Figure 4a,b), there are some precipitate particles resting inside the micro-voids, indicating that those precipitate particles contributed to the micro-void nucleation and growth (indicated by the yellow arrows in the inset of Figure 4a). A significant number of casting voids were also observable in Figure 4a (indicated by the green arrows). In certain regions depicted in Figure 4a, the fracture surface exhibits a smooth texture devoid of river-like patterns or dimples; furthermore, micro-cracks and instances of crack bifurcation are discernible; these features indicate characteristics of a brittle fracture. In addition, some cracks were initiated at the location of large-sized precipitates (indicated by the white arrows in Figure 4b), which extended to the matrix during subsequent tensile loading. Therefore, the main type of fracture at room temperature is a brittle fracture, but the micro-dimples induced by carbides provide certain plasticity to the alloy.

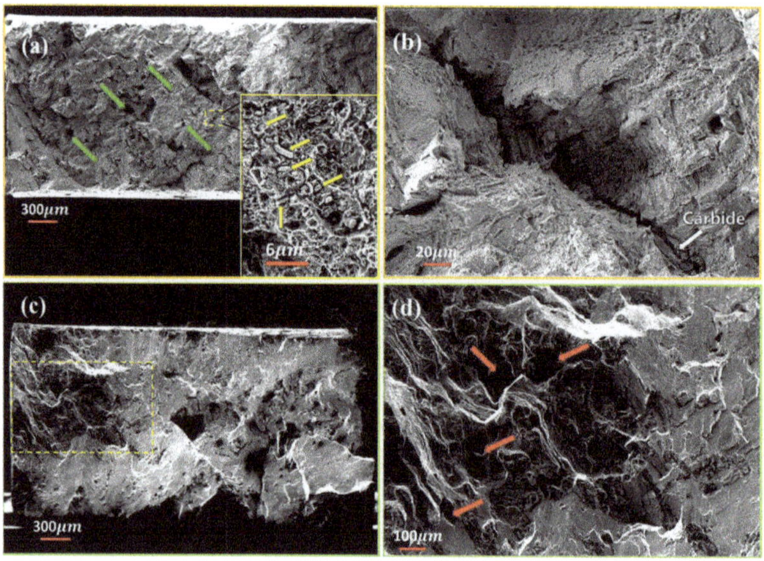

Figure 4. Fracture surface of the samples tested at (**a**,**b**) RT and (**c**,**d**) 850 °C. Note that the images are presented at different magnifications.

As the temperature increased to 850 °C (Figure 4c,d), the fracture surface became less flat (Figure 4c) and not as faceted as in that from the RT test, and the local magnified view (yellow dashed box in Figure 4c, which is shown in Figure 4d) indicated some tearing ledges and cleavage steps together with a dimpled zone showing evidence of relatively large void growth (indicated by the red arrows). It can be inferred that the fracture surface is composed of multiple (111) planes (i.e., the dislocation pile-up plane) separated by tearing ledges due to the intersection of dislocation pile-ups from adjacent slip planes. Dimples are a characteristic of ductile fracture and are formed due to the casting voids and precipitates. The enlargements of the dimples were facilitated by the tensile loading at the higher temperature. The contrast of void growth at the HT to that at RT is consistent with the larger sample elongation observed at the HT (Figure 3b). The mechanism behind the formation of these dimples is subjected to further study but can be related to the more active dislocation slip at the elevated temperature which provides the prismatic dislocation loops and the passage for vacancy diffusion, both of which facilitate void growth.

3.3. Deformed Microstructure After Tensile Fracture

The bright-field TEM images of the deformed microstructure after tensile fracture at room temperature are shown in Figure 5. At first sight, high density dislocations are tangled in the narrow γ matrix, and some dislocation pairs sheared into the γ' precipitates along the <110> directions (indicated by the red triangles in Figure 5c). Occasionally, some γ' demonstrated elevated dislocation punch-in and entanglement, as highlighted by the white dashed boxes in Figure 5a,b. More frequently, stacking faults are found in the γ' phase (marked white triangle in Figure 5c). The chemical compositions in the dislocation-enriched γ' were analyzed by scanning transmission electron microscopy (STEM) and the corresponding energy dispersive x-ray spectroscopy (EDS) elemental mapping (at.%) is shown in Figure 6. Interestingly, the heterogeneously deformed regions where dislocations shear into the γ' precipitates (in the white dashed boxes) exhibit significant enrichments in Co, Cr, and Re, which are the γ stabilizing elements. In contrast, Al and Ta, which are the γ' stabilizers, are depleted in those regions.

Figure 5. TEM bright-field images after fracture at RT. (**a**,**b**) Dislocations accumulated in the γ-channel and sheared into the γ' along the <110> direction. (**c**) Anti-phase boundaries formed by partial dislocation pairs (red arrow) and stacking faults in the γ' phases.

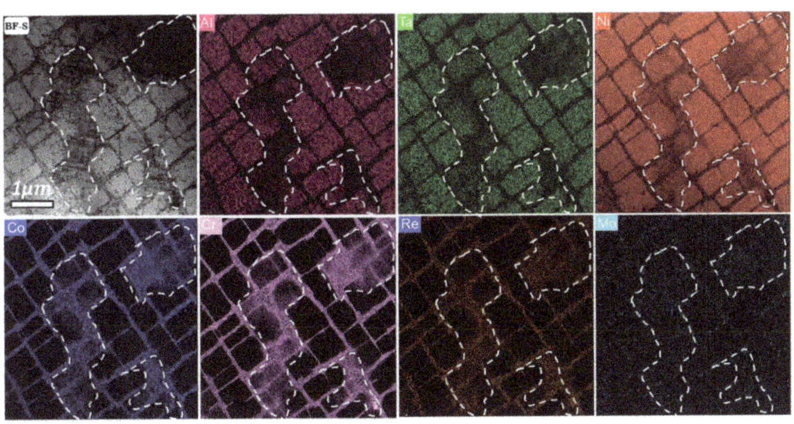

Figure 6. STEM images and corresponding EDS elemental mapping (at. %) at RT. The encircled regions indicate local segregation of γ' stabilizers: Co, Cr, and Re.

The bright-field TEM images of the deformed microstructure after tensile fracture at 850 °C are shown in Figure 7. The dislocation configurations exhibit significant differences compared to those at RT. At 850 °C, the dislocations are more homogeneously distributed across both the γ and γ' phases (Figure 7a). Notably and in contrast to room temperature deformation, stacking faults are absent, which may be attributed to an increase in stacking fault energy at higher temperatures [23,24], although the anti-phase boundary (APB) dislocation pairs (indicated by the white triangles in Figure 7b) are present, similar to the RT case. Furthermore, the misfit between γ and γ' increases at high temperatures, prompting a formation of irregular dislocation networks at the γ/γ' interfaces due to the release of misfit strain, as indicated by the arrow in Figure 7b. Additionally, some dislocations enter the γ' phase through cross-slip under thermal activation, as demonstrated by dislocations CD and EF in Figure 7c. During deformation, certain dislocations extended into the γ' precipitate parallel to the gliding direction in the γ phase, exemplified by dislocation AB in Figure 7c. In the micrograph shown in Figure 7d, some [110] curved dislocation pairs glide on (111) planes (indicated by the black triangles), while another portion of curved dislocation pairs initiated cross-slip (indicated by the red triangles). Importantly, unlike the RT case, there is no significant dislocation pile-up in both γ and γ' at the HT. Dislocations sheared into the γ' precipitates by a mixture of slip transmission and cross slip.

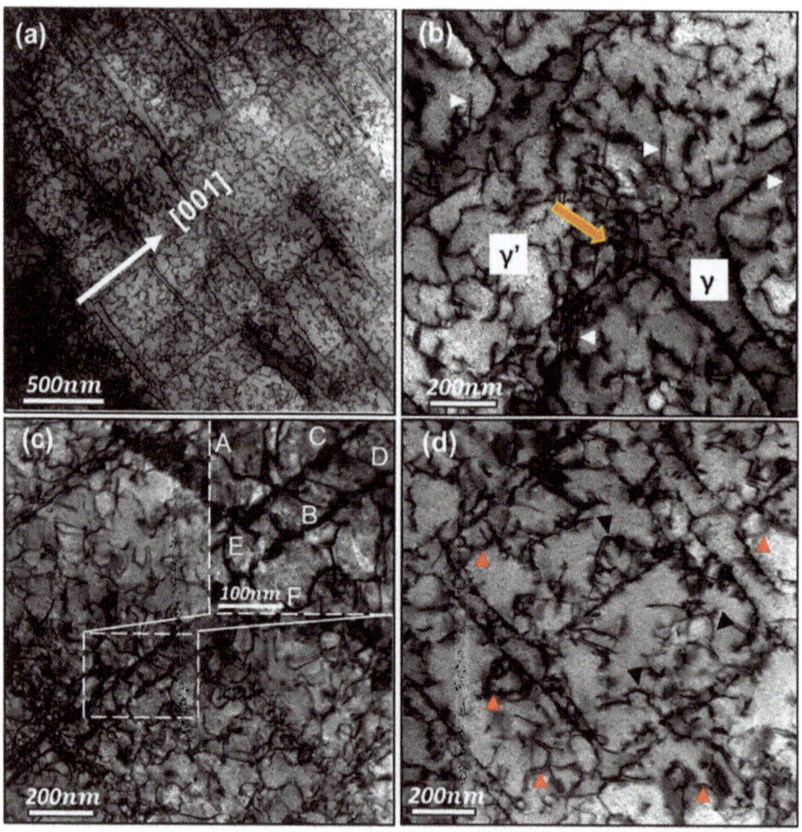

Figure 7. TEM bright-field images after fracture at 850 °C. (**a**) Curved dislocations distributed uniformly in the γ and γ' phases, (**b**) irregular dislocation network at the γ/γ' interface. The white triangles indicate partial dislocation pairs, (**c**) screw dislocations cross slip into γ', and (**d**) partial dislocation pairs cut into γ' by cross-slip, indicated by the red triangles.

The composition of the region where dislocations sheared into γ' was also analyzed by STEM and the corresponding EDS elemental mapping (at.%) is shown in Figure 8. It seems that the element distribution is more homogeneous in the sample tested at the HT. The Co, Cr, and Re preferential segregation into some of the γ' phases, as observed in the RT test, is absent at the HT.

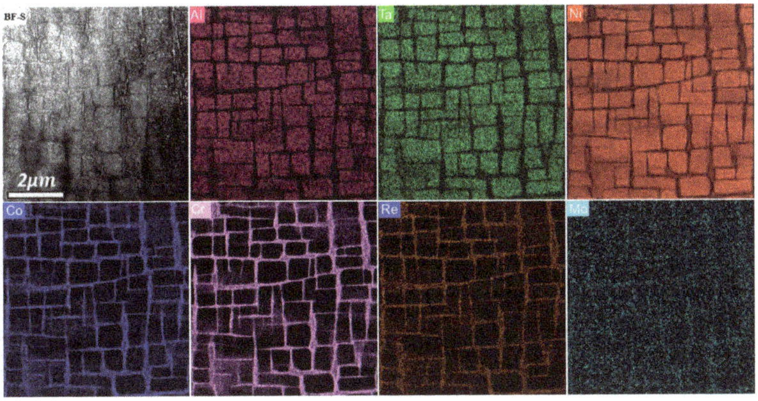

Figure 8. STEM images and corresponding EDS elemental mapping (at.%) at 850 °C.

3.4. Quasi In Situ EBSD with Different Orientations

Figure 9 shows the meso-scale geometrically necessary dislocation (GND) density distributions during room temperature tensile tests for different secondary orientations, investigated at strains of 3%, 8%, and 12%. Following the concepts proposed by Hughes [25] and Gao [26], the additional dislocation storage required for lattice rotation arising from non-uniform plastic deformation in the presence of strain gradients is termed geometrically necessary dislocation. The local GND density changes indicated the degree of strain localization in the alloy. As depicted in Figure 9, although the total GND density increased with increasing tensile strain, the variation in GND density for each orientation was uneven.

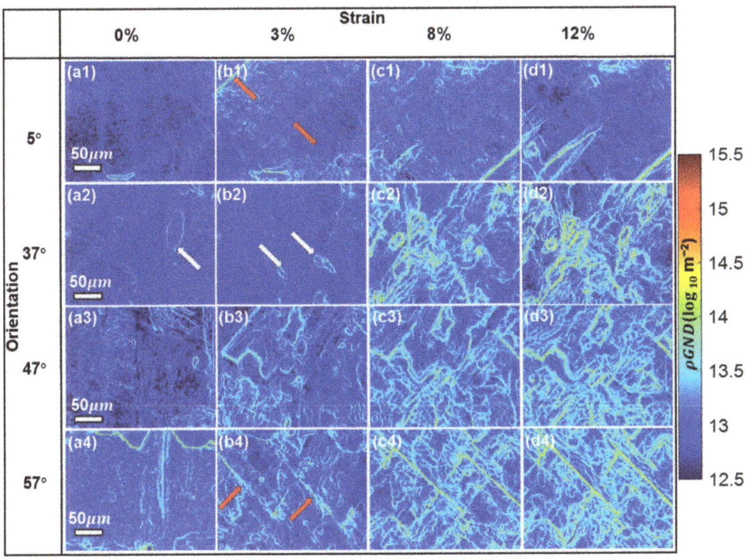

Figure 9. Quasi in situ observation of the GND density distribution of different secondary orientations during RT tensile tests at strains of 0%, 3%, 8%, and 12%.

As illustrated in Figure 9(a1–a4), the GND density (ρGND) started low before loading. Around yielding at ~3% strain (Figure 9(b1–b4)), the 47° and 57° orientations accumulated higher dislocation densities compared to the other two orientations. When the strain reached 8% (Figure 9(c1–c4)), the dislocation accumulation on the 37° sample started to catch up. The features of slip bands were not clearly observed in the GND density maps because dislocations in the slip bands are theoretically not "geometrically necessary", meaning they do not induce lattice rotations apart from a relative lattice shift. A few slip trace features can be observed on the maps and are indicated by the red arrows in Figure 9(b1,b4). Those features are better named "shear bands" rather than slip bands, as the interactions between multiple dislocation types within those bands gives rise to local lattice curvature, hence the observed GND hot spots. A few cases of GND density concentration were observed around the precipitates, as indicated by the white arrows in Figure 9(a2,b2). The overall strain accumulation for the 5° orientation sample was not as significant as those of the other orientations as indicated by the lack of dislocation cell build-up evidenced in Figure 9(d1) compared to Figure 9(d2–d4). Moreover, the rectangular-shaped dislocation cells possibly indicate that the cell formation at RT is primarily influenced by the deformation within the γ channel.

Figure 10 illustrated the GND density distribution during tensile testing at 850 °C for the same sets of secondary orientation samples, obtained at 3%, 8%, and 18% strains. Irregularly high dislocations are seen on the undeformed samples, which are likely due to the residual scratches from mechanical polishing. These scratches are less likely to affect other GND maps at higher strain levels as they are subsequently removed by electro polishing (to remove oxides from HT testing). The total GND density also increased with strain, but, in sharp contrast to the RT distributions in Figure 9, the GNDs from HT testing are more uniformly distributed. Yielding occurs at approximately 3% strain; a notable feature was the presence of precipitates in the 5° sample which accelerated the GND build-up at 3% strain (Figure 9(b1)) compared to the precipitate-free cases. A dendrite boundary (indicated by the white arrow) was captured in the HT cases, which serves as a strain concentrator as indicated by the elevated GND densities along the boundary. The accumulation of GND near precipitates could be attributed to the pile up of dislocations at the interface together with strain gradient plasticity due to the differing elastic properties between the precipitate and the matrix. An abrupt increase in GNDs was observed when the strain went beyond 8%. The GNDs at 18% strain are more uniformly distributed in the matrix with a smaller dislocation cell size (compared to the RT case) and local GND hot spots at the dendrite boundary and precipitates.

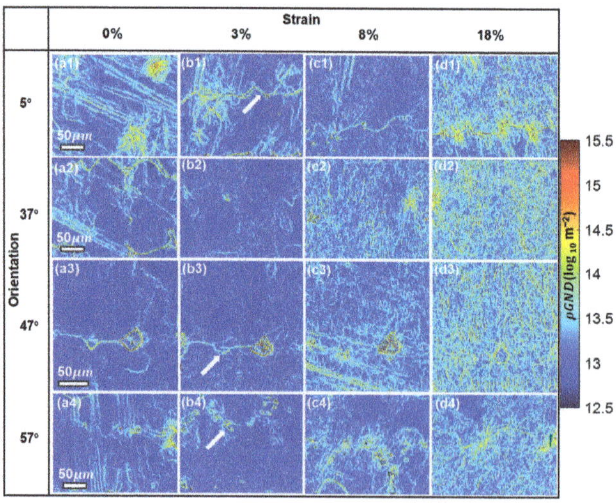

Figure 10. Quasi in-situ observation of the GND density distribution of different secondary orientations during 850 °C tensile tests at strains of 0%, 3%, 8%, and 18%.

4. Discussion

4.1. Deformation Mechanism at Different Temperatures

Based on the experimental results, Figure 5 depicts the complete structural defects observed at room temperature. Since the higher intrinsic strength of γ', plastic deformation primarily occurs in the γ channels and results in local dislocation multiplications. The confined dislocation multiplication together with the higher RT shear strength of the γ' phase resulted in pronounced work hardening as shown in Figure 3a. Stacking faults are usually difficult to observe in the γ channels attributed to the high density of dislocations [27]. With the applied stress further increased, some dislocations are able to shear into the γ' phase in the form of partial dislocation pairs.

It is well known that the formation of APB and a stacking fault (SF) is related to their formation energies, which in turn are highly dependent on the deformation temperature [23,24,28,29]. APBs and SFs are identified as the main features in the γ' phase from the fractured specimens deformed at RT. On the one hand, due to the low stacking fault energy at RT [23,24,30], the super-dislocations $[10\overline{1}]$ shear into γ' precipitates and then decompose into a couple of super-Shockley partial dislocations on the (111) plane, corresponding to the Burgers vectors of $1/3[11\overline{2}]$ and $1/3[2\overline{1}\overline{1}]$, with SFs distributed between the two super-partials dislocations. This reaction in the γ' can be expressed as follows:

$$\mathrm{a}\,[10\overline{1}] \rightarrow (\mathrm{a}/3)[11\overline{2}] + (\mathrm{SISF}) + (\mathrm{a}/3)[2\overline{1}\overline{1}] \tag{1}$$

where a is the lattice constant.

On the other hand, considering the highly heterogeneous plastic deformation, dislocations with the same type of Burgers vector tend to pile up at the γ/γ' interface. When the local shear stress in front of the pile up becomes high, super-dislocations shear into the γ' phase and may decompose into APB pairs, such as the red triangles indicated in Figure 5c. As the APB pairs cut into the γ' phase, the reaction can be expressed as follows:

$$\mathrm{a}\,[\overline{1}10] \rightarrow (\mathrm{a}/2)[\overline{1}10] + (\mathrm{APB}) + (\mathrm{a}/2)[\overline{1}10] \tag{2}$$

Although different APBs may possess the same scalar Burgers vector, b = [110], the activation of different slip systems during tension causes APBs to slip and decompose on various slip planes. Additionally, the a/2[110] dislocations of different slip systems shear into the γ' phase and interact with each other to form a dislocation network within the γ' phase, as indicated by the white dotted box in Figure 5a,b. These dislocation networks hinder the movement of subsequently punched-in dislocations, contributing to a certain level of work hardening.

Compared to room temperature deformation, deformations at 850 °C exhibit two notable characteristics. Firstly, there is a higher dislocation density within the γ' precipitates. Notice that the APB energy decreases with temperature [23,28]; hence, as the external stress increases, more super-dislocations break through the γ/γ' interface and decompose into APB couple pairs, which explains the drop in work hardening in the later stages of the tensile testing. The initial working hardening of the HT test samples may be similar to that of those tested at RT, i.e., due to dislocation multiplication in the γ phase and pile up at the γ/γ' interface, but due to the easier transmission into the γ' phase, the overall work hardening rate is much lower. The dislocations participated uniformly in both the γ and γ' phases as can be seen from Figure 7, indicating a more homogeneous deformation at 850 °C. This dislocation patterning characteristic may explain the improved HT ductility. The other notable feature is the presence of curved dislocations as mentioned in Section 3.3. This phenomenon has been observed in other studies [8,30], and is potentially related to the dislocation self-arrangement into a lower energy state. Referring to the linear dislocation energies formula [31]: $E_s = (1-\nu)E_e$, where v is the Poisson's ratio, E_s represents the elastic energy of screw dislocations and E_e denotes that of the edge dislocations. It follows that the creation of curved dislocations from an initially straight edge dislocation is energetically favorable. By comparing the dislocation configurations between Figure 5

(RT, more straight dislocations) and Figure 7 (HT, more curved dislocations) it may be the case that thermal activation plays an important role in this process. The formation of a higher proportion of curved dislocations can also be seen by contrasting the mesoscale dislocation configuration between Figures 9 and 10. When considering the cross-slip behavior of certain mixed dislocation pairs such as those indicated by the red triangles in Figure 7d, they initiate cross slip from the octahedral (111) planes to the cubic (100) planes due to the higher APB energy on the former [32,33]. This process occurs under thermal activation. However, the (100) planes are not easy glide planes for FCC nickel-based superalloys, making further slip of APB pairs on the (100) planes difficult and thus hindering subsequent dislocation motion, thereby forming Kear–Wilsdorf (K-W) locks [34]. This effect may have countered the strength decrease due to the easier dislocation activation at 850 °C and therefore maintained a similar yield strength level as that of the RT tests.

As mentioned above, γ', in which there is γ stabilizer (Co, Cr, and Re) enrichment, is more heavily deformed at RT. Two possibilities can be considered here. Firstly, the diffusion rate is higher along the dislocation line compared to the bulk material. This results in element segregation in the heavily deformed region, commonly referred to as pipe diffusion [35,36]. The second possibility is that elemental segregation results from the casting process, where the γ stabilizer-enriched γ' precipitates have lower stacking fault energy and therefore make dislocation generation and dislocation transmission easier from the neighboring γ phase.

4.2. Effects of Temperature and Secondary Orientations on Deformation Behavior

The true stress–strain curves during tensile testing at RT and 850 °C show minimal variation in yield strength, a phenomenon commonly observed in other superalloys as well [30,37]. Abnormal yield strengths of superalloys are observed at 600–800 °C, due to thermally activated dislocations that cross-slip from {111} planes to {001} planes, forming K-W locks. These locks tend to be loosened at higher temperatures, returning the yield strength to the room temperature level; therefore, our test at 850 °C did not capture the abnormal yield phenomenon.

To quantify the extent of work hardening at the two temperatures, the work hardening rate θ was calculated and plotted as a function of true plastic strain in Figure 11a,b, i.e., $\theta = d\sigma/d\epsilon$, and the work hardening response can be divided into three stages. In Stage 1, dislocations are impeded by the γ' precipitate, leading to an increased hardening rate at RT. In contrast, as dislocations are relatively easy to shear into γ' at 850 °C, a drop in the work hardening rate is observed in Stage 1. Stage 2 involves initial softening followed by hardening at RT, attributed to the localized strain concentration [38], as depicted in the distribution of GNDs in Figure 9. Hardening at high temperatures is due to the formation of K-W locks. When it comes to Stage 3, the higher applied load causes the unlocking of the K-W locks, resulting in softening at 850 °C. The capability of dislocations shearing into the γ' phase at 850 °C facilitates uniform plastic deformation, as shown in Figure 10, and ultimately leads to a gradual decrease in the work hardening rate until fracture.

The effect of the secondary orientation on mechanical behavior remain unclear throughout the literature. According to some research in the literature, it seems that the effect of secondary orientation takes effect only through local structural irregularities. For example, Guo et al. [21] discovered that in cylindrical hole specimens, the secondary orientation influences the location of creep crack initiation at hole edges. Zhou et al. [20] confirmed that the presence of a circular hole induces a multi-axil stress condition that is favorable for anisotropic plastic deformation around the hole. Furthermore, secondary orientation influences the activation of slip systems adjacent to the hole, with a greater number of activated slip systems leading to more uniform plastic deformation and higher fracture strains. Zhou et al. [17], by conducting tensile testing on a specimen with circular holes, concluded that the three-dimensional tensile stress parallel to the stress axis around the holes promotes stress concentration. Under these conditions, the <100> slip systems experience higher resolved shear stresses, leading to greater activation of slip systems, enhanced work hardening, and higher strength.

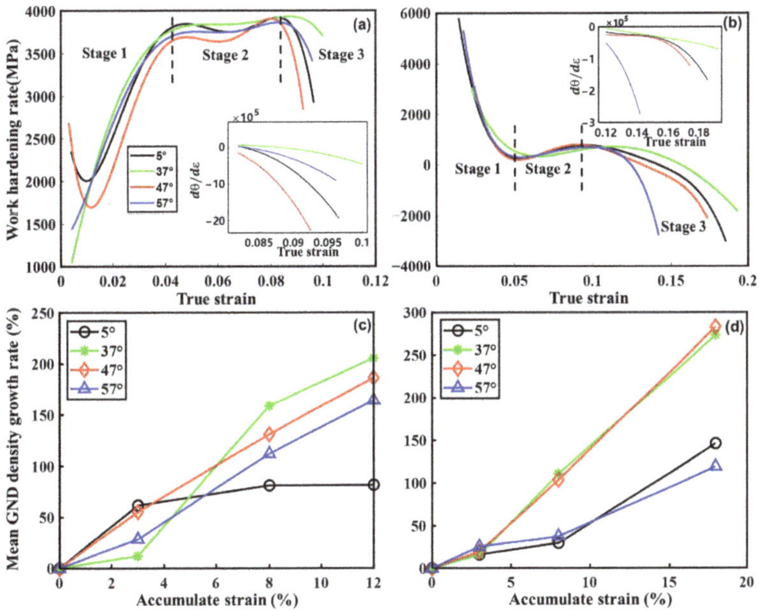

Figure 11. (**a**,**b**) The work hardening rate variations among secondary orientations at (**a**) RT and (**b**) 850 °C. (**c**,**d**) indicate the evolution of the GND density growth rate vs. strain for different secondary orientations at (**c**) RT and (**d**) 850 °C.

On the other hand, some experiments have also demonstrated that the secondary orientation has minimal impact on the mechanical properties of single-crystal superalloys. Guo et al. [39] found that during RT tensile testing of specimens with circular holes, the secondary orientation had no influence on the stress–strain curve. The only difference observed was in the initiation of slip bands at the edges of the holes. The lack of influence on mechanical properties may be attributed to the rapid initiation and propagation of cracks in the large circular holes, which could mask any secondary orientation effects. Zhou et al. [17], through the in situ tensile testing of smooth specimens, discovered that the secondary orientation has a minor impact on yield stress and ultimate tensile strength. However, it notably influences the activation of slip systems, as well as the initiation and propagation of edge cracks in the samples. Sabnis et al. [15], through three-dimensional finite element simulations and experimental data, found that the tensile curves of edge U-notch specimens with varying secondary orientations exhibit striking similarity.

The lack of consistency in data in the literature on the effect of secondary orientations may be attributable to the specific testing conditions and chemical compositions of the specimen used, but it is clear that the most prominent effect of secondary orientations is that they lead to differences in the activations of slip systems. However, despite the distinct activation patterns of slip systems across different secondary orientations, this variability does not always translate into differences in mechanical properties. Secondary orientation effects may be amplified through methods such as reducing hole size, pre-notching, conducting tests at elevated temperatures, extending test durations, and increasing specimen thickness.

The results in the current study indicated minimal variation in strength among different secondary orientations, but the fracture plasticity exhibited greater dispersion across different secondary orientations at 850 °C compared to RT. Specifically, the 37° orientation demonstrated superior plasticity at both temperatures. Additionally, all four orientations exhibited similar work hardening behavior at the same temperatures, but all the samples showed a continuous decrease in the work hardening rate θ (obtained by taking the derivative of stress to strain; the derivative of θ with respect to strain therefore indicates changes in the work hardening rate) during Stage 3 prior to fracture. The inset in Figure 11a,b depicts the derivative of θ with respect to strain in Stage 3. Notably, the 37° samples exhibited the lowest decrease in θ at both temperatures. This slower reduction in strength suggests reduced susceptibility to fracture under high-stress conditions, resulting in higher fracture strain. Moreover, to accurately depict the change in GND in the matrix during tensile deformation, the average growth rate of GND density in the matrix, excluding carbides, was calculated using data from Figures 9 and 10, and presented in Figure 11c,d. It is evident that the 37° orientation generally shows a higher growth rate in GND density compared to the other orientations, which is consistent with the slower drop of the working hardening rate. The number of activated slip systems and their sustained multiplication capability play a crucial role in the continuous accumulation of GND density and correlate closely with plasticity [38,40,41]. Among the batch of secondary orientation samples tested, the 37° orientation meet those conditions and therefore demonstrates higher fracture strain. Notice from Figure 2b that the 37° sample also has the highest carbide volume fraction, which may be counterintuitive to the higher ductility, but the local uniformity of slip activation may override the reduced ductility due to the higher carbide area fraction in this specific orientation, which is consistent with the results indicated by Zhou et al. [20].

5. Conclusions

The influences of temperature and secondary orientations on the plastic deformation of Ni-based SC superalloys were investigated using quasi in situ tensile tests and multiscale characterizations. The main conclusions are summarized as follows:

1. Deformation at 850 °C leads to higher tensile ductility and higher variance in tensile ductility among secondary orientations. Apart from this variation in ductility at higher temperatures, we found a weak influence of secondary orientation on mechanical properties.
2. The fracture mode at RT is brittle fracture, but the micro-dimples formed at 850 °C provide certain plasticity to the alloy, leading to a mixture of cleavage and ductile fractures mode.
3. At RT, dislocations accumulate mainly at the γ phase with limited dislocation transmission into the γ' phase and the γ' phase deform by stacking faults formation. At 850 °C, stacking faults are absent in the γ' phase, but it is relatively easy for dislocation transmission to occur through paired partial dislocation mechanisms, leading to a more uniform dislocation distribution, resulting in a lower work hardening rate but improved fracture strain.
4. Local entanglements of high density dislocations were found in some of the γ' phase. This local high deformation zone is correlated with the local segregation of γ stabilizing elements.
5. The secondary orientation sample with the [100] direction rotated 37° demonstrated superior tensile ductility at 850 °C, which was supported by the high and sustained accumulation of GNDs even though this sample contained the highest area fraction of precipitates.

Author Contributions: Conceptualization, C.Z., X.C. and Y.G.; methodology, C.Z.; software, Y.G.; validation, G.M., Y.Z. and J.H.; formal analysis, S.L.; investigation, C.Z.; resources, Y.G.; data curation, S.L.; writing—original draft preparation, S.L.; writing—review and editing, Y.G.; visualization, S.L.; supervision, X.C. and Y.G.; funding acquisition, X.C. All authors have read and agreed to the published version of the manuscript.

Funding: This research was funded by National Natural Science Foundation of China No. 52201149, National Science and Technology Major Project No. J2019-VI-0023-0139, the National Science and Technology Major Project, grant number J2019-VI-0019-0134, and the Special Projects of the Central Government in Guidance of Local Science and Technology Development, grant number 2024010859-JH6/1006.

Data Availability Statement: The raw data supporting the conclusions of this article will be made available by the authors on request.

Conflicts of Interest: The authors declare no conflicts of interest.

References

1. Wu, R.H.; Yue, Z.F.; Wang, M. Effect of initial γ/γ' microstructure on creep of single crystal nickel-based superalloys: A phase-field simulation incorporating dislocation dynamics. *J. Alloys Compd.* **2019**, *779*, 326–334. [CrossRef]
2. Zhou, S.Y.; Hu, M.H.; Li, C.; Guo, Q.Y.; Yu, L.M.; Ding, H.M.; Liu, Y.C. Microstructure-performance relationships in Ni-based superalloy with coprecipitation of γ' and γ'' phases. *Mater. Sci. Eng. A* **2022**, *855*, 143954. [CrossRef]
3. Zhu, Z.X.; Luo, D.L.; Zhang, J.J.; Qu, D.F.; Zheng, M.; Chen, W.Y.; Cheng, J. Nano-cutting deformation characteristics and atomic-scale behavior of two-phase γ/γ' nickel-based single crystal superalloy. *Intermetallics* **2023**, *161*, 107985. [CrossRef]
4. Li, Y.W.; Wang, L.; He, Y.F.; Zheng, W.; Lou, L.H.; Zhang, J. Role of interfacial dislocation networks during secondary creep at elevated temperatures in a single crystal Ni-based superalloy. *Scr. Mater.* **2022**, *217*, 114769. [CrossRef]
5. Wang, S.J.; He, J.J.; Li, W.P.; Gong, Z.H.; Zhou, L.B.; Yang, J.G.; Cai, Y.H.; Du, Y.X. Microstructure analysis and cracking mechanism of aero-engine hot-end component K4169 superalloy based on in-situ EBSD test. *J. Alloys Compd.* **2023**, *960*, 170781. [CrossRef]
6. Xu, Z.C.; Britton, B.; Guo, Y. Casting voids in nickel superalloy and the mechanical behavior under room temperature tensile deformation. *Mater. Sci. Eng. A* **2021**, *806*, 140800. [CrossRef]
7. Wang, G.L.; Liu, J.L.; Liu, J.D.; Wang, X.G.; Zhou, Y.Z.; Sun, X.D.; Zhang, H.F.; Jin, T. Temperature dependence of tensile behavior and deformation microstructure of a Re-containing Ni-base single crystal superalloy. *Mater. Des.* **2017**, *130*, 131. [CrossRef]
8. Tan, Z.H.; Wang, X.G.; Du, Y.L.; Duan, T.F.; Yang, Y.H.; Liu, J.L.; Liu, J.D.; Yang, L.; Li, J.G.; Zhou, Y.Z. Temperature dependence on tensile deformation mechanisms in a novel Nickel-based single crystal superalloy. *Mater. Sci. Eng.* **2020**, *776*, 138997. [CrossRef]
9. Cheng, T.W.; Wang, Y.W.; Zhao, Y.X.; Lv, B.B.; Ma, D.X. Effect of solidification direction on microstructure and mechanical property of single crystal superalloy CMSX-4. *Mater. Charact.* **2023**, *202*, 112992. [CrossRef]
10. Liu, J.L.; Jin, T.; Sun, X.F.; Zhang, J.H.; Guan, H.R.; Hu, Z.Q. Anisotropy of stress rupture properties of a Ni base single crystal superalloy at two temperatures. *Mater. Sci. Eng. A* **2008**, *479*, 277–284. [CrossRef]
11. Wang, L.N.; Liu, Y.; Yu, J.J.; Xu, Y.; Sun, X.F.; Guan, H.R.; Hu, Z.Q. Orientation and temperature dependence of yielding and deformation behavior of a nickel-base single crystal superalloy. *Mater. Sci. Eng. A* **2009**, *505*, 144–150. [CrossRef]
12. Wu, X.; Zhang, J.H.; Liu, J.L.; Jin, T.; Xu, Y.B.; Hu, Z.Q. Plastic deformation inhomogeneity in a single crystal nickel-base superalloy. *Mater. Sci. Eng. A* **2002**, *325*, 478–483. [CrossRef]
13. Li, Y.M.; Tan, Z.H.; Wang, X.G.; Mu, Y.; Zhao, H.C.; Tan, H.B.; Liu, J.L.; Wang, B.; Li, J.G.; Zhou, Y.Z.; et al. Stress rupture anisotropy of a Ru-containing fourth-generation single crystal superalloy at 760 °C and 1100 °C. *Mater. Sci. Eng. A* **2022**, *856*, 144006. [CrossRef]
14. Arakere, N.K.; Swanson, G. Effect of Crystal Orientation on Fatigue Failure of Single Crystal Nickel Base Turbine Blade Superalloys. *J. Eng. Gas Turbines Power* **2002**, *124*, 161–176. [CrossRef]
15. Sabnis, P.A.; Mazière, M.; Forest, S.; Arakere, N.K.; Ebrahimi, F. Effect of secondary orientation on notch-tip plasticity in superalloy single crystals. *Int. J. Plast.* **2012**, *28*, 102–123. [CrossRef]
16. Qiu, W.H.; He, Z.W.; Fan, Y.N.; Shi, H.J.; Gu, J.L. Effects of secondary orientation on crack closure behavior of nickel-based single crystal superalloys. *Int. J. Fatigue* **2016**, *83*, 335–343. [CrossRef]
17. Zhou, Z.J.; Wang, L.; Wang, D.; Lou, L.H.; Zhang, J. Effect of secondary orientation on room temperature tensile behaviors of Ni-base single crystal superalloys. *Mater. Sci. Eng. A* **2016**, *659*, 130–142. [CrossRef]
18. Qin, L.; Pei, Y.; Li, S.; Gong, S.; Xu, H. Influence of stress and secondary orientation on the oxidationinduced dynamic recrystallization behavior of a Ni-based single crystal superalloy. *J. Alloys Compd.* **2017**, *706*, 455–460. [CrossRef]
19. Suzuki, S.; Sakaguchi, M.; Inoue, H. Temperature dependent fatigue crack propagation in a single crystal Ni-base superalloy affected by primary and secondary orientations. *Mater. Sci. Eng. A* **2018**, *724*, 559–565. [CrossRef]
20. Zhou, H.; Zhang, X.; Wang, P.; Lu, S.P. Crystal plasticity analysis of cylindrical holes and their effects on the deformation behavior of Ni-based single-crystal superalloys with different secondary orientations. *Int. J. Plast.* **2019**, *119*, 249–272. [CrossRef]
21. Guo, Z.X.; Song, Z.Y.; Liu, H.H.; Hu, D.J.; Huang, D.W.; Yan, X.J.; Yan, W.T. A dislocation-based damage-coupled constitutive model for single crystal superalloy: Unveiling the effect of secondary orientation on creep life of circular hole. *Int. J. Plast.* **2024**, *173*, 103874. [CrossRef]
22. Rohaizat, N.I.; Alharbi, K.; Pinna, C.; Ghadbeigi, H.; Hanlon, D.N.; Azid, I.A. Using a Correction Factor to Remove Machine Compliance in a Tensile Test on DP1000 Steel Validated with 2D Digital Image Correlation Technique. In *Progress in Engineering Technology III. Advanced Structured Materials*; Bakar, M.H.A., Zahelem, M.N., Öchsner, A., Eds.; Springer: Cham, Switzerland, 2021; pp. 113–123. [CrossRef]
23. Lv, P.S.; Liu, L.R.; Zhao, G.Q.; Guo, S.D.; Zhou, Z.R.; Zhao, Y.S.; Zhang, J. Temperature effects on tensile behaviors and relevant deformation mechanisms of a low-cost nickel-based single crystal superalloy containing 1.5% Re. *J. Alloys Compd.* **2022**, *926*, 166819. [CrossRef]
24. Tian, S.G.; Zhu, X.J.; Wu, J.; Yu, H.C.; Shu, D.L.; Qian, B.J. Influence of Temperature on Stacking Fault Energy and Creep Mechanism of a Single Crystal Nickel-based Superalloy. *J. Mater. Sci. Technol.* **2016**, *32*, 790–798. [CrossRef]

25. Hughes, D.A.; Hansen, N.; Bammann, D.J. Geometrically necessary boundaries, incidental dislocation boundaries and geometrically necessary dislocations. *Scr. Mater.* **2003**, *48*, 147–153. [CrossRef]
26. Gao, H.J.; Huang, Y.G. Geometrically necessary dislocation and size-dependent plasticity. *Scr. Mater.* **2003**, *48*, 113–118. [CrossRef]
27. Song, W.; Wang, X.G.; Li, J.G.; Meng, J.; Yang, Y.H.; Liu, J.L.; Liu, J.D.; Zhou, Y.Z.; Sun, X.F. Effect of Ru on tensile behavior and deformation mechanism of a nickel-based single crystal superalloy. *Mater. Sci. Eng. A* **2021**, *802*, 140430. [CrossRef]
28. Kruml, T.; Viguier, B.; Bonneville, J.; Martin, J.L. Temperature dependence of dislocation microstructure in $Ni_3(Al,Hf)$. *Mater. Sci. Eng. A* **1997**, *234–236*, 755–757. [CrossRef]
29. Li, W.G.; Ma, J.Z.; Kou, H.B.; Shao, J.X.; Zhang, X.Y.; Deng, Y.; Tao, Y.; Fang, D.N. Modeling the effect of temperature on the yield strength of precipitation strengthening Ni-base superalloys. *Int. J. Plast.* **2019**, *116*, 143–158. [CrossRef]
30. Wang, X.G.; Liu, J.L.; Jin, T.; Sun, X.F. The effects of ruthenium additions on tensile deformation mechanisms of single crystal superalloys at different temperatures. *Mater. Des.* **2014**, *63*, 286–293. [CrossRef]
31. Cai, W.; Nix, W.D. *Imperfections in Crystalline Solids;* Cambridge University Press: Cambridge, UK, 2016.
32. Liu, S.H.; Wang, C.Y.; Yan, P.; Yu, T. The effect of Ta, W, and Re additions on the tensile-deformation behavior of model Ni-based single-crystal superalloys at intermediate temperature. *Mater. Sci. Eng. A* **2022**, *850*, 143594. [CrossRef]
33. Qiu, D.; Feng, L.S.; Zhao, P.Y. Modulating superdislocation cores and planar faults of Ni3Al through applied stresses. *Comput. Mater. Sci.* **2024**, *237*, 112865. [CrossRef]
34. Hirsch, P.B. Kear-Wilsdorf locks, jogs and the formation of antiphase-boundary tubes in Ni_3Al. *Philos. Mag. A* **1996**, *74*, 1019–1040. [CrossRef]
35. Love, G.R. Dislocation pipe diffusion. *Acta Metall.* **1964**, *12*, 731–737. [CrossRef]
36. Wu, X.X.; Makineni, S.K.; Kontis, P.; Dehm, G.; Raabe, D.; Gault, B.; Eggeler, G. On the segregation of Re at dislocations in the γ' phase of Ni-based single crystal superalloys. *Materialia* **2018**, *4*, 109–114. [CrossRef]
37. Shi, Z.X.; Liu, S.Z.; Yu, J.; Rong, L.J. Tensile Behavior of the Second Generation Single Crystal Superalloy DD6. *J. Iron Steel Res. Int.* **2015**, *22*, 738–742. [CrossRef]
38. Jiang, W.X.; Lu, J.X.; Li, F.Q.; Wang, J.; Zhang, Y.F.; Zhang, Z.; Zhao, Y.S.; Zhang, J. In-situ EBSD investigation of the effect of orientation on plastic deformation behavior of a single crystal superalloy. *Mater. Sci. Eng. A* **2022**, *849*, 143453. [CrossRef]
39. Guo, Z.X.; Song, Z.Y.; Ding, X.; Guo, K.M.; Liu, H.Z.; Yan, H.; Huang, D.W.; Yan, X.J. Slip Band Evolution Behavior near Circular Hole on Single Crystal Superalloy: Experiment and Simulation. *Int. J. Plast.* **2023**, *165*, 103600. [CrossRef]
40. Rui, S.S.; He, Z.W.; Guo, Y.Y.; Su, Y.; Han, Q.N.; Ma, X.F.; Shi, H.J. Secondary orientation effects on the low cycle fatigue behaviors of rectangular-sectional Ni-based single crystal superalloys at medium and high temperatures. *Fatigue Fract. Eng. Mater. Struct.* **2023**, *46*, 3290–3305. [CrossRef]
41. Wang, L.; Liu, J.; Zheng, M.R.; Min, S.L.; Wang, D.; Zhang, G.; Dong, J.S.; Lou, L.H. Combined effect of secondary dendrite orientation and wall thickness on creep behavior of a Ni-base single crystal superalloy. *J. Alloys Compd.* **2023**, *968*, 171968. [CrossRef]

crystals

Review

Advancements and Perspectives in Additive Manufacturing of Tungsten Alloys and Composites: Challenges and Solutions

Mehrdad Zarinejad [1,2,*], Yunxiang Tong [3,*], Mojtaba Salehi [4,*], Chengfa Mu [1], Nian Wang [2], Yonglong Xu [2], Sajjad Rimaz [5], Lintao Tian [1], Kai Xiang Kuah [6] and Xiaotong Chen [1]

1 Research and Development Institute, Wenzhou Hongfeng Electrical Alloy Co., Ltd., No. 5600, Oujin Avenue, Oujiangkou Industry Cluster, Wenzhou 325700, China; muchengfa@wzhf.com (C.M.); tianlintao@wzhf.com (L.T.); chenxiaotong@wzhf.com (X.C.)
2 Wenzhou Hongfeng Alloy Co., Ltd., No. 1633, Binhai First Avenue, Wenzhou Economic and Technology Development Zone, Wenzhou 325603, China; wangnian@wzhf.com (N.W.); xuyonglong@wzhf.com (Y.X.)
3 Institute of Materials Processing and Intelligent Manufacturing, College of Materials Science and Chemical Engineering, Harbin Engineering University, Harbin 150001, China
4 Additive Manufacturing Division, Singapore Institute of Manufacturing Technology (SIMTech), Agency for Science, Technology and Research (A*STAR), 5 Cleantech Loop, Singapore 636732, Singapore
5 Department of Chemical and Bimolecular Engineering, National University of Singapore, 4 Engineering Drive 4, Singapore 117585, Singapore
6 Department of Materials Science and Engineering, National University of Singapore, 9 Engineering Drive 1, Singapore 117575, Singapore; kxkuah@u.nus.edu
* Correspondence: mehrdad@wzhf.com (M.Z.); tongyx@hrbeu.edu.cn (Y.T.); mojtaba_salehi@simtech.a-star.edu.sg (M.S.)

Abstract: This review explores additive manufacturing (AM) for refractory tungsten (W) and its alloys, highlighting the primary challenges and determining factors in the AM of pure W, W alloys and composites. The challenges mainly arise from W's high melting point, low laser absorptivity, high thermal conductivity, high melt viscosity, high oxygen affinity, high ductile-to-brittle transition temperature, and inherent embrittlement, which lead to defects and anomalies in AM-produced parts. This review focuses on both processes and alloying strategies to address the issues related to densification, micro-cracking, and the resultant properties in W-based components. Cracking in additively manufactured W remains a persistent issue due to thermal stress, embrittlement, and oxide formation. Powder characteristics, process parameters, and thermal management strategies are crucial for W densification. Throughout the review, existing knowledge and insights are organized into comprehensive tables, serving as valuable resources for researchers delving deeper into this topic. Future research in W-AM should focus on understanding the interaction between AM process parameters and microstructural and material design. Advances in atomic-level understanding, thermodynamic modeling, and data analytics have the potential to significantly enhance the precision, sustainability, and applicability of W-AM.

Keywords: refractory alloys; tungsten alloys and composites; additive manufacturing; laser and electron beam melting; challenges and mitigation strategies

1. Introduction

Refractory alloys, including tungsten (W) and its alloys, are integral to aerospace and energy-generation industries [1]. W is well known for its high melting point (approximately 3422 °C), good mechanical properties, and low thermal expansion coefficient of less than 4 μm/m·K [2]. Additionally, its impressive thermal conductivity, surpassing 150 W/m·K [2], exceptional resistance to heat and wear [3], notable chemical stability, and remarkable hardness [3] make W suitable for high-temperature and radiation-resilient applications. Examples of its applications include collimators, heat sinks [4], integrated circuits [5,6], switch contacts [7], rocket nozzles [5,8], turbine blades [5], and nuclear reactor parts [5,9].

Harnessing the full potential of refractory alloys, particularly those of W-based materials, has been challenging owing to their high melting points and complex metallurgical behavior [10].

Conventional production methods for W-based alloys primarily rely on powder metallurgy techniques, which require high-temperature sintering. Achieving full densification is particularly challenging. Additionally, the hardness and brittleness of W alloys complicate formative and subtractive manufacturing processes, such as forming and machining methods. Furthermore, W oxidizes at relatively low temperatures (beginning around 400–500 °C), which increases the complexity and cost of manufacturing. Historically, the manufacturability of W-based alloys has been hindered by high costs, scalability issues, and challenges in producing intricate structures [11]. Additive manufacturing (AM) of refractory alloys offers unparalleled design freedom. Unlike traditional subtractive techniques that can waste up to 70% of material [12–14], AM's layer-by-layer approach ensures material efficiency, especially with costly refractory alloys [15,16].

AM could address some of the traditional challenges associated with the production of W-based alloys and provide significant advantages in terms of design flexibility, material efficiency, and the capability to create complex geometries. For instance, AM facilitates the production of intricate and custom molds for high-temperature tooling. As another example, AM allows for the creation of sophisticated channels and intricate geometries, enhancing the cooling efficiency of plasma-facing components in nuclear fusion reactors. Although W-AM can mitigate many of the issues associated with traditional techniques, the inherent properties of W alloys present significant challenges for AM processes. The high melting points and elevated ductile-to-brittle transition temperatures (DBTTs) pose concerns for all refractory alloys but are amplified for W because of its high viscosity (8×10^{-3} Pa·s) [17], surface tension (2.361 N/m) [17], and rapid solidification rate [18]. In the solid state, W's body-centered cubic structure adds complexity to thermal processes, particularly given its DBTT [19,20] and the notable influence of impurities on its DBTT. Issues such as interstitial contamination, balling [21], cracking [22], and the limited understanding of process-structure-property relationships [18,21,23] represent substantial hurdles to overcome.

Over the past decade, extensive research has been undertaken on the AM of unalloyed and alloyed W, as well as W-matrix composites [24,25]. Most studies on unalloyed W have utilized Powder Bed Fusion (PBF) processes, namely, Selective Laser Melting (SLM) [8,17–19,21,24,26–61], and Electron Beam Melting (EBM) [33,34,47,59–69] of W. Other notable methods employed include Direct Energy Deposition (DED) processes, namely laser-based DED [2,34,47,59–61,70–77] and electric arc-based DED [78,79], and binder jetting (BJ) which shows promise for W-based materials [80–84]. Other novel methods such as ultrashort-time liquid phase sintering (LPS) [85], and bound metal deposition (BMD) have also been introduced [86].

In most AM processes, W materials initially melt, and then solidify, leading to nucleation and subsequent grain growth. This thermal history contrasts with traditional processing techniques like sintering, which primarily deal with powder densification followed by sintering and grain growth. Post-solidification grain growth in AM shares similarities with sintering and is guided by kinetic laws [60,87]. However, in AM of nano-W, surface diffusion prevails [60,88], unlike traditional sintering's grain boundary diffusion. Recognizing these differences and understanding the influence of process parameters and alloy chemistry on molten pool behavior, solidification, and further solid-state microstructural evolution and strengthening effects are essential for advancing the field of W-AM [24,60,74,89–91].

Building on the foundation of previous review articles on W-AM [58,60,61], our review delves deeper into understanding process parameters' design and recent advancements in materials design strategies aimed at mitigating W-AM challenges. We begin by providing an overview of the current status of AM techniques for W alloys, focusing on the PBF, DED, and BJ processes. We will then look at the challenges that are often associated with their AM processes, such as densification, microcracking, and limited material properties, followed by mitigation strategies that researchers have developed from both design and material perspectives to address these issues. Finally, our review concludes with an exploration of promising future directions for W-AM, aimed at paving the way for broader adoption and commercialization.

2. Techniques for Additive Manufacturing of W-Based Materials

W and W-based metal systems demand substantial energy input during AM. With technological advancements, AM systems have been equipped with high-energy sources like lasers and electrons to provide the energy required. Table 1 provides a brief comparison of process parameters and relative densities achieved via different AM techniques for W-based materials (pure, alloys, and composites). Figure 1 depicts the schematic setups of these methods [92–96]. The choice of an AM method for W materials depends on several factors including the material type, post-processing needs, geometric accuracy, cost, surface finish, and property requirements [33,61].

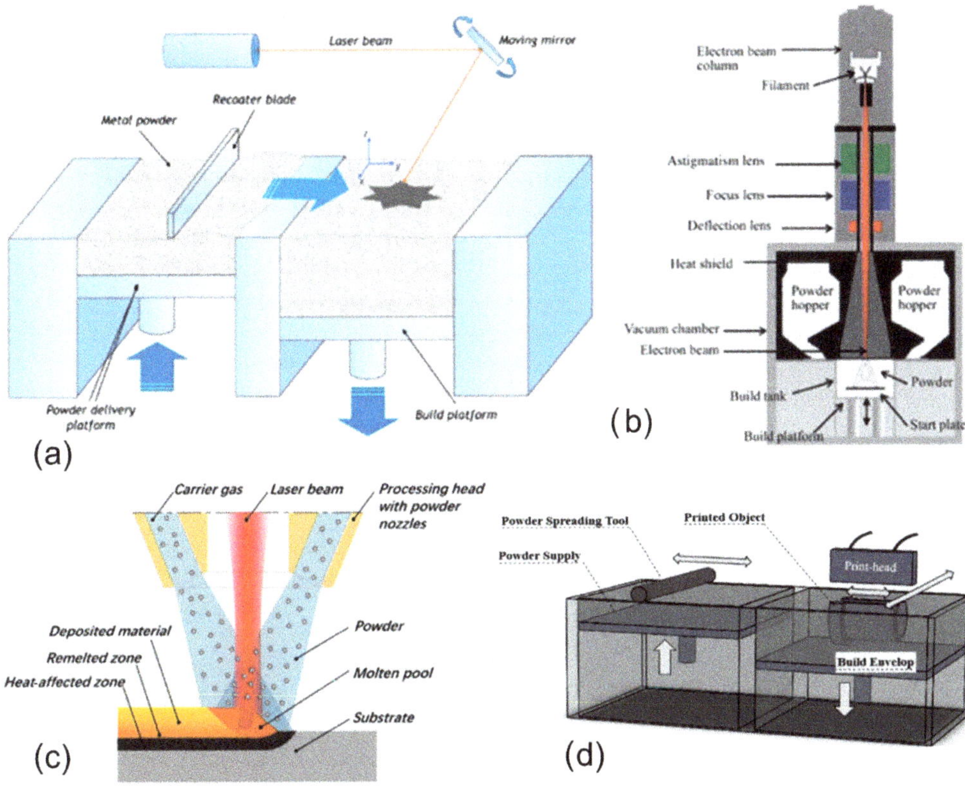

Figure 1. Schematic setups of AM techniques used in W materials fabrication: (**a**) SLM [92], (**b**) EBM [94], (**c**) laser-based DED [93], and (**d**) BJ [96].

SLM (Figure 1a) has been extensively studied for fabricating W and its alloys [97]. Crack forming, especially at grain boundaries, is a major issue in SLM-produced W parts. This happens because of factors such as rapid cooling creating thermal stresses exceeding DBTT, and oxygen trapped at the boundaries during the coalescence of the particles [19,29,63]. Nevertheless, SLM-fabricated W materials can achieve relative densities of 95% to 99.2% [30,47,54,98]. Advancements like femtosecond fiber lasers can potentially improve this technique [18].

Using the double melt strategy also improves W densification [46]. SLM has the potential for integration with other techniques (such as infiltration) to achieve higher density in W parts [19,60].

EBM (Figure 1b) is another prominent AM technique for W [94]. This method employs an electron beam as its primary heat source, which is manipulated through an electromagnetic lens, as detailed in Galati [62]. EBM requires a high vacuum level post He/N_2 purge due to electron travel [32,63]. In comparison to SLM, EBM offers a higher power level of powder [61,99], with typical values between 1000 and 3000 W [32,60,63]. This increased power is responsible for a relative density close to 99.5% in W materials [32]. With appropriate preheating, EBM can address cracking issues in AM for W materials [32,61,62], with many promising results from EBM-fabricated W being reported [32,60,61]. EBM has also been adapted for W surface treatment to enhance quality, due to its controlled beam size and thermal influence [69]. Despite the advantages of EBM, challenges in its use that include complexities associated with electron focusing and the need for maintaining a high vacuum remain [32].

Laser-based DED (Figure 1c) has the distinction of directly adding energy during processing [71–77,93]. In laser-based DED, the relationship between molten pool size and powder flow is crucial [4]. Although increased laser energy can widen the molten pool, when the molten pool exceeds the flow of the powder, it can result in a reduction in the percentage of W entering [2]. The high-energy laser beam rapidly heats and cools powders, which reduces the microstructural coarsening duration [71,76]. Laser-based DED, with its larger melt pools (about several mm) and reduced cooling rates [61,71,76,99], promotes W particle rearrangement and reduces porosity [72–75], while SLM-processed W materials show fusion deficiencies and non-uniform W particle distribution (despite powder preheating) that lead to brittleness. There remains a growing interest in laser-based DED for W alloys because it can produce materials with higher relative density compared to SLM. This improved performance is largely credited to the technique's larger laser spot size [6,31]. Challenges arise in laser-based DED such as managing surface roughness [70] and the formation of intermetallic compounds due to high energy inputs [31].

In BJ (Figure 1d), a flowable powder is laid onto a build platen, followed by the selective deposition of binder material in the shape of the intended part [96]. After each layer, the platen descends, a new powder layer is spread, and the procedure is repeated. These layers are adhered together to create the targeted component. Once the printing is finished, the build box undergoes a curing process in an oven to solidify the binder material [80–84]. The green body is subsequently separated from the surplus powder for further processing. Then, the part undergoes sintering, similar to a standard compacted W component. Tungsten Heavy Alloys (WHAs) with a relative density of 99.7% have been obtained via this method [80]. While the BJ method offers the significant advantage of stacking parts for mass production, achieving fully dense W parts is extremely challenging due to the necessity of high-temperature sintering.

In subsequent sections, we will discuss the AM of W-based materials, which includes pure W, W alloys, and W composites. In this review, we consider all combinations of W with other elements as alloys and refer to combinations of W with ceramic materials as composites. We begin by examining the factors and processing strategies involved in the AM of unalloyed W. We next highlight methods to address the challenges related to processing and properties. Following this, we discuss additively manufactured W alloys and composites, presenting alloying strategies and the role of dispersed ceramic phases (in the W composites) on AM processing, and the resulting properties of W alloys and composites.

Table 1. A brief comparison of AM processes for W-based materials.

AM Process	Energy (W)	Beam Size (μm)	Scan Speed (mm/s)	Temperature Gradient (°C/mm)	Cooling Rate (°C/s)	Relative Density (%)
SLM	10^2–10^3 (normally below 1000 [60,61,99])	30–200 [61,99]	10–1000 [99]	10^6–10^7 [61,99]	10^5–10^7 [61,99]	99.2 [98]
EBM	10^2–10^3 [61,99] (normally more than 1000 W, up to 3000 W [32,60,63], but can be as high as 6000)	100–1000 [61,99]	10–1000 [99]	-	10^3–10^4 [61,71,99]	~100 [100]
Laser-based DED	10^2–10^4 [71,99] (normally between 600 and 4000 [61])	100–1000 [61,71,76,99]	10–1000 [61,76,99]	10^4–10^6 [61,71,99]	10^2–10^5 [99], mostly on the order of 10^4 °C/s [19]	More than 99 for W [2]: even higher for W alloys [31]
BJ (Binder Jetting) [80–84]	Not applicable	Not applicable	Not applicable	Not applicable	Not applicable	99.7 for WHA [80]

3. Additively Manufactured Unalloyed W: Challenges, Determining Factors, and Mitigating Strategies

3.1. Problems and Resolutions in Brief

The remarkable properties of unalloyed W, including its exceptional melting point and high thermal conductivity, make it highly desirable in different applications. However, problems associated with the AM of W still need to be overcome. This section will briefly introduce key challenges and look at their potential solutions (Table 2). The mechanical properties of additive-manufactured unalloyed W, particularly its elongation, tensile strength, and fracture toughness, often fall short of expectations. Researchers have found ways to improve these properties through targeted alloying, the inclusion of ceramic reinforcements, and optimizing the parameters used during the AM process itself [49].

W is inherently brittle due to its high Peierls stress that stems from its BCC crystal structure and weak grain boundary cohesion. The ductility of W can be improved through appropriate alloying, which improves the stacking fault energy of the microstructure to enhance the slip mechanism [60]. The inherent brittleness of W is further exacerbated by its high DBTT, thermal stresses, and oxygen-induced grain boundary issues, all of which led to cracking during the AM process. While strategies like alloying, incorporating carbide/oxide reinforcements, and process adjustments have shown some promise in improving the ductility of additively manufactured W, further studies are needed to completely eradicate these cracking problems [21,38,49].

Another problem stems from the confluence of W's high melting point, low laser absorptivity (<70% in SLM), and high viscosity, making it difficult for powders of W and alloying elements to properly melt and fuse during the printing process. Furthermore, the AM process can be plagued by phenomena like balling, where molten material forms spheres instead of adhering properly. The selection of W powders with improved laser absorption and optimization of processing parameters can mitigate these issues [30,38,40,41,43,51,53,101]. Optimizing process control parameters, particularly laser power and scan variables, has been shown to be instrumental in achieving better densification of W parts [30,38,41,102].

W's propensity for oxygen poses another challenge, as it leads to the formation and incorporation of W oxide and subsequent hot cracking during AM. There are limited options for mitigating the oxidation of W during AM, including the use of high-quality starting materials, establishing better processing atmospheres, and using specific alloying combinations [49].

Table 2. An overview of the primary challenges, governing factors, and potential or attempted mitigation strategies in the AM of W.

Challenges	Possible Strategy/Developments to Tackle the Challenge	Reference
The inherent brittleness and low ductility of W stem from (1) the high Peierls stress of its BCC crystal structure and (2) weak grain boundary cohesion.	Requires appropriate alloying to increase stacking fault energy for easier slip, and microstructural optimization.	[60]
High melting point, significant laser absorptivity (<70% for SLM), pronounced viscosity, densification, porosities, balling, and deformation.	Use of smaller, spherically shaped W powders to enhance laser absorptivity; adjusting processing parameters to influence outcomes like defects, densification, and grain structures. Optimizing parameters like power and volumetric energy density have been found pivotal for improved W densification.	[30,38,40,41,43,53,71,102]
Oxygen affinity and W oxide segregation lead to hot cracking.	Better quality starting material (powder, wire) production, optimized processing atmosphere, alloying.	[49]
Cracks due to W embrittlement and high DBTT; thermal stress-induced, difficult-to-control crystallization; and oxygen-induced cracking along the grain boundaries.	Adjusting laser speeds and reducing grain sizes; techniques to adjust laser speeds and reduce grain sizes showed some positive results but did not completely resolve the issue. Scanning strategy optimization: speed, rotation, remelting.	[8,19,21,38,45,49]
Mechanical property issues (tensile strength and limited fracture toughness) especially in SLM-fabricated W.	Alloying, introduction of reinforcements, process adjustments.	[49]

3.2. *High Melting Point, Related Challenges, and Controlling Strategies*

W's intrinsic high melting point can result in pronounced thermal gradients during AM. Rapid heating and cooling cycles lead to thermal distortion, significant residual stresses, and crack initiation. This compromises the mechanical integrity of the printed components. Furthermore, W exhibits high thermal conductivity. In AM, this causes rapid heat dissipation, complicating the sustenance of the molten state, leading to partial melting and inconsistent fusion. These characteristics can influence the resultant component's microstructure and mechanical properties [29,41,42,53,67]. In the case of alloyed W, the evaporation of low melting point elements can also be a challenge. Working at elevated temperatures also demands more of AM machinery, leading to heightened maintenance needs and increased costs [61,103]. Concurrently, the susceptibility of W and its alloys to oxidation increases at high temperatures, often leading to surface oxide formations that undermine material performance [49].

3.2.1. Melting

Melting behavior is integral to W-AM, often serving as a key factor that defines its AM processing window. The direct melting of W presents significant challenges due to the practical constraints associated with high laser power sources and the large beam sizes required. The very high melting point of W necessitates powerful laser systems capable of delivering sufficient energy for melting. These systems often require power sources that are not commonly available in standard AM machines. Additionally, the large beam sizes needed to ensure adequate energy distribution across the W powder bed can lead to significant thermal gradients, causing residual stresses and potential cracking. The combination of high laser power and large beam sizes also limits the precision and resolution of the final parts. Such strict criteria lead to many partial melting situations, making it difficult to fabricate high-quality parts [52]. The particle size and morphology of W powder, layer thickness, heat source power, volumetric energy density controlled by scanning variables and strategies, melting behavior, and interfacial characteristics are all important in the melting of W during AM processing.

Particle Size and Morphology

Several powder characteristics are decisive for W-AM, especially for PFB processes [104,105]. The primary factors include powder particle morphology, particle size, and purity. Particle size and morphology significantly influence the flowability of the powder and the resulting powder bed density, which in turn influences the density and mechanical properties of the final component. Smaller particle sizes generally lead to better packing density and surface finish but may increase the risk of oxidation and require higher laser energy due to an increased surface area. The shape of the powder particles (spherical vs. irregular) directly impacts the flowability such that spherical particles tend to flow more easily and pack more densely than irregularly shaped particles, resulting in a consistent layer formation during the built cycle. The purity of the powder is critical in avoiding contamination that can lead to defects in the final product. Impurities can cause the formation of brittle phases, reduce ductility, and introduce cracks or other anomalies in AM-fabricated parts.

W's inherent properties necessitate the modification of its powders. These adjustments are needed to improve morphology, ensure uniform size, maximize laser absorption [26], and enhance surface conditions. Techniques such as the utilization of radio frequency plasma have been employed for these modifications [26,27]. The modifications are critical for powder flowability and to achieve a desired density [27]. Zhang et al. studied the influence of particle size on laser absorptivity [53], finding absorptivity to be higher when the powder bed was on a substrate because of incident laser ray entrapment. As an example, a 5 μm powder natively had 0.53 absorptivity but this increased to 0.603 on a stainless steel substrate. Finer particles increase absorptivity, which is the ability of a material to absorb energy from the laser or electron beam, due to their higher surface area-to-volume ratio. Fine powders can also enhance printability by improving detail resolution, packing density, and surface finish. Notably, printability is a comprehensive measure of how effectively a material can be used in a given AM process. It is influenced by a variety of factors, including powder material properties, machine capabilities, process parameters, and design considerations.

Given the cost and limited availability of fine powder sizes, Wang et al. [26] explored an alternative approach to enhance W powder absorptivity. They managed to transform irregularly shaped powder particles into near-perfect spheres (close to 100% sphericity) using radio frequency (RF) induction plasma. These spherical particles were significantly smaller (average size of 31.5 μm) compared to the raw powder (average size of 56.8 μm). This transformation resulted in a significant increase in absorptivity, exceedingly not only in the raw powder (as shown in Figure 2) but also exceeding the values predicted by Zhang et al. [53]. Very fine particles with an average size of 17.3 μm were used to successfully obtain continuous tracks without cracks [29]. However, it is important to note that studies have not established a direct correlation between particle size and achieving crack-free tracks of pure W.

Power, Volumetric Energy Density, Layer Thickness

Volumetric energy density (*VED*) is a crucial parameter in AM design, as it plays a decisive role in defining processing parameters. This can be mathematically represented in Equation (1), where the *VED* is expressed as follows [106,107]:

$$VED = P/(v \times s \times t) \quad (1)$$

where P is the laser power, v is the laser scanning speed, s is the hatch spacing, and t is the layer thickness. The energy input required for AM varies depending on the specific AM processes, machine parameters, material characteristics, and part geometries. Due to the interplay of these factors, a single, universal value for energy input in the AM of a material cannot be established. These variations are reflected in the reported literature, where a wide range of energy input values is observed for W-based materials (e.g., 400 J/mm^3,

8.595 J/mm^3, 1000 J/mm^3). These variations highlight the importance of considering the specific experimental conditions under which such values were determined. Table 3 offers an overview of the VED's effect on W-AM.

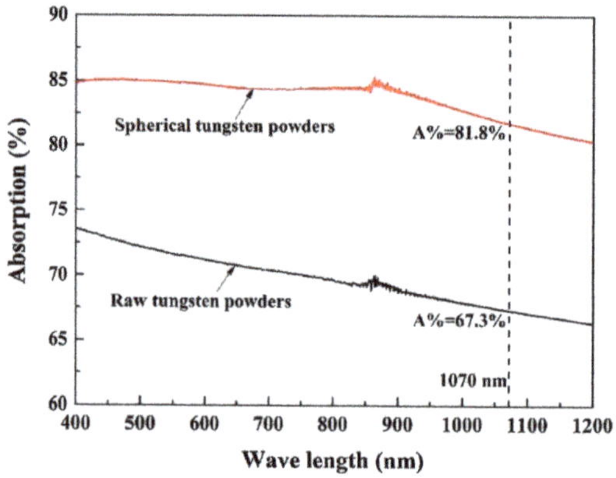

Figure 2. Comparison of absorption properties between non-spherical (raw) and spherical W powders [26]. Reproduced with permission from Elsevier.

Sidambe and Fox [108] as well as Sidambe et al. [36] investigated the impact of VED on the relative density of W parts. Both studies found relative density to increase with higher VED, although the latter reported even higher relative density values at the same VED (Figure 3a). Guo et al. [41] and Hu et al. [42] examined the impact of VED on relative density and hardness using spherical particles. They found that increasing VED led to increased relative density and decreased pore size to achieve a maximum relative density of approximately 98.3%. Microhardness also increased with higher VED, reaching values of 474 Hv [41] and 430 Hv [42]. Other studies on SLM, however, reveal the complexity of the influence of VED and laser power. Yamamoto et al. investigated how laser power, scanning speed, and hatch spacing affect the density of W cylinders [50]. The parts printed from pure polyhedral W powder at a preheating temperature of 35 °C showed that higher laser power and wider hatch spacing led to denser parts, matching the results of those from Rebesan et al. [37]. Plotting the VED against relative density showed the highest relative density reached was 98.58% at a VED of 411 J/mm^3 (P = 370 W, v = 500 mm/s, h = 90 μm, t = 20 μm; Figure 3b). This VED trend differed from those of Guo et al. [41] and Enneti et al. [51], who reported increasing relative density within the VED range. This difference highlights the uncertain role of VED in relative density estimation. Hatch spacing and scanning strategy affect part integrity, porosity type, and thermal history for each layer. Thus, interpreting these trends requires a full understanding of each study's parameters and specific porosity characteristics [50,58]. Wen et al. [30] studied the fabrication of W cubes with varying laser power, scanning speed, and hatch spacing. They found the use of low laser power to result in dense, high-relative-density cubes due to the laser's impact on melt pool temperature. Increasing the line (laser) energy density (LED) led to warped morphologies, which were eliminated with a further increase in LED due to stronger interlayer bonding. Cracks were observed but reduced with higher scanning speed, reaching a maximum relative density of 98.71% at 200 mm/s scanning speed.

For W materials, a higher VED and subsequent higher processing temperature may reduce the resultant residual stresses due to enhanced annealing effects [60,109]. This reduction in residual stresses might be why using EBM generally leads to a reduced cracking probability in W materials when compared to using SLM [63]. Extremely high VEDs (e.g., >300 J/mm^3) are not necessarily beneficial (Figure 3c) [19,50,110]. Density in AM-produced W parts is influenced by several factors. One issue is unstable laser melting. The intense laser beam can create deep holes, called keyholes, that do not fully solidify, leaving air pockets (porosity) throughout a part [19]. Another problem arises from using too much laser power. Excessive melting disrupts the molten metal pool, trapping air bubbles and reducing density [32]. Powder characteristics can also be problematic. If particles do not flow smoothly into the melt pool but instead stick to the edges, they create voids and hinder densification [32]. Finally, W's high melting point presents another hurdle. Some elements that evaporate easily (volatile elements) may vaporize during the printing process, leaving behind tiny holes that contribute to lower density [60,61].

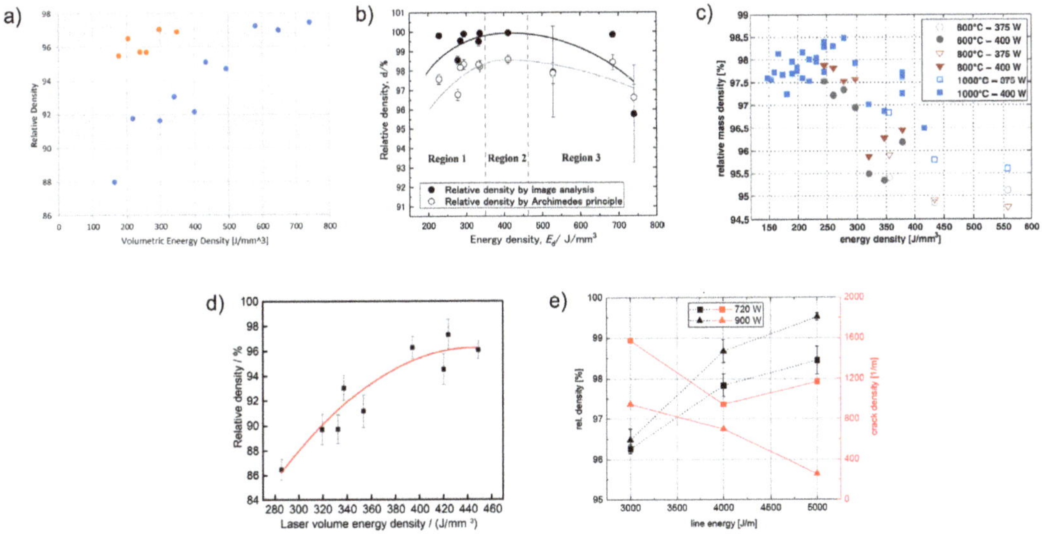

Figure 3. (**a**) The relative density of SLM-fabricated W cubes [36,108], (**b**) influence of VED on the density of pure W parts made with SLM [50], (**c**) influence of substrate preheating temperatures on the relative density of W samples [19], (**d**) relationship between VED and density in W-nickel alloy (W-15Ni) parts fabricated using DED [74], and (**e**) influence of LED and electron beam power on the relative density of W parts made with EBM [58]. Data and figures were reproduced with permission from Elsevier.

The work of Wang et al. on the laser-based DED processing of W found the relative density of fabricated parts increases with VED (Figure 3d) [74]. In their study on EBM processing of unalloyed W, Dorow-Gerspach et al. investigated the impact of LED variations by altering the power and speed of the electron beam to assess its effect on cubic samples [65]. These experiments were carried out at a substrate preheating temperature of 1000 °C. They achieved a high relative density of 99.5% but observed cracking in all fabricated cubes. While increasing electron beam energy led to a decrease in relative density, it also resulted in a decrease in crack density (Figure 3e), highlighting the trade-off between densification and structural integrity in this AM process [65].

High energy inputs in AM processes that include SLM affect the molten pool due to the Marangoni effect. The severity of this effect is directly proportional to the laser energy input [38,111,112]. This effect can drag gas into the molten pool, leading to pore formation upon solidification. Excessive laser energy density, especially at lower scan speeds, can also result in molten pool instabilities, yielding a rough track [29]. As SEM images showing the typical surface morphologies of scanning tracks at different laser powers in the AM processing of W (Figure 4a–e [29]) indicate, an optimal laser power can result in a regular flow front and no cracking (Figure 4b).

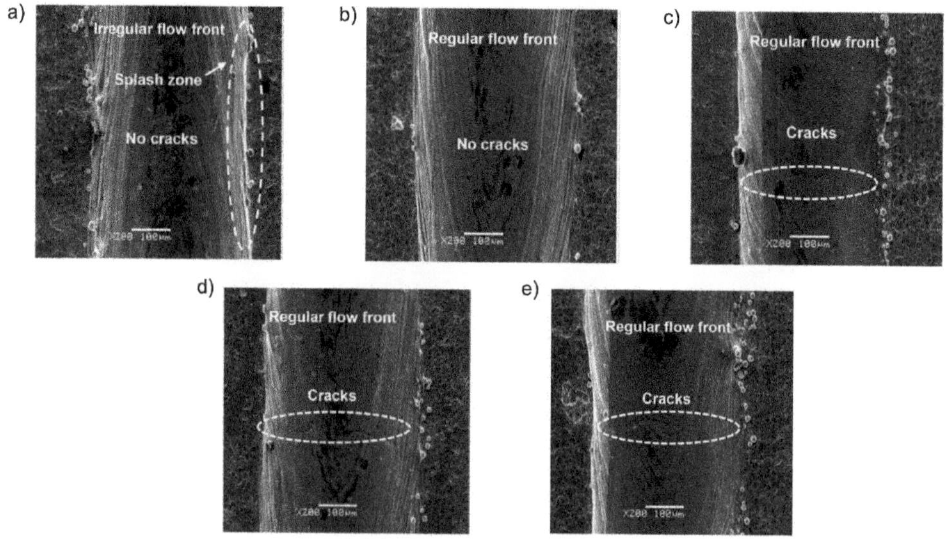

Figure 4. SEM images showing surface morphologies of scanning tracks at a laser power of (**a**) 250 W, (**b**) 300 W, (**c**) 350 W, (**d**) 400 W, and (**e**) 450 W; the scanning speed is 200 mm/s [29].

Properly controlling the melting behavior during AM can yield several significant advantages. High energy input facilitates the mixing of elements with lower melting points, enhancing interfacial bonding through Marangoni convection [29,97]. Stable molten pools promote the formation of complete columnar crystals in additively manufactured W parts [32]. Additionally, techniques such as EBM can introduce remelting during AM, annealing the built portion of the W parts, and reducing their susceptibility to cracking [63]. Another parameter affecting melting in AM processing is the layer thickness, as indicated in Equation (1). The comprehensive review by Marcos et al. indicates that reducing the layer thickness in PBF processes requires less energy to melt the powder, resulting in continuous conduction mode tracks at lower energy input levels [58].

Scan Speed

Scanning speed is a dominant factor, accounting for a ~75% variation in W densification within the SLM process [51,60]. Ren et al. [17] examined how the morphology of W single tracks can be influenced by laser power and scanning speed. These single tracks were printed onto a stainless steel substrate preheated to 200 °C. Establishing a processing map based on the surface characteristics of the single tracks (Figure 5a) allows different regions to be delineated, where one can observe regular tracks, irregular tracks, or instances of balling. The scanning speed plays a critical role in determining the quality of the fabricated tracks. At lower scanning speeds, the laser imparts sufficient energy to the powder, facilitating the formation of regular and continuous tracks. Conversely, as the scanning speed increases, the reduced time available for melting and consolidation leads to the formation of irregular tracks and balling, where molten material forms spherical particles instead of continuous lines.

Rebesan et al. [37] explored how adjusting the laser scanning speed and spacing between hatch spacing affects the density of SLM-fabricated W parts under Ar atmosphere at a power of 170 W and spherical particles (Figure 5b). This approach resulted in achieving high relative densities of 96.4% to 99.6%. Enneti et al. took a different approach, with the laser power kept at 90 watts, a VED of 1000 J/mm^3, and the base plate preheated to 80 °C [51,102]. With these settings, densities ranging from 60% to 75% were achieved. Li et al. [113] examined the impact of process parameters (laser power, scanning speed, hatch spacing) on the density of as-SLM bulk W parts, achieving a maximum relative density of 98.31%, and creating process maps (Figure 6). Here, there was a direct correlation between laser power and density (Figure 6a), with hatch spacing and scanning speed having minor effects (Figure 6b,c). High densities (>98%) were consistently attained with 250–300 W power, 0.08–0.1 mm hatch spacing, and 0.03 mm layer thickness, offering valuable insights into process optimization. Xiong et al. [45] used image analysis to measure density and established process parameters for achieving high-density W coupons. They achieved densities of 97.3% to 98.1% with laser powers between 250 and 300 W and scanning speeds of 400 to 500 mm/s. Dong et al. [114] conducted a study to investigate how various factors influence the quality of AM-produced W parts. They utilized a preheating temperature of 100 °C and employed irregularly shaped (polyhedral) particles for the powder material. Furthermore, they observed that higher laser power and slower scanning speeds produced denser parts. However, when the scanning speed was excessively high, it caused the material to ball up and led to the formation of surface pores. Specific studies on the effect of nitrogen on W-AM are limited. Dong et al. [114] found that W parts printed under a nitrogen atmosphere achieved higher density, hardness, and overall strength compared to those printed under an argon atmosphere.

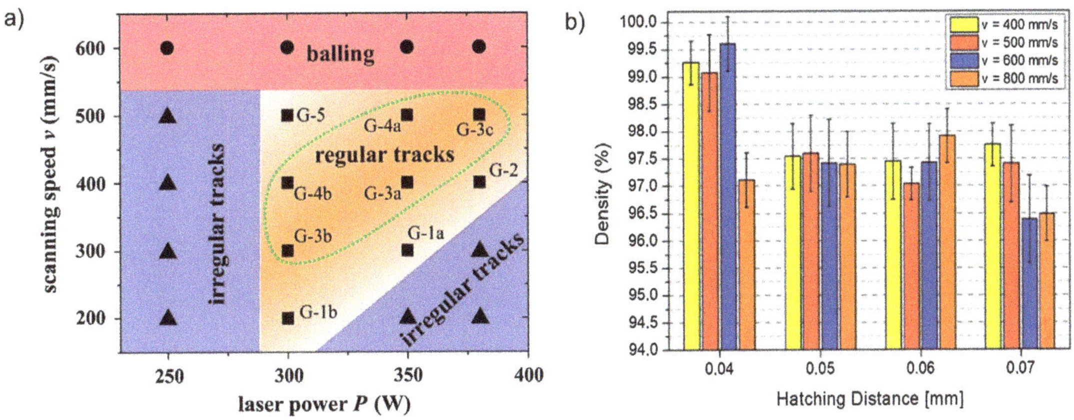

Figure 5. (**a**) Processing map for SLM of unalloyed W [17], and (**b**) the influence of hatch spacing and scanning speed on the relative density. The laser was maintained at a power of 170 W and the substrate was preheated to 80 °C [37]. Reproduced with permission from Elsevier.

Recent findings by Sharma et al. [98] have demonstrated a maximum relative density of 99.2% for pure W, which is the highest value reported in the SLM of pure W. They employed a multi-scale thermo-kinetic and thermo-mechanical computational model to fine-tune the laser parameters and understand the thermal behavior during the melting and solidification phases. Their work demonstrates the potential for further density improvement through controlling laser power, scan speed, and hatch spacing.

Other Scanning Variables and Strategies

The scanning pattern plays a critical role in determining the thermal history of processed W materials, offering flexibility suitable for a variety of applications. This flexibility represents a unique advantage of AM over traditional manufacturing techniques. Table 3 offers an overview of scanning variables and strategies employed in W-AM. The localized differences in material properties can be ascribed to the focused energy input from electron or laser beams or other energy sources that determine the thermal history experiences at a point [60]. In the EBM process, for example, one can witness multiple instances of melting and re-melting at a single scanning point [63]. EBM utilizes flexible bidirectional scan patterns to optimize densification and surface finish, leading to a high print quality [32].

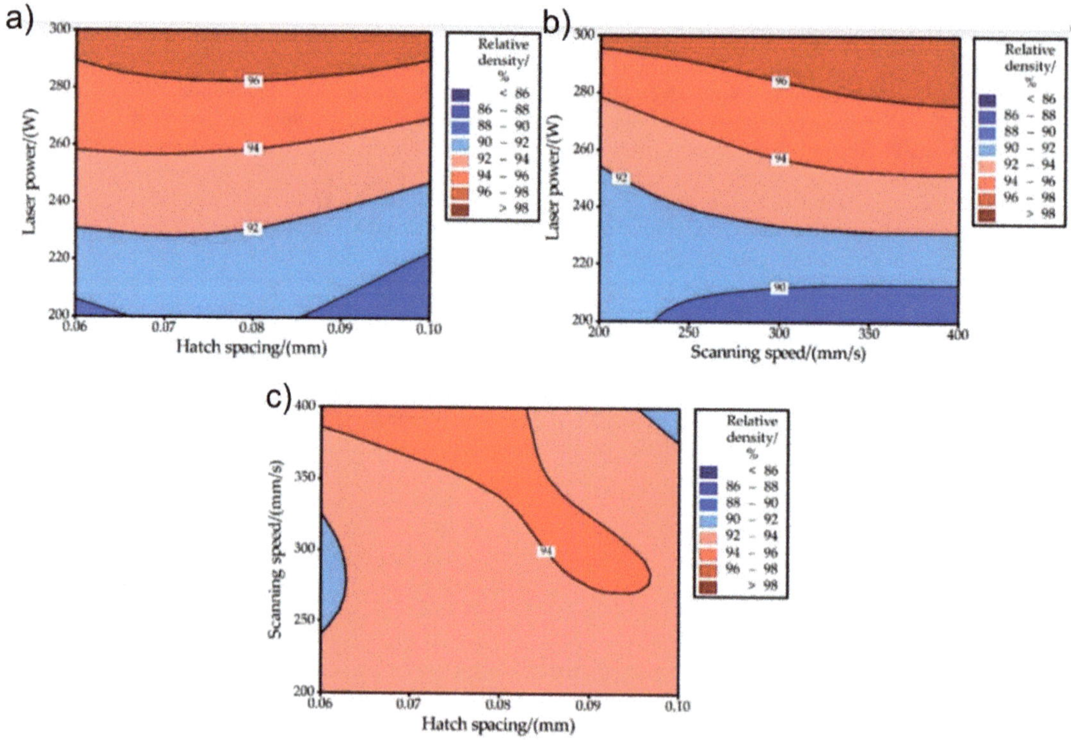

Figure 6. Processing map for SLM of W. (**a**) laser power vs. hatch spacing, (**b**) laser power vs. scanning speed, and (**c**) scanning speed vs. hatch spacing, [113].

In PBF for W materials, different scanning patterns have been explored. A 67° rotation between layers randomizes grain orientation, reducing the ladder-shaped grain structures which are potential crack-formation sites [21]. While many studies adopt a 67° rotation to minimize scan alignment, others use rotational angles of 45° or 90° [8,19,27]. Remelting, which is scanning a track multiple times before recoating, combined with rotation, curtails the formation of columnar grains and reduces longitudinal cracks, resulting in smaller grain sizes and reduced surface roughness [21,45,61]. Wang et al. [21] investigated crack development in SLM of unalloyed W. They studied three scanning strategies: parallel, 67° rotation, and remelting with 67° rotation. Matching scan tracks between layers resulted in consistent crack growth along the build direction. Rotation between layers reduced crack propagation, while remelting did not effectively reduce cracks but instead promoted thermal fatigue crack growth [21].

A growing body of research explores the relationship between scan pattern design and the properties of additively manufactured W materials. These studies have quantified the effects of specific pattern parameters. Hatch spacing, interestingly, is responsible for a mere ~7% fluctuation in the final W densification [60]. Point distance is another crucial factor in the AM relative density for W materials. Wang et al. [43] and Huang et al. [115] studied the influences of point distance and hatch spacing on the relative density of W achieved when pulsed (Figure 7a), and continuous lasers were used (Figure 7b,c), respectively. By adjusting these parameters, different melting strategies can be deployed, presenting an opportunity to oversee and regulate re-melting sequences to ensure that they sidestep randomness [46].

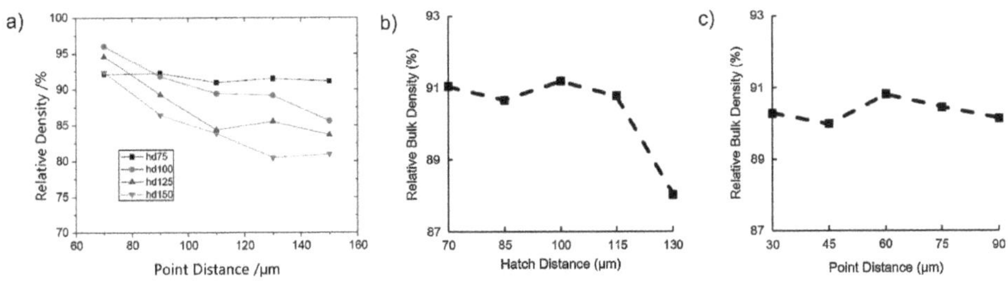

Figure 7. Influence on the relative density of manufactured W cubes due to the interaction between hatch spacing and point distance with input parameters of (**a**) a pulsed laser power of 400 W and exposure time of 100 μs [43], and (**b**,**c**) a laser power of 40 W, layer thickness of 50 μm, and exposure time of 300 μs [115]. Reproduced with permission from Elsevier.

In examining the role of scan track design, we find it plays an important role in pattern design, especially in upholding the quality of SLM-fabricated parts [60]. Parameters such as the single scan track width and hatch spacing can be adjusted. Within SLM, the overlap rate is devised in line with these two parameters, serving as an analytical tool to gauge conditions across multiple scan tracks. Previous studies have underscored a non-linear relationship between scan track characteristics, which is dictated by overlap design choices, and the eventual densification of W material. This points to the existence of an optimal overlap rate that balances these parameters [17].

Table 3. An overview of scanning variables and strategies employed in the AM of W.

Parameters	Details	Reference
VED	- **VED:** - ***Increased power*** in AM affects the molten pool via the Marangoni effect, with the severity proportional to laser energy. - Can introduce gas, causing pore formation and pool instabilities, resulting in rough tracks. - Enhances the densification of manufactured parts. - Reduces residual stresses via annealing. - Relates to pattern design. - ***Extremely high energy densities*** (>300 J/mm^3) have drawbacks: more porosity, excessive fusion risks, disrupted melt pools with stuck powder particles, and the vaporization of volatile elements. These challenges impact printed part quality. - ***Controlled melting has benefits,*** allows the mixing of low melting point elements, and improves interfacial bonding. It promotes the growth of full columnar crystals in W-AM.	[19,29,32,38,50,60,61,63,74,97,109–112]
Scanning Variables	- **Direction and Angle:** - EBM process allows multiple instances of melting and re-melting at a single scanning point. - 90° alternating bidirectional scanning in SLM. - 67° alternating bidirectional scan. - **Speed:** scanning speed affects ~75% of W densification variation in SLM. - **Hatch Spacing:** affects ~7% of W densification variation. - **Overlap Rate:** The relationship between scan track characteristics and final W material densification is non-linear. An optimal overlap rate exists. - **Scan Track Width**	[17,19,27,32,46,51,60,63]

Wetting Behavior and Interfacial Characteristics

For producing high-quality components via AM, interfacial design and W wetting behavior at multiple scales are vital due to their well-documented influence on reducing crack initiation and propagation, and on mechanical properties (Table 4). Microscopic wetting behavior involves the internal heterogeneous boundaries within the AM-produced W parts. W is not easily wettable by most metals at low temperatures (below 2000 °C). The temperature-dependent wettability of W has been a focal point for many researchers [60,116]. Muramatsu et al. [97,117] underscored the significance of the heterogeneous interface, especially when considering the broad temperature range covered in AM processes [97,116,118]. To enhance the AM quality of W-based materials, researchers have introduced small quantities of transition metals like Ni and Fe. These elements reduce the activation energy required during the sintering of W, accelerate W diffusion, and provide improved wettability to facilitate an AM process [60,97].

Table 4. An overview of W wetting behavior and interfacial phenomena at micro as well as macro scale in AM processes.

Category	Description	Examples and Developments	References
Microscopic Wetting Behavior	Concerns internal heterogeneous boundaries in W-AM.	- Temperature-dependent wettability of W - Introduction of Ni and Fe to improve wettability and AM quality	[60,97,116–118]
Macroscopic W Material/Substrate Interfacial Wetting and Reaction	Relates to interfaces between the substrate and W material, affecting parts quality.	- Use of stainless steel interlayers containing nickel - Substituting substrates with Ti-based alloys, e.g., Ti–6Al–4V - Sandwich-structural printing sequences - Risks of diffusion-induced secondary phases and unwanted reactions	[32,73,97,119]

On the other hand, macroscopic W material/substrate interfacial wetting and reaction involve the interfaces between the substrate and the W material to influence the final quality of an AM part. Stainless steel interlayers, containing nickel, have been employed to improve the adhesion between W and other alloying elements because of an enhanced solid-state diffusion and grain boundary diffusion [97]. For better outcomes, substrates can be substituted with Ti-based alloys such as Ti–6Al–4V due to the enhanced thermodynamic compatibility between W and Ti [32]. Novel approaches, like sandwich-structural printing sequences, have been developed for W materials, leading to better quality AM parts [119]. However, using such interlayers can lead to diffusion-induced secondary phases. Additionally, interactions with other elements might bring about unwanted reactions, such as the peritectic reaction seen between Fe and W during non-equilibrium solidification [73,97].

3.3. *Oxygen, DBTT, and Micro-Cracking*

Oxygen, along with other interstitial contaminants like nitrogen and hydrogen, presents serious challenges in the AM of refractory alloys (including W materials). Such contaminants, particularly oxygen, critically affect properties like the DBTT value in W materials, with molten W being especially sensitive to oxygen [17,120–122]. This heightened sensitivity leads to embrittlement, increasing DBTT, and decreased strength. Table 5 summarizes the current understanding of oxygen contamination for W in AM processing. Common sources of oxygen contamination include oxidized powder surfaces, residual oxygen in the chamber [49], and oxygen present in the substrate due to high thermal energy inputs which suggests that substrates may function as oxygen reservoirs [19]. The powders utilized in AM, owing to their vast surface-to-volume ratio, are notably prone to absorbing oxygen from their surroundings. Moreover, when W is exposed to oxygen at room temperature, it promptly forms an oxide layer within an hour [123]. Nagy and Humphry-Baker [122] conducted a study on the oxidation of W and developed a W oxidation mechanism map to reveal the kinetics of W oxidation at the 600–1600 °C temperature range (Figure 8).

This can be useful in W-AM. Under AM's elevated temperatures, oxygen can not only diffuse swiftly into the alloy but also react with W, resulting in intergranular cracking upon cooling [48,63,124].

Table 5. Current understanding of interstitial contamination in W during AM processing.

Factor/Parameter	Details/Effects	References
Contaminants	Oxygen, nitrogen, hydrogen	[17,120–122]
Adverse Effects	- Embrittlement - Shifts in DBTT - Strength reduction - Cause cracking especially at HAGBs	[17,21,120–122]
Sources of Oxygen in W Materials	- Oxidized powder surface - Residual oxygen in the chamber - Substrates due to high thermal energy inputs	[19,49]
Oxygen Sensitivity in W Materials	- Molten W is sensitive to oxygen - Oxide layer formation (1 nm thick within an hour) - Potential for intergranular cracking upon cooling	[48,63,123,124]
Oxygen Control Techniques in AM Processes	- Higher beam power - Enhanced VED - Radio frequency induction plasma treatment	[26,63]
Material Design for Reduced Oxygen Sensitivity	Introduction of alloying elements such as Ta to W	[48]
Important Studies and Findings	Oxidation mechanism map for AM applications exists	[122]

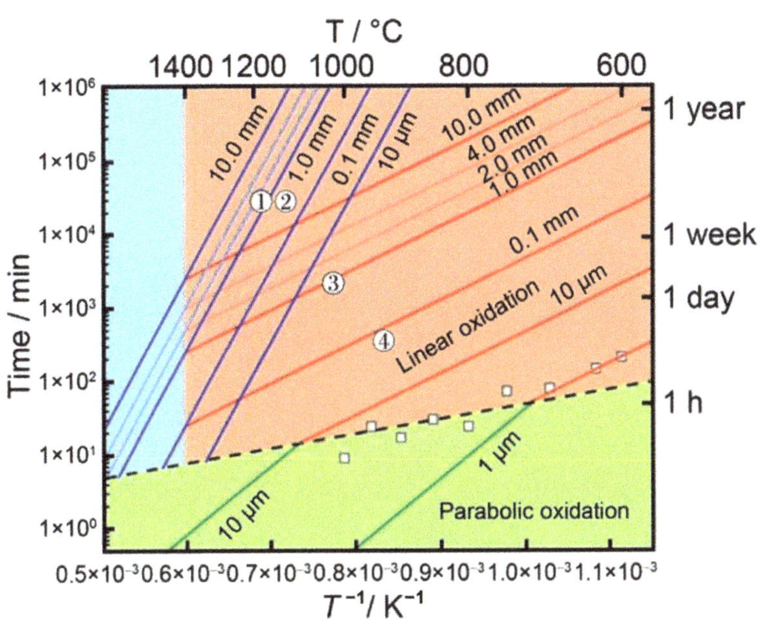

Figure 8. W oxidation map in the temperature range of 600–1600 °C and up to 2 years. Contours indicate the thickness of the metal oxidized. The parabolic, linear, and sublimation regions are shaded in green, red, and blue respectively [122]. Reproduced with permission from Elsevier.

Reducing oxygen levels during the AM process is paramount, as this will improve part quality by reducing the occurrence of cracks [49]. W oxides can remain stable at up to 2000 °C, potentially initiating cracks during rapid solidification [120]. To confront these challenges, researchers can turn to several strategies in many metallic systems. Gas atomization under inert conditions delivers powders with appreciably lower oxygen and nitrogen contents. The use of inert gas environments, such as argon or helium, during AM offers a practical way to minimize contamination by isolating the alloy from potential contaminants. Additionally, enhancing powder handling and storage protocols—like the vacuum packaging of powders and handling under inert atmospheres—is another avenue being explored to eliminate contamination. Implementing techniques like using higher beam power and enhancing volumetric energy density can diminish the oxygen content and curb cracking [63]. The role of oxygen is not just restricted to the fabrication process but extends to preheating, underscoring the necessity for an oxygen-regulated environment [32]. Innovatively, as pointed out earlier, methods such as radio frequency induction plasma treatment have been deployed to manage oxygen levels effectively [26]. Furthermore, material design advancements, like introducing Ta to W, offer a promising avenue to counteract oxygen sensitivity [48]. To facilitate deeper insights, detailed studies to understand W oxidation have been undertaken, resulting in the creation of an oxidation mechanism map tailored for AM [122].

Crack nucleation and propagation in the AM of W correlate with its high DBTT of approximately 200–600 °C. Support for this comes from an observed delay between solidification and crack appearance using in situ high-speed cameras reported by Vrancken et al. [24,35]. Cracks often propagate along high-angle grain boundaries (HAGBs) [21], where the formation of cracks in the intergranular regions of SLM-fabricated W has been mostly observed. Such behavior is linked to W's grain boundary sensitivity to impurities, notably oxygen, found in amounts between 30 and 370 ppm in W powder [24,38]. While some researchers attribute cracks to W oxide aggregation during solidification [21,37,41,42], persistent cracks at low oxygen levels indicate other possible influencing factors as well [24]. The role of impurities, including oxygen and hydrogen, in SLM on W brittleness has not been fully explored. High-speed in situ SLM videos show cracking across W's DBTT, likely due to the increased stress [24]. Such stresses from SLM may only be offset by cracking at impurity-contaminated grain boundaries [42,125]. A study by Rebesan et al. on SLM-fabricated W specimens revealed a high cracking tendency [37]. Two types of cracks were identified: longitudinal (along the melt-pool center) and transverse (perpendicular to laser-induced surface ripples). Longitudinal cracks are linear and 30–100 μm long, while transverse cracks are shorter with an "S"-shape along grain boundaries (Figure 9). The extensive cracking results from thermal stresses during rapid solidification or recrystallization (Figure 10a), particularly below the DBTT. Micro-cracks that resemble intergranular hot cracking form due to nanopore aggregation at grain boundaries induced by boiling W oxides (WO_x) in the melting pool (Figure 10b) [37].

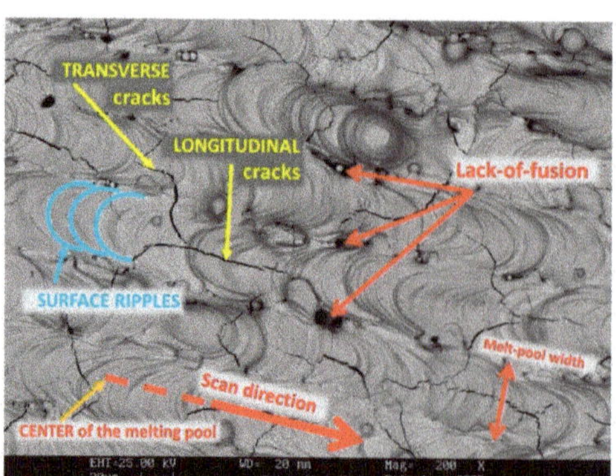

Figure 9. SEM image of surface morphology and crack type in SLM-fabricated W [37].

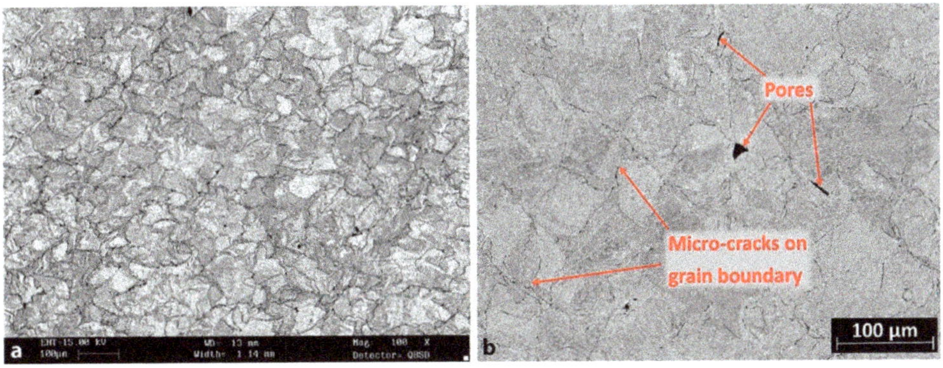

Figure 10. SEM images of SLM-fabricated W sample, (**a**) electrolytically etched displaying a crack network; (**b**) a higher magnification of grain boundaries displaying micro-cracks and pores [37].

Cracking in EBM-produced W is comparable to in an SLM one, with their mechanisms still under debate. Solid-state cracks may arise from inelastic deformation near low-angle grain boundaries (LAGBs) (Figure 11a), which is supported by EBSD data (Figure 11b) [63]. In laser-based DED, residual stress-induced cracking due to high thermal gradients and porosity are common challenges [61]. Oxygen contamination and its contribution to microcracking is another challenge.

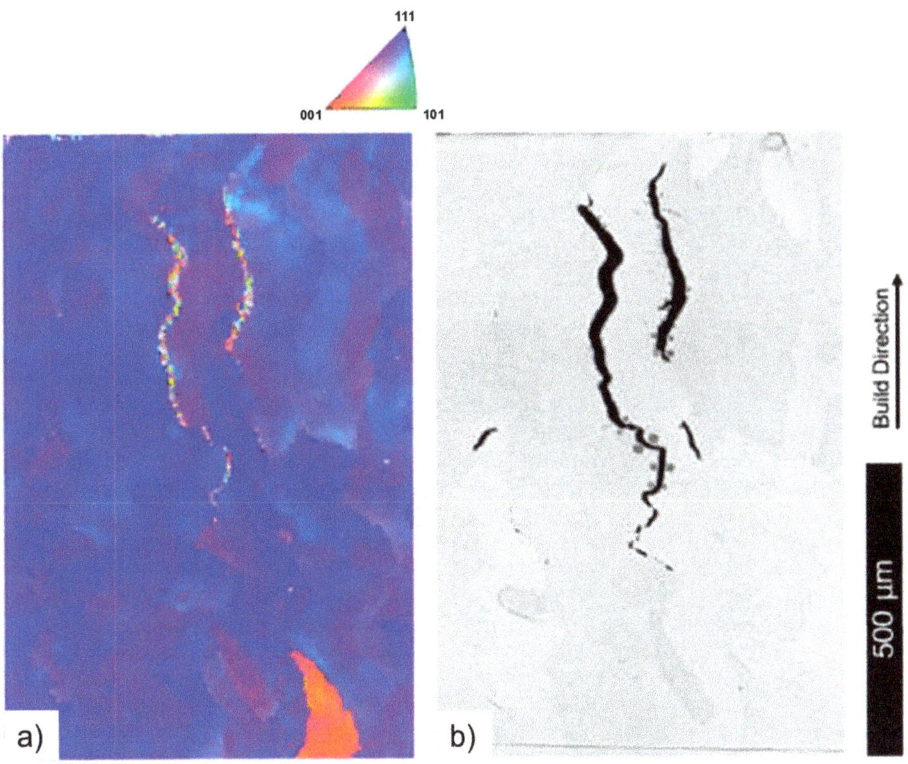

Figure 11. A vertical crack at grain boundaries of EBM-fabricated W samples: (**a**) EBSD in which the crack appears as random unindexed pixels, and (**b**) an optical image [63].

Overall, micro-cracking in W-based alloys arises due to several mechanisms, which are as follows. Thermal stress: The high melting point and low thermal expansion coefficient of W lead to substantial thermal gradients during the rapid heating and cooling cycles in AM processes. These thermal gradients induce thermal stresses that exceed the material's fracture toughness, resulting in micro-cracks. Solidification cracking: During the rapid solidification in AM, the difference in cooling rates can cause uneven shrinkage and create tensile stresses in the material. If the material solidifies with significant residual stresses, it can lead to solidification cracking. The presence of impurities and alloying elements can intensify this issue by creating brittle phases or weakening grain boundaries. Phase transformations: Certain tungsten-based alloys undergo phase transformations during the cooling process. These transformations can involve volume changes that induce additional stresses within the microstructure, contributing to micro-cracking. The incompatibility between different phases in terms of their mechanical properties can further complicate this problem. Porosity and defects: Incomplete melting and impurities can introduce porosity and other defects within the material. These defects act as stress concentrators and can initiate micro-cracks under the applied thermal and mechanical loads during the AM process and subsequent cooling. Grain boundary embrittlement: Alloying elements or impurities that segregate to grain boundaries can weaken these boundaries and make them more susceptible to cracking under thermal and mechanical stresses. This embrittlement can significantly contribute to the formation and propagation of micro-cracks. In addition to powder and process parameter-related solutions, post-processing techniques like hot isostatic pressing (HIP) can be explored as viable techniques for closing microcracks in W-based alloys to enhance the overall mechanical properties of AM-fabricated W components.

Substrate Preheating

Utilizing a heated substrate plate in PBF processes can potentially mitigate the problems posed by thermal gradients [19,38]. Preheating is an essential preparatory technique implemented before the actual PBF processes to optimize the material's structure and improve the resultant properties. This method confers several advantages (Table 6). Firstly, preheating aids in the mitigation of built-up thermal residual stresses within the material to enhance its stability [109]. Moreover, by controlling the melt pool's cooling rate and the thermal gradient during solidification, preheating ensures improved material formation and properties. This is particularly critical for W which has a high DBTT; preheating can potentially delay or even avert this transition [44]. Additionally, preheating can emulate the annealing process, refining W's microstructure to produce parts with increased density and fewer microcracks [109]. In some cases, preheating the substrate to temperatures close to 1000 °C has proven effective in achieving those outcomes [19].

Elevating substrate temperature beyond W's DBTT decreases cracking likelihood, thereby improving the relative density of parts manufactured using AM with W. SLM processing at increased temperatures from 200 °C to 1000 °C reduced cracking. Such findings have been visually represented in the works of Muller et al. (Figure 12) [19], where they compare the microstructure of SLM-fabricated W with a substrate temperature of 200 °C that includes grain boundary cracks (Figure 12a) to that of a SLM-fabricated W using a substrate temperature of 1000 °C without cracks (Figure 12b). In another study, using ~850 °C substrate heating led to minor cracking [32]. Other observations indicate that temperatures below 400 °C are inadequate for preventing microcrack formation [8]. Vrancken et al. also showed that the exact temperature above which cracks in SLM-fabricated W are eliminated ranged from 500 °C to 600 °C. The crack-mitigating effect of preheating persists when scaling up from a single scan track to a hatched area. Without preheating, a large crack network is formed (Figure 13a), whereas only small, isolated cracks are found using 500 °C preheating (Figure 13b) [44].

Table 6. An overview of the effects and benefits of preheating in the AM of W materials.

Preheating Effects	Information/Description	Reference
Purpose	Used before the process to enhance obtained properties	
Benefits		
- Mitigation of Stress	Reduces built-up thermal stresses	[102]
- Melt Pool Dynamics	Controls cooling rate and thermal gradient during solidification	
- Delay/Avoidance of DBTT	Delays or prevents DBTT	[44]
- Annealing Effect	Simulates the annealing process, refining W microstructure	[109]
- Enhanced Density and Reduced Microcracks	Increases part density and reduces microcrack formation, especially with preheating up to 1000 °C in SLM, and more in EBM	[19,63]
Optimal Preheating Range		
- Observation	Low preheating temperatures below 400 °C are insufficient for microcrack mitigation of W materials	[8,32]
- General Findings	Higher preheating temperatures minimize high DBTT effects and enhance the density of W parts	[19]
Drawbacks of Excessive Preheating		
- Recrystallization	Extremely high temperatures trigger recrystallization in material	[63]
- Grain Vulnerabilities	Larger recrystallized grains are prone to intergranular cracking, acting as pathways for cracks near grain boundaries	[60,121]

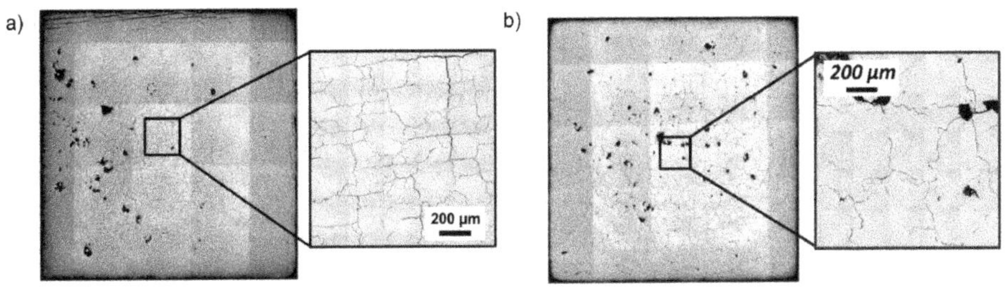

Figure 12. Effect of temperature on SLM-fabricated pure W cubes: (**a**) optical image of a cross-section at 200 °C, (**b**) optical image of a cross-section at 1000 °C [19]. Reproduced with permission from Elsevier.

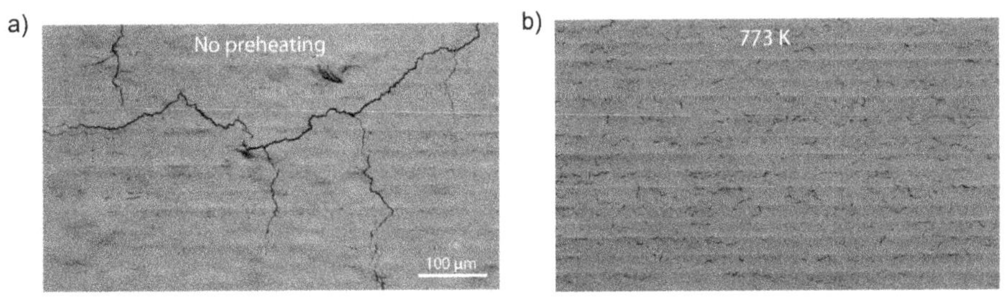

Figure 13. Crack formations in SLM-fabricated W samples: (**a**) crack pattern of a 1 by 2 mm area scanned using hatch spacing h = 25 μm, and (**b**) absence of cracks in a 1 by 2 mm area scanned using 500 °C preheating [44]. Reproduced with permission from Elsevier.

In EBM processing of W, preheating each layer up to 1800 °C may be required to fabricate high density parts [100]. Ellis et al. observed a lower crack density at 1500 °C (corresponding to a preheating input energy of 735 J/mm^2) than at 1100 °C (corresponding to a preheating energy input of 325 J/mm^2), as shown in Figure 14a,b [63]. This led them to postulate that the higher preheat energy suppresses cracking by raising the ambient temperature of the build [63]. While preheating is beneficial, overdoing it can have detrimental effects. Excessive preheating temperatures can induce undesirable recrystallization processes within W materials [63]. The formation of larger recrystallized grains, a consequence of intense preheating, makes the material susceptible to intergranular cracking [60,121]. These grains can then become crack-initiation points, especially in regions close to or on the grain boundaries [60,121]. As a result, extremely high preheating temperatures can be counterintuitive and diminish the quality of the product [63].

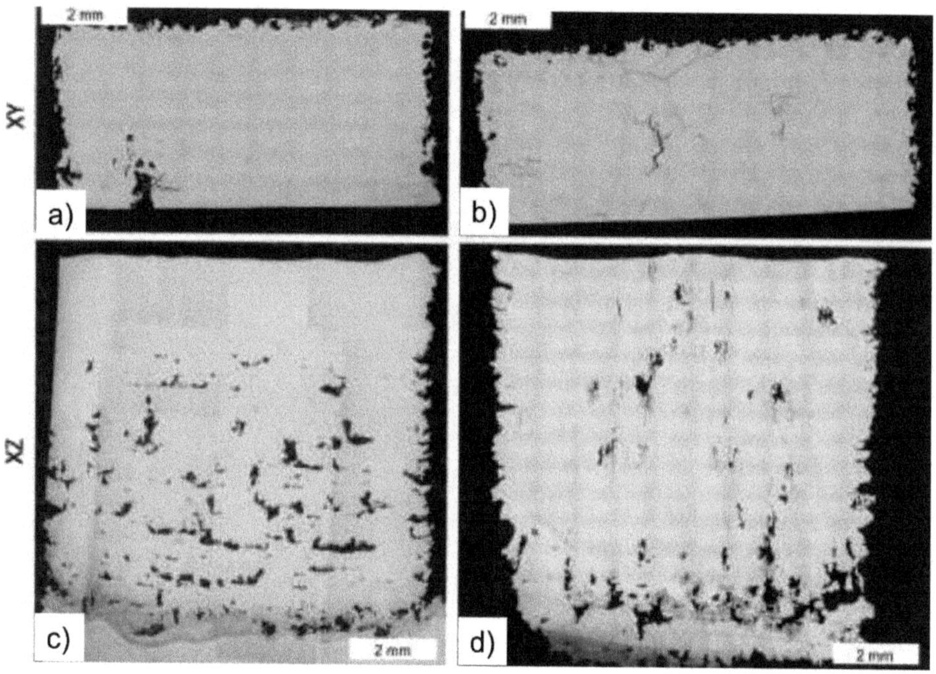

Figure 14. Cracking in EBM-fabricated W blocks built with nominally identical melt parameters: (**a**,**b**) XY sections, approximate elevations 9 mm, (**c**,**d**) XZ sections, approximately at the midline. The build shown in (**a**,**c**) did not crack; the build in (**b**,**d**) exhibited cracking [63].

Residual stresses in W-AM arise from W's high melting point and low thermal expansion coefficient. These stresses are generated by the complex thermal cycles experienced during AM processes, including thermal gradients, solidification stresses, phase transformations, and mechanical constraints. W parts undergo rapid heating and cooling, leading to steep thermal gradients that cause differential expansion and contraction. This results in tensile and compressive stresses within different regions of the part. Furthermore, as the molten W solidifies, the material shrinks. Non-uniform solidification and cooling rates across the part introduce residual stresses due to uneven shrinkage. Additionally, W-based materials may experience phase transformations during cooling, involving volume changes that generate further stresses. The build platform and surrounding material can also constrain the part, leading to residual stresses as the material contracts upon cooling. Mitigation strategies involve controlled cooling post-deposition to minimize thermal stresses through preheating, optimizing scan strategies, and designing effective support structures.

Thermal management is crucial in W-AM and can be addressed through various strategies during the design, AM processing, and post-AM processes. Enhanced support structures specifically designed for W-AM can effectively dissipate heat and stabilize the part. Preheating the build platform, optimizing scan strategies, and adjusting process parameters to control cooling rates help maintain a more uniform temperature distribution, thereby reducing thermal gradients during the build cycle. Post-processing heat treatments, such as annealing or HIP, are employed to relieve residual stresses and improve the microstructural integrity of AM-produced W components, significantly reducing internal stresses and enhancing mechanical properties.

Ledford et al. [100] employed a comprehensive experimental design to establish an optimal processing window for achieving high-density, crack-free pure W using the EBM method. They successfully attained full density in select fabricated samples by preheating the substrate to 1800 °C and optimizing the other parameters such as scan speed and layer thickness. The as-fabricated W samples exhibited subgrains approximately 10–20 μm in size, with low-angle grain boundaries characterized by high dislocation densities ranging from approximately 1015 to 1016 per square meter in certain instances. High-temperature tensile testing revealed significant tensile anisotropy attributed to inherent crystallographic variations. Yield strength ranged from 85 to 202 MPa, while tensile strength ranged from 121 to 274 MPa, closely resembling annealed tungsten properties.

Similarly, Antusch et al. [126] utilized EBM to manufacture pure W components for nuclear fusion reactors, achieving 99.8% density and a crack-free microstructure characterized by elongated columnar grains. The absence of microcracks was attributed to reduced thermal stress facilitated by optimized pre-heating strategies during EBM and the absence of oxide precipitates due to manufacturing in a vacuum environment. This resulted in a stable melt pool during the fabrication process. Tensile testing conducted between 600 °C and 1000 °C demonstrated ductile behavior with a total elongation of 80% along the building direction and DBTT ranging from 700 °C to 900 °C. Thermal conductivity measurements were comparable with values in the literature for conventionally produced W counterparts. These studies highlight the potential of EBM as a promising method for processing W (Figure 15).

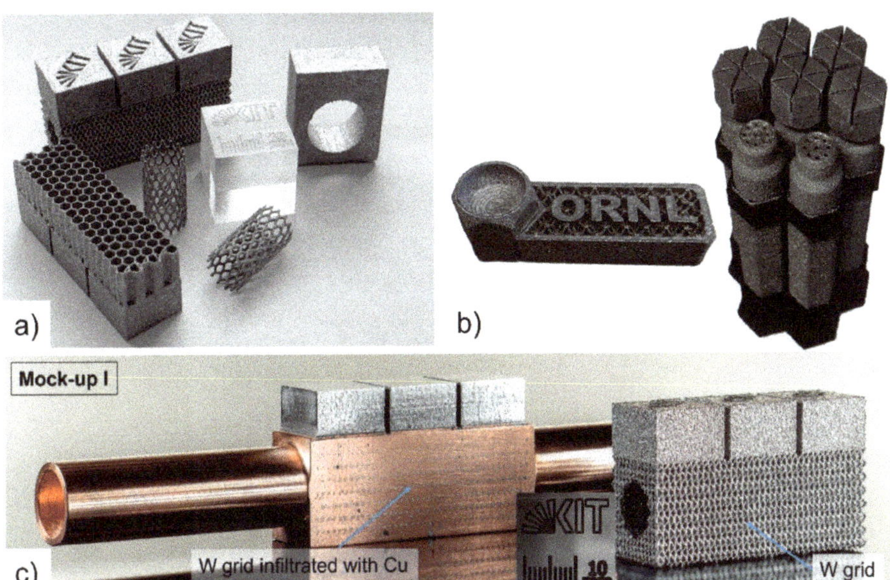

Figure 15. Pure W components fabricated using EBM: (**a**) complex geometrical parts [126], (**b**) a diverter tile and mesh demonstration piece, along with a mockup illustrating the potential diverter geometry featuring tungsten plasma-facing tiles [100], and (**c**) the integration of W-EBM with copper infiltration tailored for fusion applications [126].

4. Additively Manufactured W Alloys and Composites

Amongst the strategies to mitigate the challenges discussed in the preceding sections, alloying is a particularly useful approach. Alloying has been shown to impart multifaceted effects on the properties and behavior of W in AM contexts. Several W alloys have been studied in the realm of AM. Since W is often insoluble or immiscible with other elements, its alloyed materials with some elements, have been referred to as composites in some studies. In this review, we consider all combinations of W with other elements as alloys and refer to combinations of W with ceramic materials (such as carbides and oxides) as composites (Table 7). The AM-fabricated alloy groups that have been studied include W-Re [127–129], W-Ni [74,130], W-Ni-Fe [71,72,75,77,80,83,85,103,131–133], W-Ni-Fe-Co [134], W-Ni-Fe-Cu [80], W-Ni-Cu [4,135], W-Cu [4,6,136–138], W-Cu-Sn [139], W-Ta [8,140–143], W-Ta-Re [144,145], W-Nb [110], W-Fe [73,146], W-Mo [147], and W-Cr [86]. W-based matrix composites with different ceramic phases such as TiC [148,149], ZrC [120], TaC [150], La_2O_3 [44], CeO_2 [44], a mix of La_2O_3–ZrO_2–Y_2O_3 [44], and Y_2O_3 [42,61] have also been fabricated via AM methods.

Table 7. Additively manufactured W-based alloys and W-matrix composites.

Materials	References
W Alloys	
W-1–25%Re	[128,151]
W-0.1–40%Ni	[74,130]
W-7%Ni-3%Fe	[71,72,75,77,80,83,85,103,131,132]
W-4.6%Ni-2.4%Fe	[133]
W-6%Ni-2%Fe-2%Co	[134]
W-12%Ni-4%Fe-4%Co	[134]
W-18%Ni-6%Fe-6%Co	[134]
W-8.7%(Ni-Fe-Cu)	[80]
W-5%Ni-15%Cu	[3]
W-5%Ni-25%Cu	[135]
W-10%Ni-10%Cu	[135]
W-10–40%Cu	[4,6,136–138]
W-20%($Cu_{10}Sn$)	[139]
W-1–12%Ta	[8,140–143]
W-14.8%Ta-17.2%Re	[144,145]
W-5%Nb	[110]
W-14–79%Fe	[73,146]
W-50%Mo	[147]
W-12.85%Cr	[86]
W Matrix Composites	
W-0.5–2.5%TiC	[148,149]
W-0.5%ZrC	[120]
W-5%TaC	[150]
W-5%La_2O_3	[44]
W-2%CeO_2	[44]
W-1.75%La_2O_3-0.12%ZrO_2-0.12%Y_2O_3	[44]
Y_2O_3	[42,61]

A comprehensive review of the impact of alloying elements and ceramic particles on mitigating the challenges associated with the AM of W-based materials, as well as their microstructure and mechanical properties (Table 8), reveals several key purposes for these additions. Alloying elements and ceramic particles are introduced for the following purposes: to lower the melting point of W to enhance densification, for the modification of laser absorption, reduction in thermal conductivity of W, and reduction in melt viscosity to enhance densification, to form binding phases to enhance densification, improve grain boundary strengthening, modify grain structure, enhance grain boundary cohesion and improve pore distribution to enhance mechanical properties and to mitigate microcracking, to strengthen through secondary dispersed or precipitated phases to restrict grain boundary movement and enhance mechanical properties to mitigate microcracking and enhance low-temperature ductility, increasing stacking fault energy, boosting dislocation mobility to improve ductility and lower DBTT, and for other purposes such as self-passivating behavior and the reduction in fuzz formation in the nuclear application.

Table 8. An overview of the impact of alloying elements and dispersion particles on mitigating the challenges of the AM of W-based materials and on their microstructure and mechanical properties.

Effects on W-AM	Alloying Element(s)	Specifics/Notes	Ref
Enhanced Densification	Ni, Fe, Co, Mo, Re, Ta, etc.	Due to the lower melting points of alloying phases densification is promoted. The melting of Ni, Fe, and Co enhances the dissolution of W particles.	[19,41,74,86,130,134–138]
	Cu	Occurs due to the role of alloying elements in forming binding phases. Good wetting allows the molten Cu to spread and adhere strongly to the W particles. Surface smoothing also contributes to a denser final structure.	[4,6]
	Ni, Cu, Ta, Fe, Mo	Modifications in laser absorption, melt viscosity, and fluidity. Ni works better than Cu. Ta increases recrystallization temperature and decreases thermal conductivity of WNi and Fe for improves wettability of W.	[4,6,8,74,130,135–138,140–143]
Microcrack Mitigation	Re, Ta, Nb, Mo, Cr, Ti, Ir	Re lowers the DBTT of W alloys, significantly reducing their tendency to crack during AM processes. Ta forms a special cellular structure that keeps tiny air pockets (nanopores) separated from each other. This reduces the overall tendency of the material to crack. Ta also oxidizes more easily than W during the building process. As a result, fewer nanopores form in the final W alloy, and the risk of microcracks is reduced. Nb improves the intergranular bonding of the alloy via solid solution strengthening. Cr forms the Cr-rich Cr-W phase and causes grain refinement. Ti (in conventional manufacturing of W) gives rise to a heterogeneous chemical distribution to prevent the nanostructured microstructure from coarsening. Ir (proved in conventional manufacturing of W) strengthens grain boundary cohesion, optimizes dislocation mobility, and reduces cracking. Adding some of these elements prevents the formation of tiny cracks (microcracks) in the building process. In general, W solidifies first, causing stress within the material. Alloying elements helps lessen this stress by essentially acting as a buffer. They also result in grain refinement of W, improved grain boundary strengthening, modified grain structure, and better pore distribution.	[4,44,48,86,110,127–129,141–144,152,153]
	Re	W, when alloyed with Re, showed a reduced DBTT and increased low-temperature ductility. This improvement is attributed to Re's high solubility in W and its capability as a solution hardener. The other benefits include the enhancement of grain boundary cohesion, boosting dislocation mobility and lowering DBTT, and reducing embrittlement.	[127–129]
	TiC, ZrC, TaC, Y_2O_3, La_2O_3, etc.	Y_2O_3 leads to more LAGBs and reduces cracking. W–Y_2O_3 has a better low-temperature ductility and suppresses cracking. La_2O_3 not only bolsters radiation resistance but also provides an indirect solution to these interstitial contamination challenges. ZrC nanoparticles decrease crack density due to finer grains and increased grain-boundary length. They also capture oxygen to form ZrO_y, reducing embrittlement. TaC supports the formation of W_2C phases within the material, further increasing the overall strength and crack resistance. It also reacts with any oxygen present, forming TaO_x, effectively removing oxygen from the system. TiC enhances hardness.	[42,47,120,141,150]
Mechanical Properties Adjustments	Fe	Strengthening by the formation of secondary or precipitated phases during AM processes. The non-equilibrium solidification between Fe and W allows for the potential formation of intermetallic phases like Fe_7W_6 and Fe_2W, improving the hardness.	[73,144,146]
	Ni	Tensile strength and fracture toughness enhancement.	[74,130]
	Ti, Ir, Hf	Increase in recrystallization temperatures. Potent carbide formers mitigate the adverse effects of carbon impurities on the alloy's mechanics. Elevate the recrystallization temperatures and improve W's low-temperature ductility and high-temperature strength. Ir can improve mechanical properties by increasing grain boundary cohesion.	[152–156]
	Ta, Nb, and other group V and IV transition metals	Improvement of intrinsic ductility of W alloys. Ta can increase hardness. Solid solution strengthening by Nb.	[110,141–144]
Others	Ta	Fuzz formation reduction, especially during He plasma irradiation.	[140]
	Cr and Y	Self-passivation behavior for high-temperature applications.	[157]

In the subsequent sections, we will closely examine some of these alloying strategies.

4.1. Alloying for Enhanced Densification

Alloying with metals that have lower melting points is a strategy to decrease the overall melting point of W. But it is important for this to be carried out in moderation to ensure that the intrinsic high-temperature properties of the W alloys are retained [19,41]. W alloys behave differently when melted compared to pure W. The alloying elements can form phases with lower melting points than pure W, aiding in the binding and dissolution of W particles [4]. These elements can also affect melt viscosity and fluidity during AM, especially in SLM. Alloying elements can lead to unique solidification behaviors, leading to improved grain structures, and reduced stresses, among other benefits. Overall, W alloys generally exhibit superior AM part quality compared to the use of pure W [60,61]. A major consequence observed is the enhanced densification of the W matrix. Most alloying elements, including Ni, Fe, Co, Cu, and Re contribute to this phenomenon. The basis for this enhanced densification is the lower melting points of these alloying elements compared to W and the possibility of melting at lower energy inputs and faster scan rates. In the context of the SLM technique, the melting of Ni, Fe, and Co has been noted to expedite the dissolution of W particles [134,135], leading to a more consolidated alloy [6,130,134,135]. Cu, as an alloying element, has been studied to reveal its distinct role in W densification during AM [4,6,136–138]. The significance of Cu lies in its ability to form binding phases with W. The molten state of Cu ensures the effective wetting of W particles, while W particles concurrently undergo surface smoothing. Both these effects foster the increased densification of the W-Cu composite [4].

The alloying elements Ni, Cu, and Ta introduce significant changes in the AM process by altering parameters such as laser absorption, melt viscosity, and fluidity [4,6,8,74,130,135–138,140–143]. Among these alloying elements, Ni demonstrates superior efficacy over Cu in such modifications, primarily due to its lower thermal conductivity [42]. The use of Ta comes with its benefits, including the elevation of the recrystallization temperature and a decrement in thermal conductivity [4,130]. Re plays a multifaceted role in the AM of W alloys, one of which is its contribution to densification. The addition of Re allows for parts with higher relative density to be achieved (relative to pure W) (Figure 16a) [127]. The addition of alloying elements such as Cu, Ni, Fe, Mo, and Cr can facilitate higher densification in AM W materials (Figure 16b). Obtaining a relative density of more than 95% in SLM-fabricated unalloyed W is usually difficult, but precise scanning parameter adjustments, preheating, and the use of optimized starting materials have resulted in achieving relative densities over 98%. The EBM technique for unalloyed W fabrication has been shown to achieve relative densities up to 99.5%.

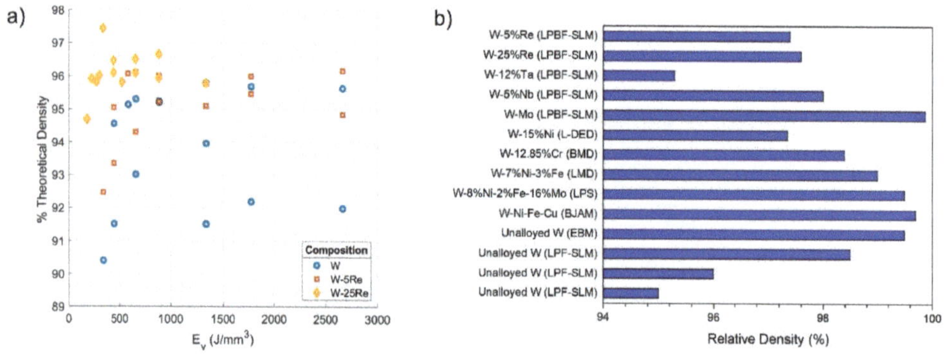

Figure 16. (**a**) Achieved relative (% of theoretical) desnity for SLM-manufactured W-5%Re, W-25%Re alloys, and unalloyed W versus energy density [127], and (**b**) obtained relative density of alloyed and unalloyed W fabricated via different AM techniques.

4.2. *Alloying and Ceramic Dispersions for Microcrack Mitigation*

4.2.1. Alloying Elements

Alloying is a very useful strategy for reducing interstitial contamination. The introduction of specific alloying elements in W during AM has been shown to substantially impact the microcrack mitigation process, which is achieved through mechanisms such as grain structure modification, improved pore distribution, grain boundary strengthening, stress reduction, a decrease in DBTT, and reduced nanopore formation. The addition of Ta induces a cellular structure that spreads the nanopore distribution, culminating in an 80% reduction in cracking [48,143] (Figure 17). Furthermore, the role of Ta in reducing oxygen sensitivity during solidification, courtesy of its in situ oxidation, is essential in alleviating microcracks in the W alloy [48,141–143].

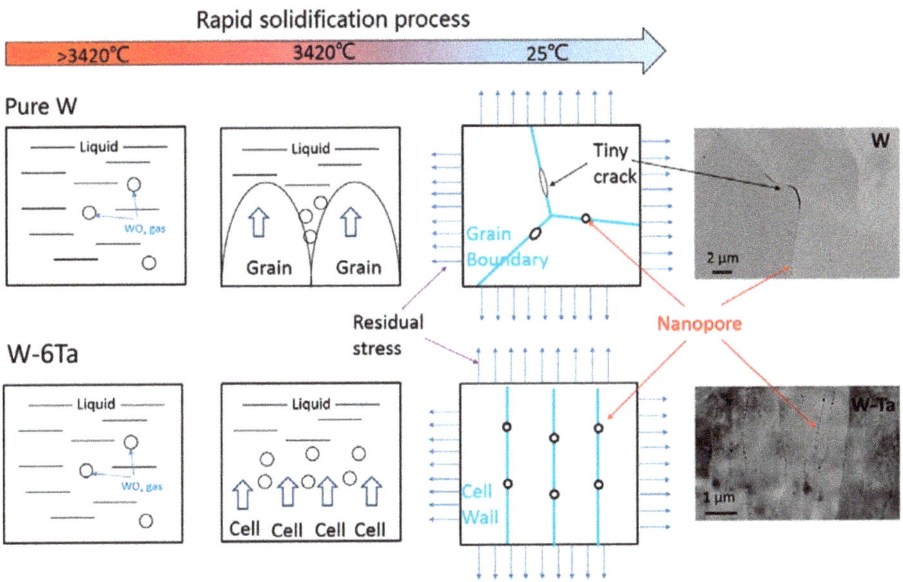

Figure 17. Schematic illustration of how adding Ta affects the formation of nanopores during SLM of W and W-Ta alloys [48].

Nb is beneficial as an alloying element due to solid solution strengthening, which strengthens intergranular bonding to provide an effective countermeasure against microcracking [110]. Alloying W with Re offers promising outcomes. Apart from its contribution to enhancing densification, the resultant alloys show enhanced grain boundary cohesion and an increased mobility of screw dislocations, significantly altering their mechanical behaviors. The addition of Re significantly reduces the ease at which W alloy cracks during the manufacturing process due to the significant lowering of the DBTT [127–129]. Eckley et al. [127] used EBSD maps and optical microscopy images to show significant cracking in pure W parts fabricated by SLM (Figure 18). Cracking increased with density and followed scan tracks, branching at 45-degree angles, and longitudinal cracking traced columnar grain boundaries in the build direction (Figure 18a,b). The introduction of 5 wt.% Re failed to effectively mitigate the observed cracking (Figure 18c,d). However, an increase in Re content to 25 wt.% resulted in a significant reduction in both surface and longitudinal cracking, and the characteristic horizontal and 45-degree angle cracking behaviors were notably minimized (Figure 18e,f). This improvement can be attributed to the lowered DBTT of W due to Re alloying, decreasing brittleness during cooling. Despite these enhancements, some cracking persisted. This study shows Re alloying to be effective in enhancing the crack resistance of W-AM. Notably, conventional fabrication has demonstrated the profound influence of Re, with W alloys containing 26 wt.% Re exhibiting a DBTT as low as −101 °C [60,155,156].

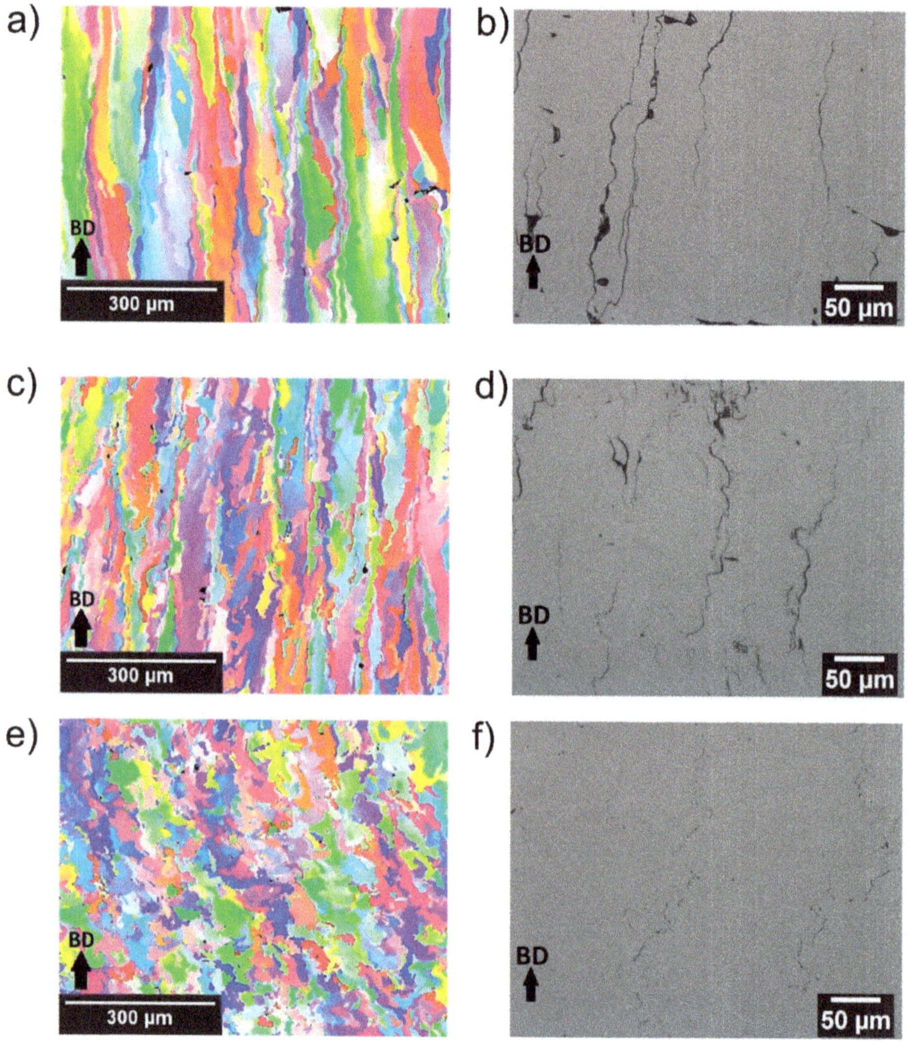

Figure 18. EBSD maps and optical microscopy images of W cubes made using SLM with different Re contents: (**a**,**b**) pure W cubes, (**c**,**d**) W with 5 wt.% Re cubes, and (**e**,**f**) W with 25 wt.% Re cubes. The laser power, hatch spacing, and scan speed were all kept constant at 200 W, 50 μm, and 400 mm/s, respectively. The build direction is also indicated in the image [127].

The inherent characteristics of alloying elements relieve stress by compensating for the shrinkage experienced as W solidifies (a direct consequence of its elevated melting point). This phenomenon plays a direct role in minimizing microcracking [110]. Further contributions from Cr result in grain refinement, indirectly aiding in crack mitigation. Cr induces a Cr-rich Cr-W phase to optimize grain refinement [86]. In conventional fabrication techniques of W-based alloys, adding Ti as an alloying element creates a heterogeneous chemical distribution, stalling the coarsening of nanostructured microstructures [152], whereas adding Ir strengthens grain boundary cohesion, optimizes dislocation mobility, and subsequently deters crack formation [153]. These alloying additions (Ti and Ir) may be beneficial in materials design for AM as well.

4.2.2. Ceramic Dispersions (W Composites)

Incorporating ceramic particles in W has been shown to enhance composite resistance to cracking. W's challenges with oxide retention and the power balling effect are clear indicators of oxidation difficulties, but solutions are on hand. The integration of La_2O_3 not only bolsters radiation resistance but also provides an indirect solution to the problem of interstitial contamination. Innovations like nanopore segregation have been introduced, inducing and identifying crack initiation. Achieving a high theoretical density range of 96–98.5%, combined with the optimal conditions characterized by elevated laser energy density and low oxygen levels, represent meaningful advancements in countering these challenges.

W-Y_2O_3 composites offer a refined grain size and commendable low-temperature ductility [42]. Based on EBSD grain size, and grain misorientation distribution studies of W-Y_2O_3 composites, the use of nano-sized Y_2O_3 particles leads to grain refinement and a greater occurrence of LAGBs. The concomitant reduction in HAGBs (which are more susceptible to cracking) markedly reduces cracking in the resulting W composites (Figure 19) [42]. Introducing micron-sized Y_2O_3 dispersions into W does not lead to grain refinement, although this increases the fraction of LAGBs (Figure 19c,d). However, when nano-sized Y_2O_3 dispersions are used in the composite, the fraction of LAGBs and grain size are both reduced (Figure 19e,f).

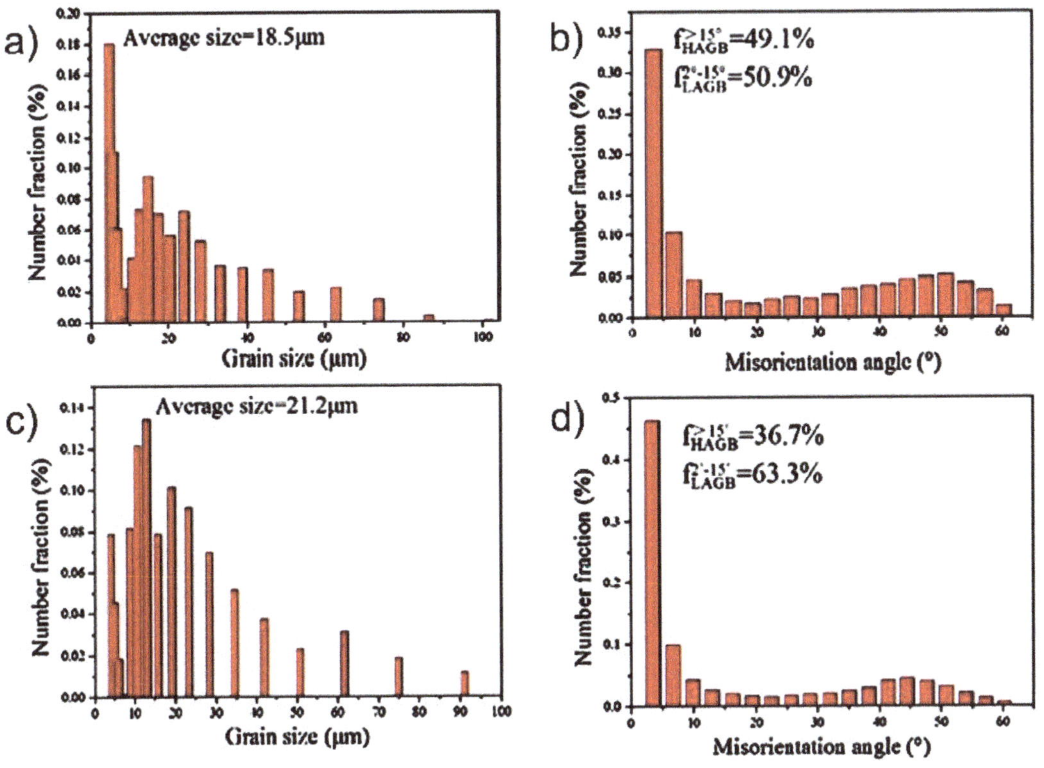

Figure 19. *Cont.*

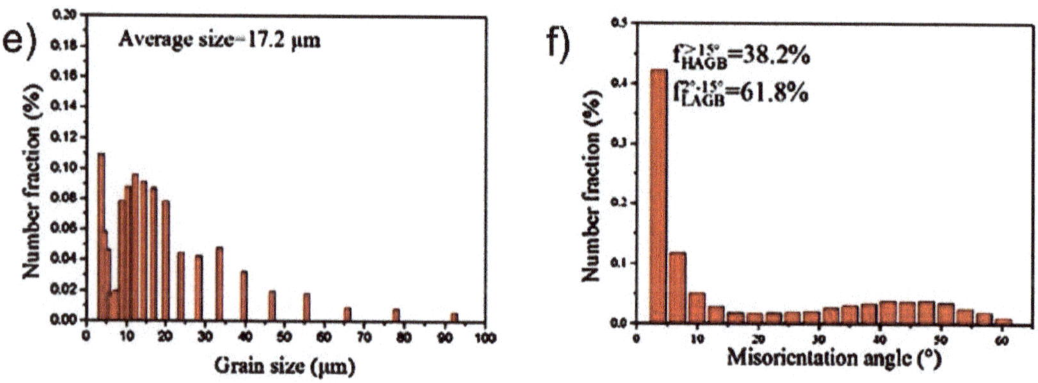

Figure 19. EBSD grain size distribution for SLM-fabricated samples: (**a**,**b**) pure W, (**c**,**d**) W-Y_2O_3 (micron-sized), and (**e**,**f**) W-Y_2O_3 (nano-sized) [42].

Li et al.'s study identified zirconium carbide (ZrC) as a crucial ceramic particle [120]. They showed that W-ZrC samples showed a remarkable reduction in crack density compared to pure W, showing fewer and more dispersed cracks in W-ZrC (Figure 20a,b). The images also reveal that, while pure W had long and densely distributed cracks, W-ZrC had shorter and sparser cracks (Figure 20c,d). This improved crack resistance is attributed to secondary-phase nanoparticles in W-ZrC, which refined the grain structure and captured oxygen impurities to hinder crack propagation. Achieving a fully uniform grain structure would require further efforts to ensure an even ZrC distribution. Li et al.'s study highlighted ZrC's role in reducing crack density, promoting finer grains, and extending grain boundaries. Their bright field TEM, HRTEM (high-resolution transmission electron microscopy), and high-angle annular dark field images (Figure 21a–c) along with EDS chemical maps (Figure 21d–h) at various magnifications, and the SAEDP (selected-area electron diffraction pattern) of selected nanoparticles (Figure 21f) provide compelling evidence of ZrC's capacity to capture and bind oxygen to form ZrO_y particles. This mechanism is a highly effective means of mitigating embrittlement [60,120].

Tantalum carbide (TaC) particles in W also play an important role in crack mitigation. This is achieved via the dual action of initiating in situ W_2C phase formations and promoting oxygen consumption through TaO_x formation, which collectively suppresses cracks [141,150]. Titanium carbide (TiC), while predominantly increasing hardness, may indirectly improve crack resistance due to the improved mechanical properties of the composite [42,47,141,150]. Hafnium carbide (HfC) dispersion in conventionally fabricated W alloys has provided excellent thermal stability and tensile properties of composites [154,155]. This can be a useful feature to have for materials design in the AM of W as well.

4.3. Alloying and Ceramic Dispersions for Mechanical Properties Adjustments

Besides mitigating the problems discussed above, alloying confers additional benefits. Fe integration in W alloys during AM strengthens the resulting material due to the formation of secondary or precipitated phases. This is due to the possible peritectic reaction between Fe and W during non-equilibrium solidification, leading to the emergence of intermetallic phases such as Fe_7W_6 and Fe_2W. This formation significantly increases the hardness of the W–Fe alloy [60,73,146]. Ni addition enhances the alloy's tensile strength and fracture toughness [74,130]. Nb provides solid solution strengthening and similarly improves the mechanical properties of the alloy [110]. The inclusion of Ta and other group V and IV transition metals improves the intrinsic ductility of W alloys. In particular, Ta not only enhances ductility but also increases the hardness of resultant alloys [141–144].

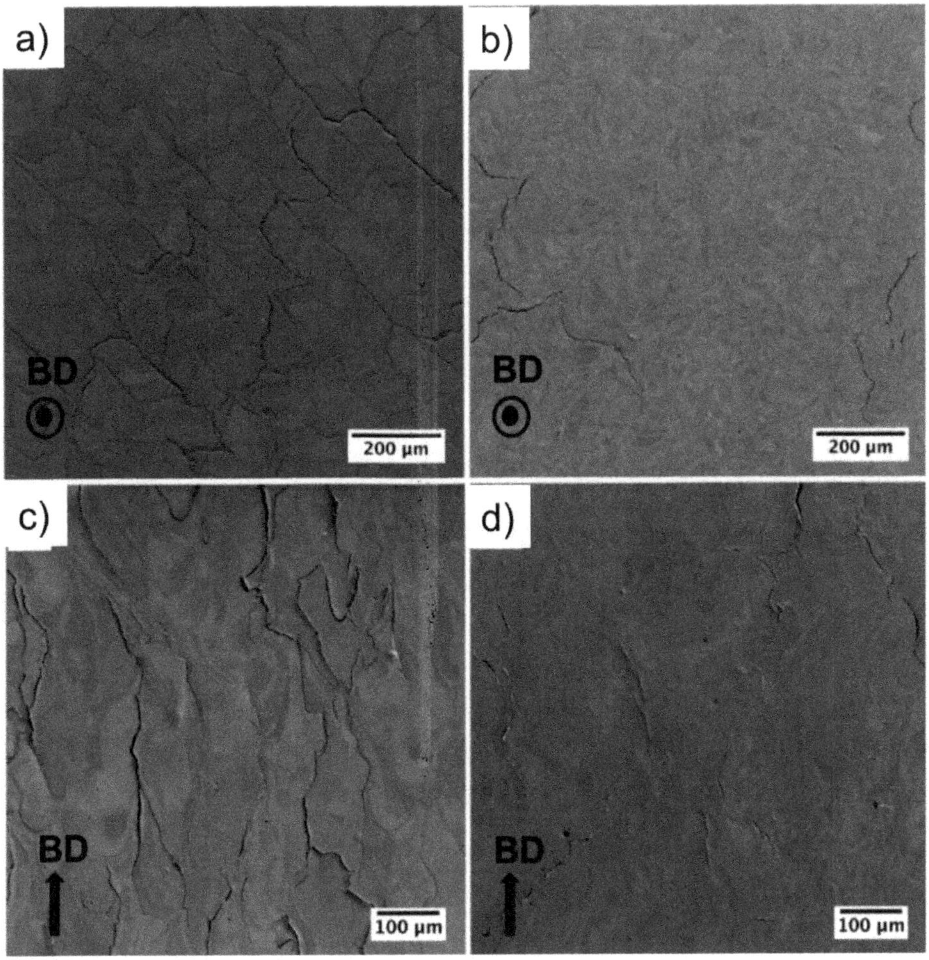

Figure 20. SEM images of SLM-fabricated pure W and W-ZrC nanocomposites: (**a**,**c**) the top and side surface of pure W cubes, and (**b**,**d**) the top and side surface of W-ZrC cubes. The build direction (BD) is also indicated in the image [120].

The incorporation of Ti, Ir, and Hf in conventionally fabricated W alloys achieves multiple objectives as they influence the mechanical properties in different ways. They increase recrystallization temperatures and, as potent carbide formers, mitigate the adverse effects of carbon impurities on the alloy's mechanics, elevate the recrystallization temperatures, and improve W's low-temperature ductility and high-temperature strength [152–156].

Since unalloyed W exhibits limited plasticity, studies have predominantly focused on reporting compression test results. Conversely, a wealth of tensile test results is available for W alloys processed using AM (Figure 22). Alloying elements play a pivotal role in enhancing the tensile properties of W, notably its elongation, which is otherwise limited or non-existent in unalloyed W. The strength and elongation of W alloys vary depending on compositions (Figure 22a,b). Wang et al. [74] reported a W–15% Ni alloy with a UTS of 560 MPa and 3.75% elongation, while W–7% Ni–3% Fe showed a UTS of 1037 MPa and 3.5% elongation [71], and W–4.6%Ni–2.4% Fe fabricated via powder extrusion 3D printing and sintering showed a UTS of 1040 MPa and remarkably high (20.7%) elongation

(Figure 22a,b) [133]. Ultrashort-time laser liquid phase sintering processing for W heavy alloy resulted in a UTS of 1374 MPa, and 8.5% elongation (Figure 22a,b) [85].

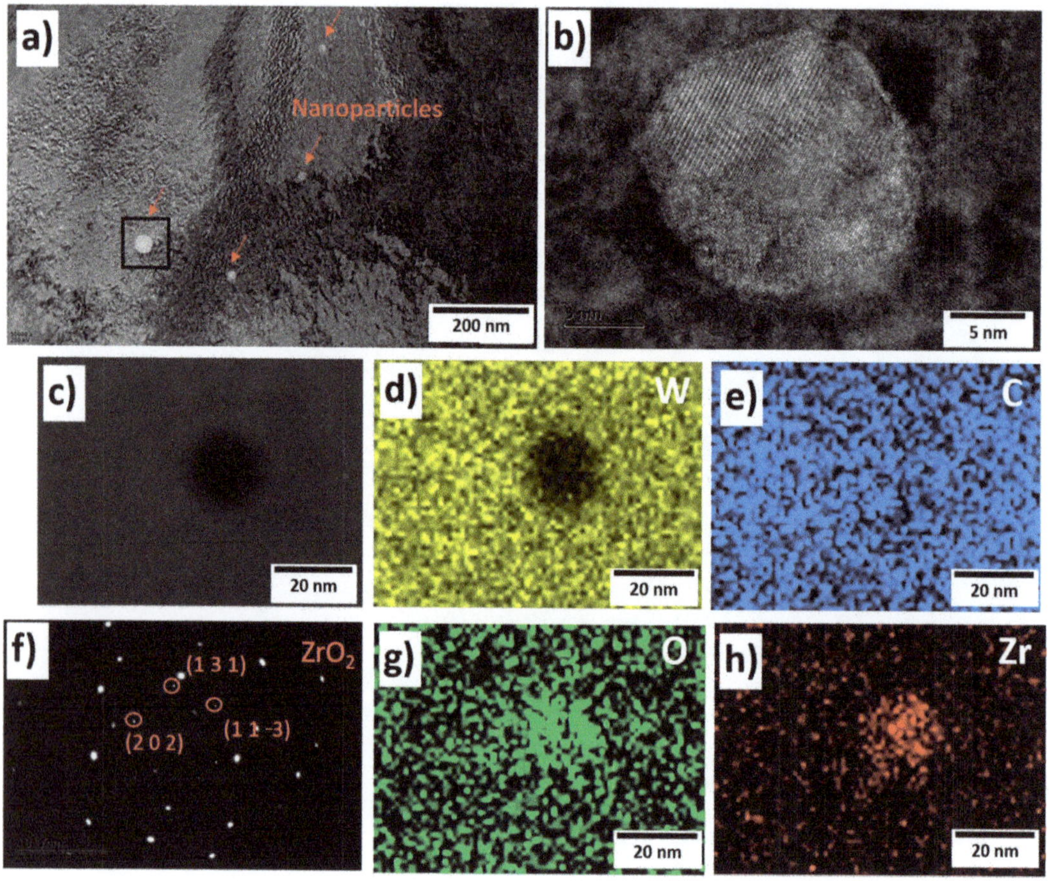

Figure 21. Analysis of the nanoparticle's dispersion in the SLM-fabricated W-ZrC nanocomposites: (**a**) distribution of nanoparticles, (**b**) the high-resolution transmission electron microscope of the selected nanoparticle, (**c**) the high-angle annular dark field image, (**d–h**) its EDS elemental maps, and (**f**) the selected-area electron diffraction pattern of the selected nanoparticle [120].

Figure 22c depicts the microhardness of AM-fabricated W-based materials. Li et al. [145] examined the variation in Vickers microhardness due to VED in the context of W-3%Ta alloy. Operating at a substrate preheating temperature of 150 °C and a 67° layer angle, they reported a range of microhardness values from 493.8 to 535.6 Hv. This range correlated with VED values spanning 192 to 1000 J/mm^3, with an average microhardness of 505.7 Hv. An upward trend in relative density was concurrently observed as VED values increased. Conversely, Bose et al. [86] reported the highest microhardness achievement of 966 Hv for a W-Cr component produced through binder jetting and sintering at 1500 °C for 1 h, demonstrating a striking contrast to the above trends. Zhang et al. [130] and Yan et al. [136] showed a contrasting narrative, showcasing a decline in hardness with an incremental influx of Ni in W-Ni and W-Ni-Cu systems. In a related study, Zhang et al. [130] highlighted a gradient of hardness reduction with the distance from the base of SLM-fabricated samples, plummeting from 810 to 300 Hv for W-10%Ni and from 490 to 250 Hv for W-40Ni.

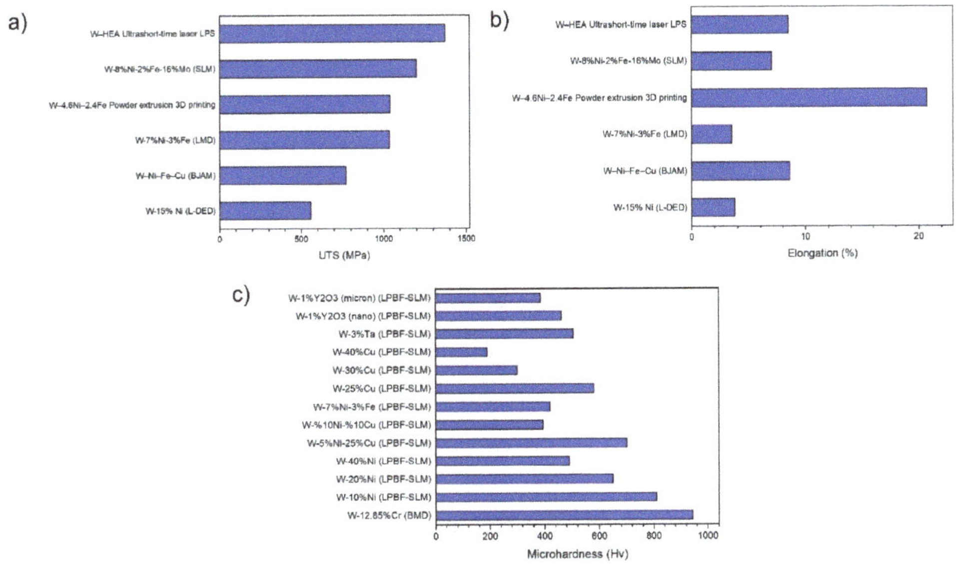

Figure 22. Mechanical properties of AM-fabricated W alloys: (**a**) ultimate tensile strength, (**b**) elongation, and (**c**) microhardness.

Ivekovic et al. [132] noted a decline in hardness post-heat treatment of SLM-fabricated W-7Ni-3Fe samples, which diverged from the augmentation in hardness expected. Conversely, Wang et al. [135] demonstrated that an escalation in the microhardness of W-Ni-Cu alloys, ranging from 360 to 395 Hv, was linked to increasing VED. This phenomenon is attributed to the formation of finer grain sizes owing to elevated melt pool temperatures. Moreover, Hu et al. [42] brought nano-structural considerations into the equation, underscoring the influence of grain size. They documented an upswing in hardness with the decreasing grain size of Y_2O_3 particles, transitioning from micro- to nano-scale dimensions.

These findings underscore the fact that enhanced and/or adjusted mechanical properties are achievable through alloying in the AM of W-based materials.

4.4. Alloying for Other Specific Purposes

The addition of Ta to W alloys is pivotal for reducing fuzz formation. Ta has demonstrated its capability to reduce fuzz during He plasma irradiation, making these alloys suitable for nuclear applications [140]. Furthermore, W alloys formulated with the Cr-Y combination exhibit self-passivation behavior in conventional fabrication. This intrinsic property makes W alloys integrated with Cr-Y especially useful in high-temperature applications [60,157].

Overall, the addition of alloying elements activates several mechanisms to enhance various properties (Table 8), which are as follows. Alloying elements such as Ta, Re, and Ni can act as grain refiners in W-based alloys. These elements reduce the grain size by inhibiting grain growth during solidification, resulting in a finer and more uniform microstructure [158]. Fine-grained microstructures typically exhibit higher strength and improved toughness due to the grain boundary strengthening mechanism. Alloying elements like Mo and Ti lead to solid solution strengthening [159]. These elements distort the W lattice and impede dislocation motion, thereby enhancing the yield strength and overall mechanical properties of the alloy. This mechanism is particularly beneficial in improving the high-temperature strength of W alloys. Certain alloying elements can form stable second phases or precipitates within the W matrix [160]. For example, carbide-forming elements such as carbon C can produce W carbides (WC) that enhance the hardness and wear resistance of the alloy. These second phases also contribute to the overall strength by hindering dislocation movement. Alloying with elements such as Cr and Si can improve the oxidation resistance of W alloys [161]. These elements form protective oxide layers on the surface, which act as barriers to further oxidation, thereby enhancing the performance of W components in high-temperature environments [162]. Other elements like Re improve the ductility of W alloys by reducing the DBTT [128]. This brings about a more ductile fracture behavior and better toughness at lower temperatures, which is crucial for applications where mechanical shock or impact resistance is required.

5. Conclusions

Overall, the proposed solutions, which include optimizing process parameters, refining powder characteristics, implementing post-processing treatments, and employing alloying elements, are essential to address the fundamental issues of densification and mechanical properties in W-based materials. Here, we provide a detailed discussion of how each approach contributes to improving densification and mechanical properties, supported by the relevant literature.

1. **Powder Characteristics**

Particle Size and morphology: Utilizing W powder with a well-controlled particle size distribution and spherical morphology enhances flowability, packing density, and uniform layer formation during the built cycle. These characteristics are crucial for achieving high relative density and consistent microstructures, as evidenced by improved mechanical properties in various studies.

Powder purity: High-purity W powder minimizes contaminants that can cause defects and degrade mechanical properties. Ensuring powder purity is fundamental for producing high-quality components with superior mechanical integrity.

2. **Process Parameter Optimization**

Laser power and scan speed: By carefully adjusting the laser power and scan speed, we can ensure adequate energy input for the complete melting of tungsten powder while avoiding defects such as keyhole porosity and excessive vaporization. Optimal control of these parameters results in improved densification and reduced residual stresses. A maximum density of 99.2% in SLM and 100% in EBM was demonstrated by adjusting process parameters.

Volumetric energy density (VED): Our study emphasizes the critical role of VED in balancing energy input to promote full melting and minimize defects. Extremely high VED can introduce porosity and fusion issues, while insufficient VED can result in incomplete melting. Optimizing VED is essential for achieving high relative density and enhancing mechanical properties.

3. **Thermal Management Strategies**

Controlled atmosphere: Using an inert gas atmosphere of argon or nitrogen during the AM process prevents oxidation and contamination, preserving the material's purity and mechanical properties. While the majority of studies used argon, processing in a nitrogen atmosphere could enhance the microstructure and mechanical performance of tungsten parts.

Substate preheat: Utilizing a heated substrate plate in PBF processes is necessary to control cooling rates and mitigate thermal gradients such as built-up thermal residual stresses within the material, and improves material formation and properties. In most cases, the substrate is preheated in the range of 500 °C to 1000 °C in SLM and up to 1800 °C in EBM to eliminate crack formation.

Optimized scan strategies: Utilizing optimized scanning strategies such as alternating scan directions, island scanning, and contour scanning can distribute heat more evenly across the build area, reducing localized thermal stresses. These strategies help in achieving uniform thermal profiles and minimizing warping and distortion.

4. Alloying Strategies

The introduction of alloying elements: Adding alloying elements such as Re, Ta, and Hf can significantly enhance the mechanical properties and processability of W. Re, for instance, is known to improve ductility and reduce the DBTT, refine the grain structure, and improve the toughness of W alloys.

The introduction of ceramic particles: carbide formers (e.g., TiC, ZrC) can be introduced, which act as grain refiners and strengthen the W matrix through dispersion strengthening to improve the hardness and wear resistance of W components.

5. Post-Processing Treatments

Heat treatment: Post-process heat treatments are vital for relieving residual stresses, promoting grain growth, and enhancing mechanical properties. Techniques such as annealing reduce micro-cracks and improve ductility.

Hot isostatic pressing (HIP): HIP applies high pressure and temperature to further densify the components, close remaining pores, and refine the microstructure. This results in significantly improved mechanical strength and integrity.

6. Future Directions

Although W-AM holds great promise for producing high-performance parts, its widespread adoption has been hindered by problems associated with W's inherent material properties and printability challenges. Research efforts into the AM of W should be prioritized through the fine-tuning of process parameters, as well as microstructural and material design. We believe in the possibilities of bringing W-AM into a new era of precision, sustainability, and widespread applicability through advancements in atomic-level understanding, and thermodynamic modeling, as well as through harnessing the power of data analytics.

1. Atomic-Level Understanding: The Key to Tailored Properties
   Further studies are needed to understand W at the atomic level, focusing on the local atomic structure and short-range order chemistry. Improving our understanding of the inherent atomic arrangement in W from the microscopic perspective will enable us to unlock new possibilities for tailored material properties and optimized performance.
2. Thermodynamic Modeling: A Roadmap for Alloy Design
   Integrating the thermodynamic calculation of phase diagram modeling into alloy design will play a key role in predicting phase equilibria and allow researchers to design W alloys with better control over composition and phase transitions. This thermodynamic approach ensures a systematic exploration of the uncharted alloy space, allowing materials with tailored properties to be developed. This provides insights into process optimization through thermal gradients, melt pool dynamics, and solidification rates, which are essential for minimizing porosity and achieving uniform microstructures.
3. Composite Design: Improved Innovative Composite
   Researchers can seek to synergize W with other advanced ceramics, to create composites that exhibit superior mechanical, thermal, and chemical properties. This approach opens avenues for tailoring W composites for specific applications across diverse industries.
4. Harnessing the Power of Data Analytics: Machine Learning and in situ Monitoring
   Researchers can leverage machine learning and statistical techniques to identify patterns and correlations which can help in the optimization of processing parameters. Multiphysics and multi-scale computational modeling are essential for establishing process–structure–property relationships in metal AM [163]. This paves the way for the development of a digital twin for W-AM. Further studies can also develop and implement advanced in situ process monitoring techniques during AM. The data generated during the real-time monitoring of temperature, microstructure evolution, and defect formation will provide valuable insights, enabling adaptive control strategies to enhance the quality and repeatability of W components.

5. Sustainable Manufacturing: the Hidden Cost of the Oxidation of W
The high oxygen affinity of W and its influence extends beyond AM-fabricated parts. The high temperatures and heating profiles employed during AM processes can lead to the inadvertent oxidation of adjacent powder beds. This oxidation significantly affects the reusability of W powder, potentially increasing waste and negating the environmental and cost benefits of AM. Future research efforts should prioritize mitigating this hidden cost by a comprehensive understanding of W oxidation to minimize the impact of oxidation and maximize powder reusability.

Author Contributions: Conceptualization, M.Z., Y.T. and C.M.; methodology, M.Z., M.S. and N.W.; validation, Y.X., M.Z. and L.T.; formal analysis, M.Z., C.M. and M.S.; resources, M.Z., N.W. and Y.X.; data curation, S.R., X.C. and L.T.; writing—original draft preparation, M.Z., M.S., Y.T., S.R. and C.M.; writing—review and editing, M.Z., M.S., X.C., N.W., Y.X., S.R. and K.X.K.; visualization, M.Z., M.S. and K.X.K.; supervision, M.Z., Y.T. and M.S.; project administration, M.Z., C.M. and Y.T.; funding acquisition, C.M. and M.Z. All authors have read and agreed to the published version of the manuscript.

Funding: This research was funded by Zhejiang Province grant number: 2023C01SA393195, and Wenzhou City grant number: ZG2023036.

Data Availability Statement: No further data are available. All the data pertaining to this work are already included in the article.

Acknowledgments: The authors from Wenzhou Hongfong Electrical Alloy Co Ltd and its subsidiary acknowledge these generous financial supports.

Conflicts of Interest: Authors Mehrdad Zarinejad, Chengfa Mu, Lintao Tian, and Xiaotong Chen were employed by the company Wenzhou Hongfeng Electrical Alloy Co., Ltd. Authors Nian Wang, and Yonglong Xu were employed by the company Wenzhou Hongfeng Alloy Co., Ltd. The remaining authors declare that the research was conducted in the absence of any commercial or financial relationships that could be construed as a potential conflict of interest.

Abbreviations

List of abbreviations used in the manuscript along with their full forms.

Abbreviation	Term
AM	Additive manufacturing
BMD	Bound metal deposition
DED	Direct energy deposition
DBTT	Ductile-to-brittle transition temperatures
EBM	Electron beam melting
HAGBs	High-angle grain boundaries
LED	Line (laser) energy density
LPS	Liquid phase sintering
LAGBs	Low-angle grain boundaries
PBF	Powder bed fusion
W	Tungsten
VED	Volumetric energy density

References

1. Rieth, M.; Dudarev, S.L.; de Vicente, S.M.G.; Aktaa, J.; Ahlgren, T.; Antusch, S.; Armstrong, D.E.J.; Balden, M.; Baluc, N.; Barthe, M.-F.; et al. Recent progress in research on tungsten materials for nuclear fusion applications in Europe. *J. Nucl. Mater.* **2013**, *432*, 482–500. [CrossRef]
2. Xie, J.; Lu, H.; Lu, J.; Song, X.; Wu, S.; Lei, J. Additive manufacturing of tungsten using directed energy deposition for potential nuclear fusion application. *Surf. Coat. Technol.* **2021**, *409*, 126884. [CrossRef]
3. Katoh, Y.; Snead, L.L.; Garrison, L.M.; Hu, X.; Koyanagi, T.; Parish, C.M.; Edmondson, P.D.; Fukuda, M.; Hwang, T.; Tanaka, T.; et al. Response of unalloyed tungsten to mixed spectrum neutrons. *J. Nucl. Mater.* **2019**, *520*, 193–207. [CrossRef]
4. Wang, G.; Qin, Y.; Yang, S. Influence of Ni additions on the microstructure and tensile property of W-Cu composites produced by direct energy deposition. *J. Alloys Compd.* **2022**, *899*, 163272. [CrossRef]

5. Engwall, A.M.; Shin, S.J.; Bae, J.; Wang, Y.M. Enhanced properties of tungsten films by high-power impulse magnetron sputtering. *Surf. Coat. Technol.* **2019**, *363*, 191–197. [CrossRef]
6. Su, S.; Lu, Y. Densified W Cu composite fabricated via laser additive manufacturing. *Int. J. Refract. Met. Hard Mater.* **2020**, *87*, 105122. [CrossRef]
7. Oponowicz, A.; Marciszko-Wiąckowska, M.; Baczmański, A.; Klaus, M.; Genzel, C.; Wroński, S.; Kollbek, K.; Wróbel, M. Gradient of Residual Stress and Lattice Parameter in Mechanically Polished Tungsten Measured Using Classical X-rays and Synchrotron Radiation. *Metall. Mater. Trans. A* **2020**, *51*, 5945–5957. [CrossRef]
8. Iveković, A.; Omidvari, N.; Vrancken, B.; Lietaert, K.; Thijs, L.; Vanmeensel, K.; Vleugels, J.; Kruth, J.-P. Selective laser melting of tungsten and tungsten alloys. *Int. J. Refract. Met. Hard Mater.* **2018**, *72*, 27–32. [CrossRef]
9. Jia, Y.; Chang, S.; Du, X.; Guo, S. Corrosion Performance of Commercial Alloys and Refractory Metals in Conditions for Electrorefining of Spent Nuclear Fuels. *Crystals* **2023**, *13*, 817. [CrossRef]
10. Wei, Q.; Ramesh, K.T.; Schuster, B.E.; Kecskes, L.J.; Dowding, R.J. Nanoengineering opens a new era for tungsten as well. *JOM* **2006**, *58*, 40–44. [CrossRef]
11. Zinkle, S.J.; Ott, L.J.; Ingersoll, D.T.; Ellis, R.J.; Grossbeck, M.L. Overview of materials technologies for space nuclear power and propulsion. *AIP Conf. Proc.* **2002**, *608*, 1063–1073. [CrossRef]
12. Mani, M.; Madan, J.; Lee, J.H.; Lyons, K.W.; Gupta, S.K. Sustainability characterisation for manufacturing processes. *Int. J. Prod. Res.* **2014**, *52*, 5895–5912. [CrossRef]
13. Valivullah, L.; Mani, M.; Lyons, K.W.; Gupta, S.K. Manufacturing Process Information Models for Sustainable Manufacturing. In *Volume 1: Materials; Micro and Nano Technologies; Properties, Applications and Systems; Sustainable Manufacturing*; American Society of Mechanical Engineers: New York, NY, USA, 2014. [CrossRef]
14. Mani, M.; Lyons, K.W.; Gupta, S.K. Sustainability Characterization for Additive Manufacturing. *J. Res. Natl. Inst. Stand. Technol.* **2014**, *119*, 419. [CrossRef] [PubMed]
15. Moghimian, P.; Poirié, T.; Habibnejad-Korayem, M.; Zavala, J.A.; Kroeger, J.; Marion, F.; Larouche, F. Metal powders in additive manufacturing: A review on reusability and recyclability of common titanium, nickel and aluminum alloys. *Addit. Manuf.* **2021**, *43*, 102017. [CrossRef]
16. Mireles, O.; Rodriguez, O.; Gao, Y.; Philips, N. Additive Manufacture of Refractory Alloy C103 for Propulsion Applications. In *AIAA Propulsion and Energy 2020 Forum*; American Institute of Aeronautics and Astronautics: Reston, VA, USA, 2020. [CrossRef]
17. Ren, X.; Liu, H.; Lu, F.; Huang, L.; Yi, X. Effects of processing parameters on the densification, microstructure and mechanical properties of pure tungsten fabricated by optimized selective laser melting: From single and multiple scan tracks to bulk parts. *Int. J. Refract. Met. Hard Mater.* **2021**, *96*, 105490. [CrossRef]
18. Bai, S.; Liu, J.; Yang, P.; Huang, H.; Yang, L.-M. Femtosecond Fiber Laser Additive Manufacturing of Tungsten. In Proceedings of the SPIE 9738-24, San Francisco, CA, USA, 13–18 February 2016; Gu, B., Helvajian, H., Piqué, A., Eds.; SPIE: San Francisco, CA, USA, 2016. [CrossRef]
19. Müller, A.V.; Schlick, G.; Neu, R.; Anstätt, C.; Klimkait, T.; Lee, J.; Pascher, B.; Schmitt, M.; Seidel, C. Additive manufacturing of pure tungsten by means of selective laser beam melting with substrate preheating temperatures up to 1000 °C. *Nucl. Mater. Energy* **2019**, *19*, 184–188. [CrossRef]
20. Feng, F.; Lian, Y.; Wang, J.; Song, J.; Yan, B.; Liu, X. Mechanical Properties and Thermal Shock Performance of High-Energy-Rate-Forged W-1%TaC Alloy. *Crystals* **2022**, *12*, 1047. [CrossRef]
21. Wang, D.-Z.; Li, K.-L.; Yu, C.-F.; Ma, J.; Liu, W.; Shen, Z.-J. Cracking Behavior in Additively Manufactured Pure Tungsten. *Acta Metall. Sin. (Engl. Lett.)* **2019**, *32*, 127–135. [CrossRef]
22. Lassner, E.; Schubert, W.-D. *Tungsten*; Springer US: Boston, MA, USA, 1999. [CrossRef]
23. Mitteau, R.; Missiaen, J.M.; Brustolin, P.; Ozer, O.; Durocher, A.; Ruset, C.; Lungu, C.P.; Courtois, X.; Dominicy, C.; Maier, H.; et al. Recent developments toward the use of tungsten as armour material in plasma facing components. *Fusion. Eng. Des.* **2007**, *82*, 1700–1705. [CrossRef]
24. Vrancken, B.; Ganeriwala, R.K.; Matthews, M.J. Analysis of laser-induced microcracking in tungsten under additive manufacturing conditions: Experiment and simulation. *Acta Mater.* **2020**, *194*, 464–472. [CrossRef]
25. Elsayed, A.H.; Sayed, M.A.; Dawood, O.M.; Daoush, W.M. Effect of Transition Metals Oxides on the Physical and Mechanical Properties of Sintered Tungsten Heavy Alloys. *Crystals* **2020**, *10*, 825. [CrossRef]
26. Wang, L.; Wu, J.; Zhang, D. Properties evolution of additive manufacture used tungsten powders prepared by radio frequency induction plasma. *Int. J. Refract. Met. Hard Mater.* **2017**, *67*, 90–97. [CrossRef]
27. Zi, X.; Chen, C.; Wang, X.; Wang, P.; Zhang, X.; Zhou, K. Spheroidisation of tungsten powder by radio frequency plasma for selective laser melting. *Mater. Sci. Technol.* **2018**, *34*, 735–742. [CrossRef]
28. Gu, D.D.; Meiners, W.; Wissenbach, K.; Poprawe, R. Laser additive manufacturing of metallic components: Materials, processes and mechanisms. *Int. Mater. Rev.* **2012**, *57*, 133–164. [CrossRef]
29. Guo, M.; Gu, D.; Xi, L.; Du, L.; Zhang, H.; Zhang, J. Formation of scanning tracks during Selective Laser Melting (SLM) of pure tungsten powder: Morphology, geometric features and forming mechanisms. *Int. J. Refract Metals Hard Mater.* **2019**, *79*, 37–46. [CrossRef]
30. Wen, S.; Wang, C.; Zhou, Y.; Duan, L.; Wei, Q.; Yang, S.; Shi, Y. High-density tungsten fabricated by selective laser melting: Densification, microstructure, mechanical and thermal performance. *Opt. Laser Technol.* **2019**, *116*, 128–138. [CrossRef]

31. Gokcekaya, O.; Ishimoto, T.; Todo, T.; Wang, P.; Nakano, T. Influence of powder characteristics on densification via crystallographic texture formation: Pure tungsten prepared by laser powder bed fusion. *Addit. Manuf. Lett.* **2021**, *1*, 100016. [CrossRef]
32. Yang, G.; Yang, P.; Yang, K.; Liu, N.; Jia, L.; Wang, J.; Tang, H. Effect of processing parameters on the density, microstructure and strength of pure tungsten fabricated by selective electron beam melting. *Int. J. Refract. Metals Hard Mater.* **2019**, *84*, 105040. [CrossRef]
33. Gokuldoss, P.K.; Kolla, S.; Eckert, J.; Processes, A.M. Electron Beam Melting and Binder Jetting—Selection Guidelines. *Materials* **2017**, *10*, 672. [CrossRef] [PubMed]
34. Klahn, C.; Leutenecker, B.; Meboldt, M. Design Strategies for the Process of Additive Manufacturing. *Procedia CIRP* **2015**, *36*, 230–235. [CrossRef]
35. Vrancken, B.; King, W.E.; Matthews, M.J. In-situ characterization of tungsten microcracking in Selective Laser Melting. *Procedia CIRP* **2018**, *74*, 107–110. [CrossRef]
36. Sidambe, A.T.; Tian, Y.; Prangnell, P.B.; Fox, P. Effect of processing parameters on the densification, microstructure and crystallographic texture during the laser powder bed fusion of pure tungsten. *Int. J. Refract. Metals Hard Mater.* **2019**, *78*, 254–263. [CrossRef]
37. Rebesan, P.; Bonesso, M.; Gennari, C.; Dima, R.; Pepato, A.; Vedani, M. Tungsten Fabricated by Laser Powder Bed Fusion. *BHM Berg-Und Hüttenmännische Monatshefte* **2021**, *166*, 263–269. [CrossRef]
38. Tan, C.; Zhou, K.; Ma, W.; Attard, B.; Zhang, P.; Kuang, T. Selective laser melting of high-performance pure tungsten: Parameter design, densification behavior and mechanical properties. *Sci. Technol. Adv. Mater.* **2018**, *19*, 370–380. [CrossRef] [PubMed]
39. Chen, J.; Li, K.; Wang, Y.; Xing, L.; Yu, C.; Liu, H.; Ma, J.; Liu, W.; Shen, Z. The effect of hot isostatic pressing on thermal conductivity of additively manufactured pure tungsten. *Int. J. Refract. Met. Hard Mater.* **2020**, *87*, 105135. [CrossRef]
40. Field, A.C.; Carter, L.N.; Adkins, N.J.E.; Attallah, M.M.; Gorley, M.J.; Strangwood, M. The Effect of Powder Characteristics on Build Quality of High-Purity Tungsten Produced via Laser Powder Bed Fusion (LPBF). *Metall. Mater. Trans. A* **2020**, *51*, 1367–1378. [CrossRef]
41. Guo, M.; Gu, D.; Xi, L.; Zhang, H.; Zhang, J.; Yang, J.; Wang, R. Selective laser melting additive manufacturing of pure tungsten: Role of volumetric energy density on densification, microstructure and mechanical properties. *Int. J. Refract. Metals Hard Mater.* **2019**, *84*, 105025. [CrossRef]
42. Hu, Z.; Zhao, Y.; Guan, K.; Wang, Z.; Ma, Z. Pure tungsten and oxide dispersion strengthened tungsten manufactured by selective laser melting: Microstructure and cracking mechanism. *Addit. Manuf.* **2020**, *36*, 101579. [CrossRef]
43. Wang, D.; Yu, C.; Zhou, X.; Ma, J.; Liu, W.; Shen, Z. Dense Pure Tungsten Fabricated by Selective Laser Melting. *Appl. Sci.* **2017**, *7*, 430. [CrossRef]
44. Vrancken, B.; Ganeriwala, R.K.; Martin, A.A.; Matthews, M.J. Microcrack mitigation during laser scanning of tungsten via preheating and alloying strategies. *Addit. Manuf.* **2021**, *46*, 102158. [CrossRef]
45. Xiong, Z.; Zhang, P.; Tan, C.; Dong, D.; Ma, W.; Yu, K. Selective Laser Melting and Remelting of Pure Tungsten. *Adv. Eng. Mater.* **2020**, 22. [CrossRef]
46. Zhou, X.; Liu, X.; Zhang, D.; Shen, Z.; Liu, W. Balling phenomena in selective laser melted tungsten. *J. Mater. Process Technol.* **2015**, *222*, 33–42. [CrossRef]
47. Wu, Y. Manufacturing of tungsten and tungsten composites for fusion application via different routes. *Tungsten* **2019**, *1*, 80–90. [CrossRef]
48. Wang, D.; Wang, Z.; Li, K.; Ma, J.; Liu, W.; Shen, Z. Cracking in laser additively manufactured W: Initiation mechanism and a suppression approach by alloying. *Mater. Des.* **2019**, *162*, 384–393. [CrossRef]
49. Braun, J.; Kaserer, L.; Stajkovic, J.; Leitz, K.-H.; Tabernig, B.; Singer, P.; Leibenguth, P.; Gspan, C.; Kestler, H.; Leichtfried, G. Molybdenum and tungsten manufactured by selective laser melting: Analysis of defect structure and solidification mechanisms. *Int. J. Refract. Met. Hard Mater.* **2019**, *84*, 104999. [CrossRef]
50. Yamamoto, T.; Hara, M.; Hatano, Y. Effects of fabrication conditions on the microstructure, pore characteristics and gas retention of pure tungsten prepared by laser powder bed fusion. *Int. J. Refract. Metals Hard Mater.* **2021**, *95*, 105410. [CrossRef]
51. Enneti, R.K.; Morgan, R.; Atre, S.V. Effect of process parameters on the Selective Laser Melting (SLM) of tungsten. *Int. J. Refract. Met. Hard Mater.* **2018**, *71*, 315–319. [CrossRef]
52. Zhang, D.; Cai, Q.; Liu, J. Formation of Nanocrystalline Tungsten by Selective Laser Melting of Tungsten Powder. *Mater. Manuf. Process.* **2012**, *27*, 1267–1270. [CrossRef]
53. Zhang, J.; Gu, D.; Yang, Y.; Zhang, H.; Chen, H.; Dai, D.; Lin, K. Influence of Particle Size on Laser Absorption and Scanning Track Formation Mechanisms of Pure Tungsten Powder During Selective Laser Melting. *Engineering* **2019**, *5*, 736–745. [CrossRef]
54. Deprez, K.; Vandenberghe, S.; Van Audenhaege, K.; Van Vaerenbergh, J.; Van Holen, R. Rapid additive manufacturing of MR compatible multipinhole collimators with selective laser melting of tungsten powder. *Med. Phys.* **2013**, *40*, 012501. [CrossRef]
55. Sidambe, A.T.; Judson, D.S.; Colosimo, S.J.; Fox, P. Laser powder bed fusion of a pure tungsten ultra-fine single pinhole collimator for use in gamma ray detector characterisation. *Int. J. Refract. Met. Hard Mater.* **2019**, *84*, 104998. [CrossRef]
56. Gear, J.I.; Taprogge, J.; White, O.; Flux, G.D. Characterisation of the attenuation properties of 3D-printed tungsten for use in gamma camera collimation. *EJNMMI Phys.* **2019**, *6*, 1. [CrossRef] [PubMed]
57. Zhou, K.; Chen, W.; Yang, Y.; Li, R.; Dong, L.; Fu, Y.-Q. Microstructure and mechanical behavior of porous tungsten skeletons synthesized by selected laser melting. *Int. J. Refract. Met. Hard Mater.* **2022**, *103*, 105769. [CrossRef]

58. Morcos, P.; Elwany, A.; Karaman, I.; Arróyave, R. Review: Additive manufacturing of pure tungsten and tungsten-based alloys. *J. Mater. Sci.* **2022**, *57*, 9769–9806. [CrossRef]
59. Omole, S.; Lunt, A.; Kirk, S.; Shokrani, A. Advanced Processing and Machining of Tungsten and Its Alloys. *J. Manuf. Mater. Process.* **2022**, *6*, 15. [CrossRef]
60. Pan, S.-H.; Yao, G.-C.; Cui, Y.-N.; Meng, F.-S.; Luo, C.; Zheng, T.-Q.; Singh, G. Additive manufacturing of tungsten, tungsten-based alloys, and tungsten matrix composites. *Tungsten* **2023**, *5*, 1–31. [CrossRef]
61. Talignani, A.; Seede, R.; Whitt, A.; Zheng, S.; Ye, J.; Karaman, I.; Kirka, M.M.; Katoh, Y.; Wang, Y.M. A review on additive manufacturing of refractory tungsten and tungsten alloys. *Addit. Manuf.* **2022**, *58*, 103009. [CrossRef]
62. Galati, M. Electron beam melting process. In *Additive Manufacturing*; Elsevier: Amsterdam, The Netherlands, 2021; pp. 277–301. [CrossRef]
63. Ellis, E.A.I.; Sprayberry, M.A.; Ledford, C.; Hankwitz, J.P.; Kirka, M.M.; Rock, C.D.; Horn, T.J.; Katoh, Y.; Dehoff, R.R. Processing of tungsten through electron beam melting. *J. Nucl. Mater.* **2021**, *555*, 153041. [CrossRef]
64. Zhang, H.; Carriere, P.R.; Amoako, E.D.; Rock, C.D.; Thielk, S.U.; Fletcher, C.G.; Horn, T.J. Microstructure and Elevated Temperature Flexure Testing of Tungsten Produced by Electron Beam Additive Manufacturing. *JOM* **2023**, *75*, 4094–4107. [CrossRef]
65. Dorow-Gerspach, D.; Kirchner, A.; Loewenhoff, T.; Pintsuk, G.; Weißgärber, T.; Wirtz, M. Additive manufacturing of high density pure tungsten by electron beam melting. *Nucl. Mater. Energy* **2021**, *28*, 101046. [CrossRef]
66. Wang, J.; Yao, D.; Li, M.; An, X.; Li, S.; Hou, W.; Zhang, X.; Yang, G.; Wang, J.; Wang, L. Hierarchical effects of multi-layer powder spreading in the electron beam powder bed fusion additive manufacturing of pure tungsten material. *Addit. Manuf.* **2022**, *55*, 102835. [CrossRef]
67. Zhao, X.; An, N.; Yang, G.; Wang, J.; Tang, H.; Li, M.; Zhou, J. Enhancing standard finite element codes with POD for reduced order thermal analysis: Application to electron beam melting of pure tungsten. *Mater. Today Commun.* **2021**, *29*, 102796. [CrossRef]
68. Fernandez-Zelaia, P.; Kirka, M.; Campbell, Q.; Rojas, J.O.; Rossy, A.M.; Ledford, C. Electron Beam Powder Bed Fusion Additive Manufacturing of Refractory Metals. In *Proceedings of the Powdermet/Tungsten 2021*; Orlando, FL, USA, 20–23 June 2021. Available online: https://www.osti.gov/biblio/1832704 (accessed on 15 July 2024).
69. Liao, W.-B.; Liu, Z.-Y.; He, M.-J.; Feng, C.; Wang, F.; Huang, J. Effect of Electron Beam Remelting Treatments on the Microstructure and Properties of Atmospheric Plasma Sprayed Tungsten Coatings. *J. Therm. Spray. Technol.* **2021**, *30*, 2128–2137. [CrossRef]
70. Jeong, W.; Kwon, Y.-S.; Kim, D. Three-dimensional printing of tungsten structures by directed energy deposition. *Mater. Manuf. Process.* **2019**, *34*, 986–992. [CrossRef]
71. Wang, Y.P.; Ma, S.Y.; Yang, X.S.; Zhou, Y.Z.; Liu, X.; Li, J.F.; Zhang, J.J.; Li, C.; Wang, X.Y.; Le, G.M.; et al. Microstructure and strengthening mechanisms of 90W–7Ni–3Fe alloys prepared using laser melting deposition. *J. Alloys Compd.* **2020**, *838*, 155545. [CrossRef]
72. Wang, G.-Y.; Gu, S.-N.; Yang, S. Microstructure and properties of tungsten heavy alloys fabricated by laser direct deposition. *Mater. Sci. Technol.* **2017**, *33*, 415–420. [CrossRef]
73. Li, C.; Ma, S.; Liu, X.; Li, J.; Le, G. Microstructures and properties of 80W-20Fe alloys prepared using laser melting deposition process. *Int. J. Refract. Met. Hard Mater.* **2018**, *77*, 113–119. [CrossRef]
74. Wang, G.; Sun, X.; Huang, M.; Qin, Y.; Yao, Y.; Yang, S. Influence of processing parameters on the microstructure and tensile property of 85 W-15Ni produced by laser direct deposition. *Int. J. Refract. Met. Hard Mater.* **2019**, *82*, 227–233. [CrossRef]
75. Zhou, S.; Wang, L.; Liang, Y.-J.; Zhu, Y.; Jian, R.; Wang, B.; Wang, L.; Xue, Y.; Wang, F.; Cai, H.; et al. A strategy to achieve high-strength WNiFe composite-like alloys with low W content by laser melting deposition. *Mater. Des.* **2020**, *190*, 108554. [CrossRef]
76. DebRoy, T.; Wei, H.L.; Zuback, J.S.; Mukherjee, T.; Elmer, J.W.; Milewski, J.O.; Beese, A.M.; Wilson-Heid, A.; De, A.; Zhang, W. Additive manufacturing of metallic components—Process, structure and properties. *Prog. Mater. Sci.* **2018**, *92*, 112–224. [CrossRef]
77. Li, C.; Wang, Y.; Ma, S.; Yang, X.; Li, J.; Zhou, Y.; Liu, X.; Tang, J.; Wang, X.; Le, G. Densification, microstructural evolutions of 90W-7Ni-3Fe tungsten heavy alloys during laser melting deposition process. *Int. J. Refract. Met. Hard Mater.* **2020**, *91*, 105254. [CrossRef]
78. Marinelli, G.; Martina, F.; Lewtas, H.; Hancock, D.; Mehraban, S.; Lavery, N.; Ganguly, S.; Williams, S. Microstructure and thermal properties of unalloyed tungsten deposited by Wire + Arc Additive Manufacture. *J. Nucl. Mater.* **2019**, *522*, 45–53. [CrossRef]
79. Pixner, F.; Buzolin, R.; Warchomicka, F.; Pilz, A.; Enzinger, N. Wire-based electron beam additive manufacturing of tungsten. *Int. J. Refract. Met. Hard Mater.* **2022**, *108*, 105917. [CrossRef]
80. Stawovy, M.T.; Myers, K.; Ohm, S. Binder jet printing of tungsten heavy alloy. *Int. J. Refract. Met. Hard Mater.* **2019**, *83*, 104981. [CrossRef]
81. Doddapaneni, V.V.K.; Lee, K.; Aysal, H.E.; Paul, B.K.; Pasebani, S.; Sierros, K.A.; Okwudire, C.E.; Chang, C. A Review on Progress, Challenges, and Prospects of Material Jetting of Copper and Tungsten. *Nanomaterials* **2023**, *13*, 2303. [CrossRef]
82. Ottensmeyer, M.; Sabet, H.; Furenlid, L.; May, M.; Kupinski, M. Collimator fabrication techniques: A comparative study. *J. Nucl. Med.* **2023**, *64*, P1600. Available online: http://jnm.snmjournals.org/content/64/supplement_1/P1600.abstract (accessed on 15 July 2024).
83. Bose, A.; Reidy, J.P.; Tuncer, N.; Jorgensen, L. Processing of tungsten heavy alloy by extrusion-based additive manufacturing. *Int. J. Refract. Met. Hard Mater.* **2023**, *110*, 106021. [CrossRef]

84. Huang, N.; Cook, O.J.; Argüelles, A.P.; Beese, A.M. Review of Process–Structure–Property Relationships in Metals Fabricated Using Binder Jet Additive Manufacturing. *Metallogr. Microstruct. Anal.* **2023**, *12*, 883–905. [CrossRef]
85. Zhou, S.; Liang, Y.-J.; Zhu, Y.; Wang, B.; Wang, L.; Xue, Y. Ultrashort-time liquid phase sintering of high-performance fine-grain tungsten heavy alloys by laser additive manufacturing. *J. Mater. Sci. Technol.* **2021**, *90*, 30–36. [CrossRef]
86. Bose, A.; Schuh, C.A.; Tobia, J.C.; Tuncer, N.; Mykulowycz, N.M.; Preston, A.; Barbati, A.C.; Kernan, B.; Gibson, M.A.; Krause, D.; et al. Traditional and additive manufacturing of a new Tungsten heavy alloy alternative. *Int. J. Refract. Met. Hard Mater.* **2018**, *73*, 22–28. [CrossRef]
87. Du, Z.-Y.; Lv, Y.-Q.; Han, Y.; Fan, J.-L.; Ye, L. Sintering densification behavior and kinetic mechanism of nano-tungsten powder prepared by sol-spray drying. *Tungsten* **2020**, *2*, 371–380. [CrossRef]
88. Han, Y.; Fan, J.; Liu, T.; Cheng, H.; Tian, J. The effect of trace nickel additive and ball milling treatment on the near-full densification behavior of ultrafine tungsten powder. *Int. J. Refract. Met. Hard Mater.* **2012**, *34*, 18–26. [CrossRef]
89. Thompson, S.M.; Bian, L.; Shamsaei, N.; Yadollahi, A. An overview of Direct Laser Deposition for additive manufacturing; Part I: Transport phenomena, modeling and diagnostics. *Addit. Manuf.* **2015**, *8*, 36–62. [CrossRef]
90. Karafi, T.; Tahiri, A.; Chabba, H.; Idiri, M.; Boubeker, B. Effect of Grain-Size in Nanocrystalline Tungsten on Hardness and Dislocation Density: A Molecular Dynamics Study. *Crystals* **2023**, *13*, 469. [CrossRef]
91. Kozyrev, N.V.; Gordeev, V.V. Thermodynamic Properties and Equation of State for Tungsten. *Crystals* **2023**, *13*, 1470. [CrossRef]
92. Griffiths, V.; Scanlan, J.P.; Eres, M.H.; Martinez-Sykora, A.; Chinchapatnam, P. Cost-driven build orientation and bin packing of parts in Selective Laser Melting (SLM). *Eur. J. Oper. Res.* **2019**, *273*, 334–352. [CrossRef]
93. Ren, K.; Di, Y.; Wang, G.; Wang, L.; Wang, H.; Rong, Y. Forward calculation model for utilization of energy and mass in laser-directed energy deposition. *Addit. Manuf.* **2023**, *68*, 103512. [CrossRef]
94. Mohammadhosseini, A.; Masood, S.H.; Fraser, D.; Jahedi, M. Dynamic compressive behaviour of Ti-6Al-4V alloy processed by electron beam melting under high strain rate loading. *Adv. Manuf.* **2015**, *3*, 232–243. [CrossRef]
95. Yang, S.F.; Li, C.W.; Chen, A.Y.; Gan, B.; Gu, J.F. Microstructure and corrosion resistance of stainless steel manufactured by laser melting deposition. *J. Manuf. Process* **2021**, *65*, 418–427. [CrossRef]
96. Ziaee, M.; Crane, N.B. Binder jetting: A review of process, materials, and methods. *Addit. Manuf.* **2019**, *28*, 781–801. [CrossRef]
97. Wei, C.; Liu, L.; Gu, Y.; Huang, Y.; Chen, Q.; Li, Z.; Li, L. Multi-material additive-manufacturing of tungsten-copper alloy bimetallic structure with a stainless-steel interlayer and associated bonding mechanisms. *Addit. Manuf.* **2022**, *50*, 102574. [CrossRef]
98. Sharma, S.; Krishna, K.V.M.; Joshi, S.S.; Radhakrishnan, M.; Palaniappan, S.; Dussa, S.; Banerjee, R.; Dahotre, N.B. Laser based additive manufacturing of tungsten: Multi-scale thermo-kinetic and thermo-mechanical computational model and experiments. *Acta Mater.* **2023**, *259*, 119244. [CrossRef]
99. Mukherjee, T.; DebRoy, T. A digital twin for rapid qualification of 3D printed metallic components. *Appl. Mater. Today* **2019**, *14*, 59–65. [CrossRef]
100. Ledford, C.; Fernandez-Zelaia, P.; Graening, T.; Campbell, Q.; Rojas, J.O.; Rossy, A.M.; Kato, Y.; Kirka, M.M. Microstructure and high temperature properties of tungsten processed via electron beam melting additive manufacturing. *Int. J. Refract. Met. Hard Mater.* **2023**, *113*, 106148. [CrossRef]
101. Matthews, M.; Trapp, J.; Guss, G.; Rubenchik, A. Direct measurements of laser absorptivity during metal melt pool formation associated with powder bed fusion additive manufacturing processes. *J. Laser Appl.* **2018**, *30*, 032302. [CrossRef]
102. Enneti, R.; Morgan, R.; Wolfe, T.; Harooni, A.; Volk, S. Direct Metal Laser Sintering (DMLS) of Tungsten Powders. In Proceedings of the Additive Manufacturing with Powder Metallurgy, Las Vegas, NV, USA, 13–15 June 2017.
103. Li, J.; Wei, Z.; Zhou, B.; Wu, Y.; Chen, S.-G.; Sun, Z. Densification, Microstructure and Properties of 90W-7Ni-3Fe Fabricated by Selective Laser Melting. *Metals* **2019**, *9*, 884. [CrossRef]
104. Hu, Z.; Liu, Y.; Chen, S.; Liu, S.; Yu, L.; Liu, Y.; Ma, Z. Achieving high-performance pure tungsten by additive manufacturing: Processing, microstructural evolution and mechanical properties. *Int. J. Refract. Met. Hard Mater.* **2023**, *113*, 106211. [CrossRef]
105. Müller, A.V.; Dorow-Gerspach, D.; Balden, M.; Binder, M.; Buschmann, B.; Curzadd, B.; Loewenhoff, T.; Neu, R.; Schlick, G.; You, J.H. Progress in additive manufacturing of pure tungsten for plasma-facing component applications. *J. Nucl. Mater.* **2022**, *566*, 153760. [CrossRef]
106. Xia, Y.; Dong, Z.; Guo, X.; Tian, Q.; Liu, Y. Towards a circular metal additive manufacturing through recycling of materials: A mini review. *J. Cent. South. Univ.* **2020**, *27*, 1134–1145. [CrossRef]
107. Gradl, P.; Mireles, O.R.; Andrews, N. Introduction to Additive Manufacturing for Propulsion and Energy Systems. In Proceedings of the AIAA Propulsion and Energy Forum 2021, Denver, CO, USA, 9–11 August 2021.
108. Sidambe, A.; Fox, P. Analysis of melt pool during the laser powder bed fusion of tungsten. In Proceedings of the RAPDASA 2019 Conference and Exhibition, Gqeberha, South Africa, 28–31 October 2019.
109. Sochalski-Kolbus, L.M.; Payzant, E.A.; Cornwell, P.A.; Watkins, T.R.; Babu, S.S.; Dehoff, R.R.; Lorenz, M.; Ovchinnikova, O.; Duty, C. Comparison of Residual Stresses in Inconel 718 Simple Parts Made by Electron Beam Melting and Direct Laser Metal Sintering. *Metall. Mater. Trans. A* **2015**, *46*, 1419–1432. [CrossRef]
110. Xue, J.; Feng, Z.; Tang, J.; Tang, C.; Zhao, Z. Selective laser melting additive manufacturing of tungsten with niobium alloying: Microstructure and suppression mechanism of microcracks. *J. Alloys Compd.* **2021**, *874*, 159879. [CrossRef]
111. Ng, G.K.L.; Jarfors, A.E.W.; Bi, G.; Zheng, H.Y. Porosity formation and gas bubble retention in laser metal deposition. *Appl. Phys. A* **2009**, *97*, 641–649. [CrossRef]

112. Zhong, M.; Liu, W.; Ning, G.; Yang, L.; Chen, Y. Laser direct manufacturing of tungsten nickel collimation component. *J. Mater. Process Technol.* **2004**, *147*, 167–173. [CrossRef]
113. Li, J.; Wu, Y.; Zhou, B.; Wei, Z. Laser Powder Bed Fusion of Pure Tungsten: Effects of Process Parameters on Morphology. *Densif. Microstruct. Mater.* **2020**, *14*, 165. [CrossRef]
114. Dong, J.; Liu, S.; Chen, H.; Li, D.; Zhang, T.; Chen, C.; Zhou, K. Effect of atmosphere on the microstructure and properties of additively manufactured tungsten. *Mater. Sci. Technol.* **2020**, *36*, 1988–1996. [CrossRef]
115. Huang, J.; Li, M.; Wang, J.; Pei, Z.; Mclntyre, P.; Ma, C. Selective laser melting of tungsten: Effects of hatch distance and point distance on pore formation. *J. Manuf. Process* **2021**, *61*, 296–302. [CrossRef]
116. Adamson, A.W. Potential distortion model for contact angle and spreading. II. Temperature dependent effects. *J. Colloid. Interface Sci.* **1973**, *44*, 273–281. [CrossRef]
117. Muramatsu, Y.; Halada, K.; Dan, T.; Isoda, Y. Solid-Liquid Interfacial Tension of the W-Cu System. *J. Jpn. Inst. Met.* **1990**, *54*, 679–684. [CrossRef]
118. Pan, S.; Guan, Z.; Yao, G.; Yuan, J.; Li, X. Mo-enhanced chemical stability of TiC nanoparticles in molten Al. *J. Alloys Compd.* **2021**, *856*, 158169. [CrossRef]
119. Wei, C.; Gu, H.; Gu, Y.; Liu, L.; Huang, Y.; Cheng, D.; Li, Z.; Li, L. Abnormal interfacial bonding mechanisms of multi-material additive-manufactured tungsten–stainless steel sandwich structure. *Int. J. Extrem. Manuf.* **2022**, *4*, 025002. [CrossRef]
120. Li, K.; Wang, D.; Xing, L.; Wang, Y.; Yu, C.; Chen, J.; Zhang, T.; Ma, J.; Liu, W.; Shen, Z. Crack suppression in additively manufactured tungsten by introducing secondary-phase nanoparticles into the matrix. *Int. J. Refract. Met. Hard Mater.* **2019**, *79*, 158–163. [CrossRef]
121. Kurishita, H.; Matsuo, S.; Arakawa, H.; Sakamoto, T.; Kobayashi, S.; Nakai, K.; Okano, H.; Watanabe, H.; Yoshida, N.; Torikai, Y.; et al. Current status of nanostructured tungsten-based materials development. *Phys. Scr. T* **2014**, *159*, 014032. [CrossRef]
122. Nagy, D.; Humphry-Baker, S.A. An oxidation mechanism map for tungsten. *Scr. Mater.* **2022**, *209*, 114373. [CrossRef]
123. Warren, A.; Nylund, A.; Olefjord, I. Oxidation of tungsten and tungsten carbide in dry and humid atmospheres. *Int. J. Refract. Met. Hard Mater.* **1996**, *14*, 345–353. [CrossRef]
124. Savitskii, E.M.; Burkhanov, G.S. *Physical Metallurgy of Refractory Metals and Alloys*; Springer US: Boston, MA, USA, 1995. [CrossRef]
125. Naghshineh, B.; Ribeiro, A.; Jacinto, C.; Carvalho, H. Social impacts of additive manufacturing: A stakeholder-driven framework. *Technol. Forecast. Soc. Chang.* **2021**, *164*, 120368. [CrossRef]
126. Antusch, S.; Klein, A.; Baumgärtner, S.; Bonnekoh, C.; Böswirth, B.; Dorow-Gerspach, D.; Dietrich, S.; Ehrhardt, M.; Ghidersa, B.-E.; Greuner, H.; et al. Additive manufacturing of novel complex tungsten components via electron beam melting: Basic properties and evaluation of the high heat flux behavior. *Nucl. Mater. Energy* **2024**, *39*, 101683. [CrossRef]
127. Eckley, C.C.; Kemnitz, R.A.; Fassio, C.P.; Hartsfield, C.R.; Leonhardt, T.A. Selective Laser Melting of Tungsten-Rhenium Alloys. *JOM* **2021**, *73*, 3439–3450. [CrossRef]
128. Yamamoto, T.; Hara, M.; Hatano, Y. Cracking behavior and microstructural. mechanical and thermal characteristics of tungsten–rhenium binary alloys fabricated by laser powder bed fusion. *Int. J. Refract. Metals Hard Mater.* **2021**, *100*, 105651. [CrossRef]
129. Klopp, W.; Witzke, W.; Raffo, P. Mechanical Properties of Dilute Tungsten-Rhenium Alloys. 1966. Available online: https://www.semanticscholar.org/paper/Mechanical-properties-of-dilute-tungsten-rhenium-Klopp-Witzke/f3a5af30bb932de1537aaa486f215920185ab265 (accessed on 15 July 2024).
130. Zhang, D.Q.; Liu, Z.H.; Cai, Q.Z.; Liu, J.H.; Chua, C.K. Influence of Ni content on microstructure of W–Ni alloy produced by selective laser melting. *Int. J. Refract. Met. Hard Mater.* **2014**, *45*, 15–22. [CrossRef]
131. Zhang, D.; Cai, Q.; Liu, J.; Li, R. Research on Process and Microstructure Formation of W-Ni-Fe Alloy Fabricated by Selective Laser Melting. *J. Mater. Eng. Perform.* **2011**, *20*, 1049–1054. [CrossRef]
132. Iveković, A.; Montero-Sistiaga, M.L.; Vanmeensel, K.; Kruth, J.-P.; Vleugels, J. Effect of processing parameters on microstructure and properties of tungsten heavy alloys fabricated by SLM. *Int. J. Refract. Met. Hard Mater.* **2019**, *82*, 23–30. [CrossRef]
133. Hu, Z.; Liu, Y.; Wu, J.; Dong, J.; Ma, Z.; Liu, Y. The simultaneous improvement of strength and ductility of the 93W-4.6Ni-2.4Fe prepared by additive manufacturing via optimizing sintering post-treatment. *Addit. Manuf.* **2021**, *46*, 102216. [CrossRef]
134. Chen, H.; Zi, X.; Han, Y.; Dong, J.; Liu, S.; Chen, C. Microstructure and mechanical properties of additive manufactured W-Ni-Fe-Co composite produced by selective laser melting. *Int. J. Refract. Met. Hard Mater.* **2020**, *86*, 105111. [CrossRef]
135. Wang, M.; Li, R.; Yuan, T.; Chen, C.; Zhang, M.; Weng, Q.; Yuan, J. Selective laser melting of W-Ni-Cu composite powder: Densification, microstructure evolution and nano-crystalline formation. *Int. J. Refract. Met. Hard Mater.* **2018**, *70*, 9–18. [CrossRef]
136. Yan, A.; Wang, Z.; Yang, T.; Wang, Y.; Ma, Z. Microstructure, thermal physical property and surface morphology of W-Cu composite fabricated via selective laser melting. *Mater. Des.* **2016**, *109*, 79–87. [CrossRef]
137. Li, R.; Shi, Y.; Liu, J.; Xie, Z.; Wang, Z. Selective laser melting W–10 wt.% Cu composite powders. *Int. J. Adv. Manuf. Technol.* **2010**, *48*, 597–605. [CrossRef]
138. Wang, G.; Qin, Y.; Yang, S. Characterization of laser-powder interaction and particle transport phenomena during laser direct deposition of W–Cu composite. *Addit. Manuf.* **2021**, *37*, 101722. [CrossRef]
139. Zhou, Z.; Tan, Z.; He, D.; Zhou, Z.; Cui, L.; Wang, Y.; Shao, W.; Wang, G. Fabrication of three-dimensional connected W-Cu10Sn composites by selective laser melting. *Mater. Lett.* **2020**, *264*, 127377. [CrossRef]
140. Li, K.; Li, Y.; Chen, W.; Zhao, C.; Yuan, Y.; Cheng, L.; Morgan, T.W.; Liu, W.; Shen, Z. Effect of Ta addition on the fuzz formation of additively manufactured W-based materials. *Nucl. Fusion.* **2020**, *60*, 064004. [CrossRef]

141. Li, K.; Ma, G.; Xing, L.; Wang, Y.; Yu, C.; Chen, J.; Ma, J.; Wu, G.; Liu, W.; Shen, Z.; et al. Crack suppression via in-situ oxidation in additively manufactured W-Ta alloy. *Mater. Lett.* **2020**, *263*, 127212. [CrossRef]
142. Guo, Z.; Wang, L.; Wang, X.-Z. Additive manufacturing of W-12Ta(wt%) alloy: Processing and resulting mechanical properties. *J. Alloys Compd.* **2021**, *868*, 159193. [CrossRef]
143. Li, J.; Wei, Z.; Zhou, B.; Wu, Y.; Chen, S.-G.; Sun, Z. Preparation, microstructure, and microhardness of selective laser-melted W–3Ta sample. *J. Mater. Res.* **2020**, *35*, 2016–2024. [CrossRef]
144. Xiao, B.; Jia, W.; Tang, H.; Wang, J.; Zhou, L. Microstructure and mechanical properties of a newly developed WTaRe refractory alloy by selective electron beam melting. *Addit. Manuf.* **2022**, *54*, 102738. [CrossRef]
145. Xiao, B.; Jia, W.; Wang, J.; Zhou, L. High-temperature mechanical behaviors of a WTaRe refractory alloy manufactured by selective electron beam melting. *Int. J. Refract. Met. Hard Mater.* **2023**, *116*, 106371. [CrossRef]
146. Chen, H.; Ye, L.; Han, Y.; Chen, C.; Fan, J. Additive manufacturing of W–Fe composites using laser metal deposition: Microstructure, phase transformation, and mechanical properties. *Mater. Sci. Eng. A* **2021**, *811*, 141036. [CrossRef]
147. Tan, Z.; Zhou, Z.; Wu, X.; Wang, Y.; Shao, W.; Guo, X.; Zhou, Z.; Yang, Y.; Wang, G.; He, D. In situ synthesis of spherical W Mo alloy powder for additive manufacturing by spray granulation combined with thermal plasma spheroidization. *Int. J. Refract. Met. Hard Mater.* **2021**, *95*, 105460. [CrossRef]
148. Lang, S.; Sun, N.; Cao, J.; Yu, W.; Yang, Z.; Hou, S. Fabrication of Ultra-Fine-Grained W-TiC Alloys by a Simple Ball-Milling and Hydrogen Reduction Method. *Materials* **2021**, *14*, 5865. [CrossRef] [PubMed]
149. Gu, D.; Dai, D.; Chen, W.; Chen, H. Selective Laser Melting Additive Manufacturing of Hard-to-Process Tungsten-Based Alloy Parts with Novel Crystalline Growth Morphology and Enhanced Performance. *J. Manuf. Sci. Eng.* **2016**, *138*, 081003. [CrossRef]
150. Chen, J.; Zhao, C.; Li, K.; Li, M.; Sun, S.; Zhang, S.; Ma, J.; Liu, W. Effect of TaC addition on microstructure and microhardness of additively manufactured tungsten. *J. Alloys Compd.* **2022**, *897*, 162978. [CrossRef]
151. Cole, N.C.; Gilliland, R.G.; Slaughter, G.M. Weldability of tungsten and its alloys. In Proceedings of the American Welding Society 49th Annual Meeting, Chicago, IL, USA, 1–5 April 1968.
152. Bose, A.; German, R.M. Microstructural refinement of W-Ni-Fe heavy alloys by alloying additions. *Metall. Trans. A* **1988**, *19*, 3100–3103. [CrossRef]
153. Luo, A.; Jacobson, D.L.; Shin, K.S. Solution softening mechanism of iridium and rhenium in tungsten at room temperature. *Int. J. Refract. Met. Hard Mater.* **1991**, *10*, 107–114. [CrossRef]
154. Wang, Y.K.; Xie, Z.M.; Wang, M.M.; Deng, H.W.; Yang, J.F.; Jiang, Y.; Zhang, T.; Wang, X.P.; Fang, Q.F.; Liu, C.S. The superior thermal stability and tensile properties of hot rolled W-HfC alloys. *Int. J. Refract. Met. Hard Mater.* **2019**, *81*, 42–48. [CrossRef]
155. Leonhardt, T. Properties of tungsten-rhenium and tungsten-rhenium with hafnium carbide. *JOM* **2009**, *61*, 68–71. [CrossRef]
156. Ren, C.; Fang, Z.Z.; Koopman, M.; Butler, B.; Paramore, J.; Middlemas, S. Methods for improving ductility of tungsten—A review. *Int. J. Refract. Met. Hard Mater.* **2018**, *75*, 170–183. [CrossRef]
157. Calvo, A.; Schlueter, K.; Tejado, E.; Pintsuk, G.; Ordás, N.; Iturriza, I.; Neu, R.; Pastor, J.Y.; García-Rosales, C. Self-passivating tungsten alloys of the system W-Cr-Y for high temperature applications. *Int. J. Refract. Met. Hard Mater.* **2018**, *73*, 29–37. [CrossRef]
158. Zhang, Y.; Huang, Y.; Liu, W.; Ma, Y.; Wang, Y.; Liang, C. Effect of tantalum addition method on microstructure and mechanical properties of the 90(W 20Ta)–Ni–Fe alloy. *Int. J. Refract. Met. Hard Mater.* **2022**, *105*, 105848. [CrossRef]
159. Chen, B.; Li, Z.-B.; Liu, J.-R.; Zhang, G.-H. Effect of molybdenum addition on microstructure and mechanical properties of 90% tungsten heavy alloys. *Int. J. Refract. Met. Hard Mater.* **2022**, *106*, 105868. [CrossRef]
160. Xie, Z.M.; Liu, R.; Miao, S.; Yang, X.D.; Zhang, T.; Wang, X.P.; Fang, Q.F.; Liu, C.S.; Luo, G.N.; Lian, Y.Y.; et al. Extraordinary high ductility/strength of the interface designed bulk W-ZrC alloy plate at relatively low temperature. *Sci. Rep.* **2015**, *5*, 16014. [CrossRef] [PubMed]
161. Suman, V.; Chaira, D. Improvement of Oxidation Resistance in Tungsten Heavy Alloys through Si, Y2O3, Ni, and Co Addition. *J. Mater. Eng. Perform.* **2023**, *33*, 5279–5290. [CrossRef]
162. Yin, C.; Terentyev, D.; Zhang, T.; Nogami, S.; Antusch, S.; Chang, C.-C.; Petrov, R.H.; Pardoen, T. Ductile to brittle transition temperature of advanced tungsten alloys for nuclear fusion applications deduced by miniaturized three-point bending tests. *Int. J. Refract. Met. Hard Mater.* **2021**, *95*, 105464. [CrossRef]
163. Sharma, S.; Joshi, S.S.; Pantawane, M.V.; Radhakrishnan, M.; Mazumder, S.; Dahotre, N.B. Multiphysics multi-scale computational framework for linking process–structure–property relationships in metal additive manufacturing: A critical review. *Int. Mater. Rev.* **2023**, *68*, 943–1009. [CrossRef]

crystals

MDPI

Review

Microstructural and Textural Evolution in Hexagonal Close-Packed Metals: The Case of Zirconium, Magnesium, and Titanium

Khushahal Thool [1,†], K. U. Yazar [2,†], V. Kavimani [3,4,†], Aman Gupta [1,5,*] and Shi-Hoon Choi [1,*]

1 Department of Advanced Components and Materials Engineering, Sunchon National University, Suncheon 57922, Republic of Korea; khushahal2@gmail.com
2 Centre for Innovative Manufacturing Research, Vellore Institute of Technology, Vellore 632014, India; yazar.ku@vit.ac.in or yazar1988@gmail.com
3 Centre for Material Science, Karpagam Academy of Higher Education, Coimbatore 641021, India; manikavi03@gmail.com
4 Department of Mechanical Engineering, Karpagam Academy of Higher Education, Coimbatore 641021, India
5 Division of Research and Development, Lovely Professional University, Phagwara 144411, India
* Correspondence: amangupta@scnu.ac.kr (A.G.); shihoon@scnu.ac.kr (S.-H.C.)
† These authors contributed equally to this work.

Abstract: Hexagonal close-packed (HCP) metals, particularly Zirconium (Zr), Titanium (Ti), and Magnesium (Mg) alloys, have attracted significant attention due to their unique properties and wide-ranging applications in the aerospace, biomedical, and energy industries. This review paper provides a comprehensive analysis of the microstructural and textural evolution in these HCP materials under various conditions, including rolling, extrusion, drawing, and annealing. The focus of the present work lies on the deformed microstructure and texture development in HCP metals, thus elucidating the fundamental mechanisms that govern their response to mechanical stress. The interaction between dislocation movements, twinning, and slip systems is discussed in detail, illustrating how these factors contribute to the anisotropic behavior characteristic of low-symmetry HCP structures. Unlike high-symmetry metals, deformation in Zr alloys depends on the activation of various slips and twin deformation modes, which are sensitive to crystallographic orientation and strain. Like Zr, Ti alloys present a more complex deformation behavior, heavily influenced by their crystallographic orientation. The most common deformation textures in Ti alloys include split-transverse direction (split-TD), split-rolling direction (split-RD), and normal direction (ND) symmetric basal fiber textures. These textures emerge due to the activation of multiple slip systems and twinning, which are dependent on external factors such as temperature, strain rate, and alloy composition. For Mg alloys, the poor formability and brittleness associated with the dominance of the basal slip system under ambient conditions is a critical material development challenge. The activation of non-basal slip systems introduces complexities in controlling texture and microstructure. However, their activation is crucial for optimizing mechanical properties such as strength and fatigue resistance. The tendency for twinning in Mg alloys further complicates their deformation behavior, leading to challenges in ensuring uniform mechanical performance. Modifying the alloy composition, grain size, and texture can additionally influence the activation of these deformation mechanisms. This review further explores the roles of dynamic recrystallization and grain growth in tailoring mechanical properties, with a particular focus on microstructure and texture evolution during annealing. Through this detailed review, we aim to present a thorough understanding of the microstructural and textural evolution in HCP materials, thereby guiding future research and industrial applications.

Keywords: HCP; hot deformation; microstructure; texture evolution; anisotropy; thermomechanical processing

1. Introduction

Hexagonal close-packed (HCP) materials and their related alloys exhibit unique material characteristics, including crystallographic anisotropy, restricted deformation behavior due to limited slip systems, and a distinct atomic configuration compared to their body-centered cubic (BCC) or face-centered cubic (FCC) counterparts. These exclusive factors result in unique processing, final properties, and underlying microstructures in HCP materials. Common HCP metals and alloys such as titanium, zirconium, and magnesium are extensively utilized in various industrial and military applications. Despite significant advancements, a deeper understanding of the relationships between processing, microstructures, and material properties remains critical. This work emphasizes the intricate relationships between microstructure and texture evolution as a function of different processing parameters relevant to various HCP materials [1–7].

Zirconium (Zr) alloys are widely employed as ideal structural materials for thermal nuclear reactors due to their low neutron absorption cross-section combined with excellent mechanical and corrosion properties at reactor operating temperatures [8]. Some grades of Zr are also used widely in the biomedical field [9]. Zr alloys exist in different grades and are typically either predominantly single-phase α HCP or two-phase α+β (HCP+BCC) microstructures. Most commercially used Zr alloys, ranging from Zircaloy-1 to Zircaloy-4, Zr-Nb, are largely HCP single-phase and polycrystalline. The fabrication of Zr component, from cast ingots to final tubes, involves stages of thermo-mechanical processing (TMP) [10–12], which is essential not only for shape modification but also for achieving the correct microstructure and, consequently, the desired properties. In service, Zircaloy-4 nuclear fuel claddings are exposed to harsh environments, leading to material degradation under operating conditions. Therefore, stringent TMP control is necessary to achieve desired properties such as low neutron absorption cross-section, radiation damage resistance, superior creep resistance, and high-temperature corrosion performance. TMP involves hot extrusion and a unique fabrication technique called pilgering [7,10,11,13–15]. Hot extrusion or hot working relies on the activation of slip-twin systems [6,16–18] and the possibilities of dynamic recrystallization (DRx). Pilgering provides triaxiality of stresses, hence higher formability, but involves heterogeneous deformation [19,20], where some grains are fragmented while others remain virtually untouched [19,21]. This heterogeneous microstructure persists even after thermal annealing [22,23]. Consequently, the final microstructure-property relationship is controlled by pilgering-induced microstructural evolution. In Zr alloys, crystallographic texture, microstructure, chemical compositions, residual stresses, and dislocation density effectively control the kinetics of various degradation mechanisms, such as radiation-induced growth [24,25] and total circumferential elongation during simulated burst tests [26,27].

Titanium (Ti) alloys are among the most widely used structural materials due to their superior specific strength, ductility, high-temperature strength, creep, and fatigue properties, making them ideal for applications in the aerospace, biomedical, energy, and petrochemical sectors [4,28–30]. Ti undergoes an allotropic phase transformation at 1155 K. Below this temperature, known as the β transus, Ti exists as an HCP structure (α phase) [4,31], and above this temperature, it exists as a BCC structure (β phase) [32]. Alpha-Ti has an HCP crystal structure with a c/a ratio about 2.8% less than the ideal, resulting in the principal slip system being prismatic $\langle a \rangle$ type, $(10\bar{1}0)\langle 11\bar{2}0 \rangle$. The secondary slip system is the basal slip system, $(0002)\langle 11\bar{2}0 \rangle$, followed by pyramidal $\langle c+a \rangle$ type slip systems, $(10\bar{1}1)\langle 11\bar{2}3 \rangle$ [33]. Extension $((10\bar{1}2)\langle \bar{1}011 \rangle, (11\bar{2}1)\langle \bar{1}\bar{1}26 \rangle)$ and compression twinning $((11\bar{2}2)\langle \bar{1}\bar{1}23 \rangle)$ mechanisms are also frequently observed, especially at low strain levels [34,35]. To accommodate strain along the c-axis, the twinning mechanisms often activate before the $\langle c+a \rangle$ slip due to the high critical resolved shear stress (CRSS) of the pyramidal slip systems [4,5]. The inherent anisotropy of the hexagonal crystal and the significant disparity in CRSS values between different slip systems result in the complex interplay of deformation mechanisms [36,37] that dictate the plastic deformation and consequent microstructure and texture evolution in α-Ti. The texture and microstructure evolution during Ti deformation depends on the deformation mode, temperature, initial texture, and grain size, directly affecting the relative activities of various deformation mechanisms.

Magnesium (Mg) and its alloys (AZ31, AZ61, AZ91, WE43, ZE41), known for their exceptionally low weight, are heavily utilized for structural applications in the aerospace, automotive, and biomedical sectors. However, their practicality is limited by their HCP crystal structure, which results in poor formability at room temperature (RT). At RT, the dominant deformation mechanism is the basal slip $(0001)\langle 11\bar{2}0\rangle$, characterized by its minimal CRSS value but offering only two independent slip systems, thus limiting formability [38]. At temperatures exceeding 523 K, other non-basal slip systems, such as prismatic and pyramidal, become active [3,39]. The activity of these non-basal slip systems can also be influenced by alloy composition adjustments, grain size refinement, and modifications to the starting material's texture. Deformation processes below the recrystallization temperature (cold rolling (CR)) of Mg alloys primarily involve basal slips and the formation of mechanical twins [40,41], which tend to favor pyramidal systems, specifically $(10\bar{1}2)\ \langle 10\bar{1}1\rangle$ nd $(10\bar{1}1)\ \langle 10\bar{1}2\rangle$. However, dynamic recrystallization (DRx) occurs concurrently with deformation mechanisms above the recrystallization temperature, resulting in recrystallized grains alongside twinning and secondary phases [42]. Mechanical twins play a vital role during deformation, impacting the strengthening mechanisms, texture evolution, asymmetric stress-strain behavior, and fracture initiation. The detailed mechanisms related to the effect of deformation on the microstructure and texture of Mg alloys are discussed in the following sections.

Overall, this review discusses the development and deformation mechanisms of structural grades of HCP materials. In this regard, the main theme of the review paper is the material-specific microstructure and crystallographic texture observed during various deformation processing routes. The different material grades and various processing conditions discussed in this review are summarized in Table 1. Further, issues specific to material development, including the effects of crystal anisotropy, grain fragmentation, and the temperature dependence of CRSS leading to differences in microstructure and texture, along with various deformation mechanisms such as twinning, shear bands (SBs), and strain localizations (SLs), are thoroughly examined and discussed for Zr, Ti, and Mg alloys. The impact of temperature, strain path, and grain size on the crystallographic texture has also been addressed, alongside the softening behavior. Critical issues such as the role of static and dynamic recrystallization, primary and secondary recrystallization, and continuous and discontinuous recrystallization mechanisms in material texture evolution are explored in detail and illustrated with figures. By providing a structured analysis of the existing literature and highlighting key material-specific issues, this review aims to serve as a valuable resource for researchers, engineers, and material scientists, offering a comprehensive and accessible guide to the critical aspects of microstructure and texture evolution in Zr, Ti, and Mg alloys.

Table 1. Discussions of various deformation and annealing mechanisms for the Zr, Ti, and Mg alloys in the present review work.

S.No.	Grades/Alloys	Characteristics Discussed	Published Year/References
1	Zr-4 Fuel Tube	Pilgering, Texture anisotropy, slip activation	2014 [12], 2015 [10]
2	Zr-4 sheet	Plane strain compression, Twinning, Role of initial texture	2014 [20], 2015 [43]
3	Zr-4 Fuel Tube	Microstructure and texture developments	2008 [13]
4	Zr-1.3Sn-0.2Fe-0.1Cr	Hot extrusion, pilgering	2008 [15]
5	Clock-rolled pure Zr plate	Role of substructure, Selection of deformation modes, Coupling between slip modes, Coupling between slip and twinning modes	2008 [16]

Table 1. *Cont.*

S.No.	Grades/Alloys	Characteristics Discussed	Published Year/References
6	Clock-rolled high-purity Zr	Effects of initial texture, Effects of temperature and strain, Secondary twinning within primary twins	2006 [6]
7	Rolled Zircaloy-4 plate	Twin boundary fraction and work-hardening rate, Effect of active deformation modes on the work-hardening rate, Double twinning	2018 [17]
8	CP-Ti	Unidirectional cold rolling and split-TD texture	2017 [44], 2013 [45], 2007 [46], 2014 [47], 2005 [48], 1987 [49], 2008 [50]
9	CP-Ti	Cold cross-rolling and split-RD texture due to varying strain path	2017 [44], 2011 [51]
10	CP-Ti	Elevated temperature and formation of basal—fiber texture	2014 [47], 2016 [52]
11	CP-Ti	Shear strain and formation of Basal fiber texture	2018 [53], 2014 [54]
12	CP-Ti	Split-TD texture in submicron-grained material	2011 [51]
13	CP-Ti	Fiber texture symmetric about extrusion and drawing axis	2010 [55], 2022 [56]
14	CP-Ti	Primary recrystallization texture and microstructure	2008 [50], 2020 [57]
15	CP-Ti	Secondary recrystallization texture and microstructure	2017 [44], 2005 [58], 2020 [57]
16	AZ31	Rolling Microstructure and Texture, Strain hardening behavior, plane slip, and twinning	2019 [59], 2023 [60]
17	AZ61	Deformation mechanism, slip systems	2022 [61]
18	AZ80	Texture evolution	2020 [62]
19	AZ91	Extrusion Texture, strain behavior	2022 [63], 2023 [64]
20	ZK60, Mg-Sn	Texture evolution, deformation mechanism	2022 [65], 2021 [66]
21	Mg-Y-Zn	bimodal microstructure	2023 [67]
22	Mg-Li-Al	Anisothermal aging	2022 [68]
23	Mg-Gd-Nd(-Zn)-Zr alloys	Strain rate and texture	2021 [69]

2. Deformed Microstructure and Texture in Zirconium

As discussed in the introduction section, several mechanisms contribute to the degradation of Zr alloy structures in reactors, including irradiation growth, oxidation, wear, creep, and hydride formation. During irradiation growth, a high flux of neutrons passing through the Zr alloy cladding induces material degradation by introducing point defects, line defects, interstitial and void clusters, precipitates, and microstructural alteration [25]. This degradation can further promote creep deformation processes in Zr alloy components under favorable stress-temperature regimes present during reactor operation, influenced by dislocation glide, dislocation climb, point defect diffusion, and grain boundary sliding (GBS) [25]. The irradiation growth, suboptimal mechanical characteristics under hydride formation, and the rapid creep rate of Zr alloys at elevated temperatures restrict the selection of Zr microstructures and associated processing [70]. Consequently, it is imperative to understand the microstructural and crystallographic texture developments during various TMPs to enhance the versatility of Zr alloys.

2.1. Role of Plastic Anisotropy during Deformation

The origin of plastic anisotropy in Zr alloys lies in their HCP crystal structure, referred to as the α phase. In Miller-Bravais indices, the hexagonal structure, planes, and orientations are depicted in Figure 1a,b. The c/a ratio of the HCP structure is 1.593 at RT, but increases to 1.597 at 773 K [71,72]. This change in the c/a ratio causes variations in thermal expansion coefficients in different crystallographic directions. For a single crystal of Zircaloy-2, these coefficients are 11.4×10^{-6} K^{-1} in the $\langle a \rangle$ direction and 5.7×10^{-6} K^{-1} in the $\langle c \rangle$ direction. This disparity in thermal expansion coefficients significantly affects slip activation at different temperatures by influencing lattice slip resistance [72,73] and the hierarchy of close-packed planes. It also results in the formation of thermal residual strains/stresses in addition to deformation-induced ones. The magnitude of these residual stresses in the basal and prismatic planes is considerable (+10^{-3} for (0002) and -10^{-3} for $(10\bar{1}0)$) [74]. These residual stresses [12,75] significantly impact the alloy's deformation behavior and in-service performance [26,76].

2.2. Effect of Temperature and Orientation on Microstructure Development

For Zr alloy components, fabrication (from cast ingots to final tubes and rolled sheets) involves stages of TMP designed to achieve the necessary dimensions. Typical Zr alloy component fabrication includes hot, warm, and cold working. Unlike high symmetry metals, deformation in HCP Zr alloys is highly dependent on the activation of different slip and twin deformation modes, as shown in Figure 1a,b, which are sensitive to crystallographic orientation and strain. Crystallographic orientation becomes a dominant factor \ in slip activation and, more dominantly, in twin activation when the loading direction is considered. Generally, deformation along the c-axis is challenging, whereas deformation along the $\langle a \rangle$ direction is less difficult at and above ambient temperatures. Deformation along the c-axis requires the activation of pyramidal $\langle c + a \rangle$ slip, which inherently has a high CRSS compared to deformation along non c-axes, which only requires the activation of prismatic slip. It is noteworthy that slip and twin activation also depend on the deformation temperature, and the CRSS for slip varies significantly, as discussed in the literature and shown in Figure 1c [16–18,77–79]. However, the type of twin formation, whether primary tensile $(10\bar{1}2)\langle 10\bar{1}\bar{1} \rangle$ or compressive, depends on the deformation loading direction at both RT and sub-zero liquid nitrogen (LN) temperatures, showing little sensitivity to deformation temperature.

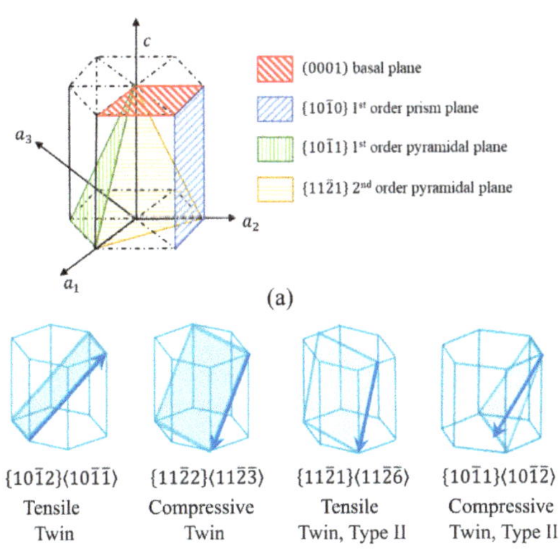

Figure 1. *Cont.*

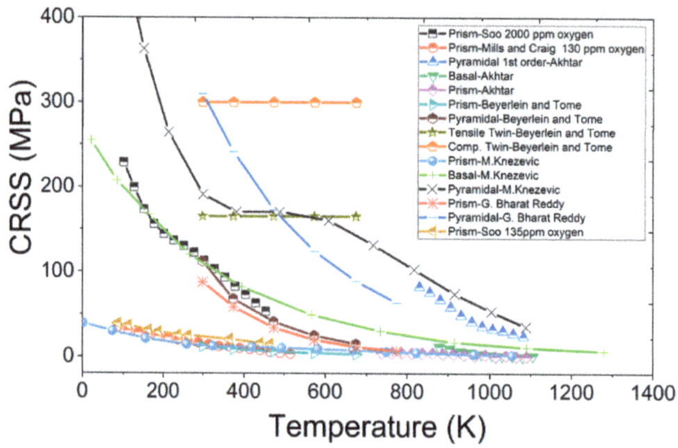

(c)

Figure 1. Illustrations showing (**a**) common slip planes and (**b**) twin deformation modes in hexagonal Zr, data from [6,72], and (**c**) combined plot showing the activation CRSS for different slip and twin mechanism as a function of working temperature, data from [16–18,80–83].

In case of deformation along non *c*-axes, i.e., in-plane (IP) compression of material with basal texture, tensile twinning, and prismatic slip consistently occurs at RT and LN temperatures. At elevated temperatures, a combination of basal and prismatic slip mechanisms accommodates the deformation. Upon twin formation, subsequent deformation of the twin domains predominantly occurs by pyramidal and $(11\bar{2}2)\langle 11\bar{2}\bar{3}\rangle$ twinning at lower temperatures, and basal slip and $(10\bar{1}1)\langle 10\bar{1}\bar{2}\rangle$ twinning at higher temperatures. Compressive twins do not form at moderate strains but appear within the primary tensile twins at higher strains, contributing substantially to plastic deformation, as shown in Figure 2g–i. The formation of twins is also evidenced by a higher work hardening rate shortly after yielding, as shown in Figure 2j, although it does not significantly contribute to plastic deformation.

In grains with *c*-axes aligned along the loading direction, tensile twinning during IP compression reorients the crystal *c*-axes by 85.2° so that they align with the compression axis near the equator. In this case, slip mechanisms facilitate the migration and spreading of pole intensities along the compression axis towards the equator. In through-thickness (TT) compression, prismatic slip and $(11\bar{2}2)\langle 11\bar{2}\bar{3}\rangle$ dominate at sub-zero temperatures, transitioning to a combination of prismatic $\langle a\rangle$ and pyramidal $\langle c+a\rangle$ slip at ambient and higher temperatures. It is noteworthy that for TT compression at higher temperatures, basal $\langle a\rangle$ slip begins to contribute alongside $(10\bar{1}1)\langle 10\bar{1}\bar{2}\rangle$ twinning [6,18]. As shown in Figure 2j, a sharp yield and a gradually increasing work hardening rate are observed for TT compression, typically characteristic of deformation twinning. Unlike IP compression, during TT compression at sub-zero conditions, the compression axes are oriented close to the *c*-axes containing $(11\bar{2}2)$ compression twins, while $(10\bar{1}2)$ tensile twins are found in grains with compression axes oriented far from the *c*-axes, as depicted in Figure 2d–f. The compressive twinning reorients the *c*-axes from near the pole (compression axis) to near the equator [6,18,84]. The loading directions (TT and IP), deformation temperature, and deformation modes contribute to distinct heterogeneous microstructural developments, including slip-induced internal misorientations and SLs features observed in the deformed samples [6,85].

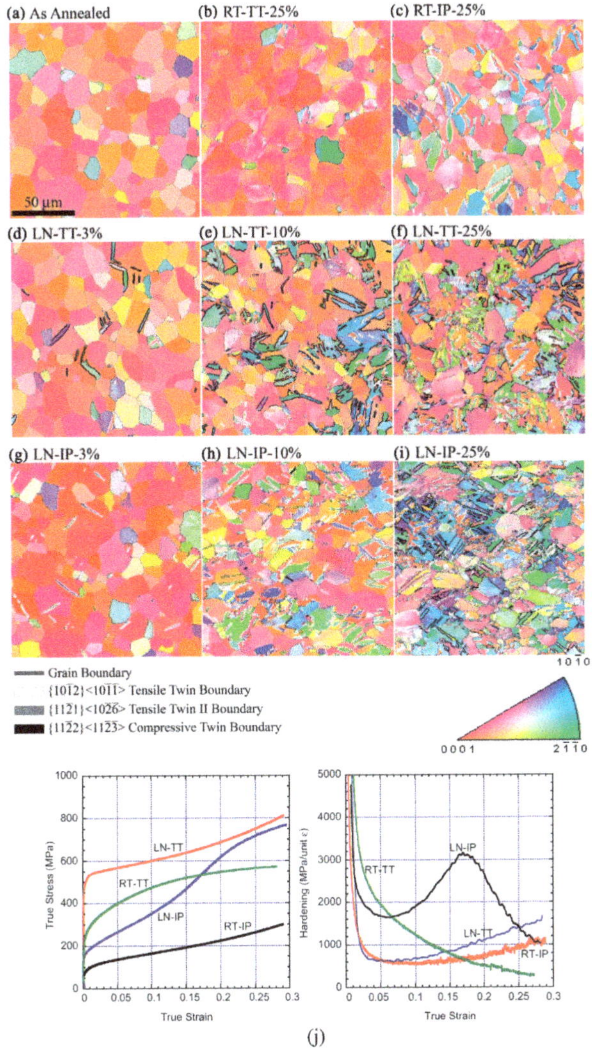

Figure 2. (**a–i**) IPF maps at various deformation stages for a clock rolled and annealed, high-purity textured polycrystalline Zirconium at 76 K and 298 K in the through-thickness (TT) and in-plane (IP) directions. Deformation conditions are labeled as room temperature (RT), and liquid nitrogen (LN). (**j**) Representative stress-strain curves for TT and IP deformation conditions at RT and LN [6]. (Reprinted with permission from ref. [6]. Copyright 2007, Taylor & Francis).

2.3. Effect of Temperature and Orientation on Texture Development

The deformation temperature and the starting crystallographic orientation, along with the deformation mode, profoundly impact the texture development of Zr. Rolling reductions of up to 70% at room temperature (RT) and liquid nitrogen temperature (LN) demonstrate significant differences in texture evolution, as analyzed by electron backscatter diffraction (EBSD), as shown in Figure 3a [86]. At RT, rolling primarily activates dislocation slip, leading to the presence of non-deforming grains with their *c*-axes oriented near the normal direction (*ND*) of the Zr sheet, rendering them unfavorable for slip. However, at LN temperatures, both slip and twinning are activated, with compressive twinning deforming these non-deforming grains and promoting more uniform microstructures. This difference in deformation behavior leads to varied texture intensity development. In RT-rolled specimens, the angle between basal pole peaks and the *ND* decreases slightly due to the activity of $(10\bar{1}1)\langle 11\bar{2}3\rangle$ pyramidal slip, resulting in the strengthening of the basal texture as compared to LN-rolled specimen, see Figure 3(a-i) [86]. However, compared to as-received, the overall texture intensity significantly drops with progressive deformation. A similar texture development is observed in commercially pure Zr and Zr-2 when uniaxially compressed along the *ND* [84]. The RT condition primarily activates dislocation slip, maintaining certain grains undeformed due to their unfavorable orientations for slip. In contrast, LN rolling not only activates slip but also induces twinning. The early initiation of twinning in LN-rolled specimens leads to crystal reorientations that help mitigate the centralization of the bimodal basal texture. Twinning-induced reorientations contribute to the strengthening of the $(10\bar{1}0)$ texture since twinning reorients the basal pole by 85° away from the *ND* [84,87], see Figure 3(a-ii). This results in a more homogeneous microstructure and improved texture uniformity compared to RT-rolled specimens. The influence of initial orientation is also critical in determining the subsequent annealing behavior, as seen in Zr-4 alloy sheets compressed along *ND* and rolling direction (*RD*) [88]. As already discussed, *ND* deformation involving extensive pyramidal $\langle c + a\rangle$ slip and minimal prismatic slip doesn't lead to significant evolution of texture, see Figure 3(b-i,b-ii), which is also confirmed by Visco-plastic self-consistent (VPSC) simulations [18,88]. On the other hand, deformation along the *RD* involves a significant amount of twining, identified to be of type $(10\bar{1}2)$ tensile twin. Since during *RD* compression the *ND* axis experiences elongation, tensile twin activation is favored over compressive twinning. The tensile twinning thus leads to enhancement of $\langle 11\bar{2}0\rangle \parallel RD$ component during *RD* deformation, see Figure 3(b-iii,b-iv). However, higher work-hardening rates and stored dislocation density in *ND* compression lead to differing annealing behaviors at higher temperatures [88].

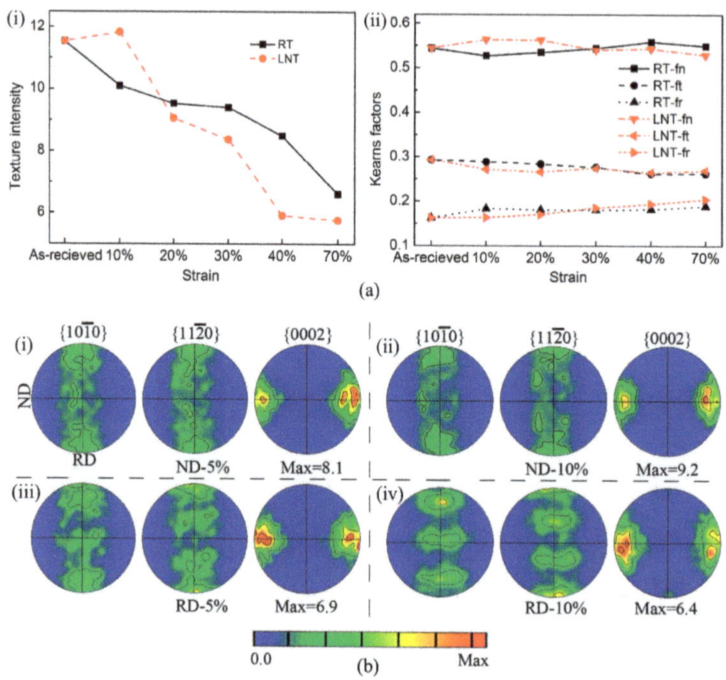

Figure 3. (**a**) Evolution of (**i**) crystallographic texture density and (**ii**) Kearns factors in the rolled specimens at RT and LN temperatures (LNT). The fn, ft, and fr represent the Kearns factor in normal,

tangential, and radial directions of the pilgered tube [86]. (**b**) Texture evolution during compressive deformation along the ND (**i**,**ii**) and RD (**iii**,**iv**) directions at 5% and 10% strains [88]. (Reprinted with permission from refs. [86,88]. Copyright 2015 and 2018 Elsevier).

2.4. Microstructure and Texture Development during Pilgering

Hot working of Zr involves hot extrusion and specialized methods such as pilgering, a deformation technique that involves the simultaneous reduction of the outside diameter, inside diameter, and wall thickness of the tubes over the working length under a pair of dies. The dies have semi-circular tapered grooves cut on them, while a tapered mandrel, with a contour matching that of the dies, controls the wall thickness during deformation [10,15]. Pilgering depends on the activation of slip-twin systems [6,7,16], providing triaxiality of stresses, thereby achieving higher formability [15]. However, the process also involves heterogeneous deformation, where some grains get fragmented while others do not [19]. Pilgering deformation in Zr alloys results in gradual texture development from an initially non-random texture with well-defined texture fibers. Pilgering enhances the $\langle 1\bar{1}00 \rangle$ and $\langle \bar{1}100 \rangle$ fiber texture intensities with an increase in the effective strain ($\bar{\varepsilon}$), as shown in Figure 4. The Kearns factors are often used to measure the texture developments in the HCP materials. The Kearns factor (f) [89,90] represents the fraction of basal poles in one of the three principal directions. In the case of pilgering, the radial Kearns factors (reference direction: *ND*) exhibit an increasing trend with effective strain, as evidenced by the increase in basal pole texture shown in Figure 4c. Conversely, axial and circumferential Kearns factors display a decreasing trend. This indicates that during pilgering, grains undergo rotation and align their *c*-axes closer to the *ND* direction, as depicted in Figure 4d [7,10]. It is also worth noting that pilgering induces texture asymmetry between the $\langle 1\bar{1}00 \rangle$ and $\langle \bar{1}100 \rangle$ fibers depending on the tube location with respect to the pilgering mandrel [10].

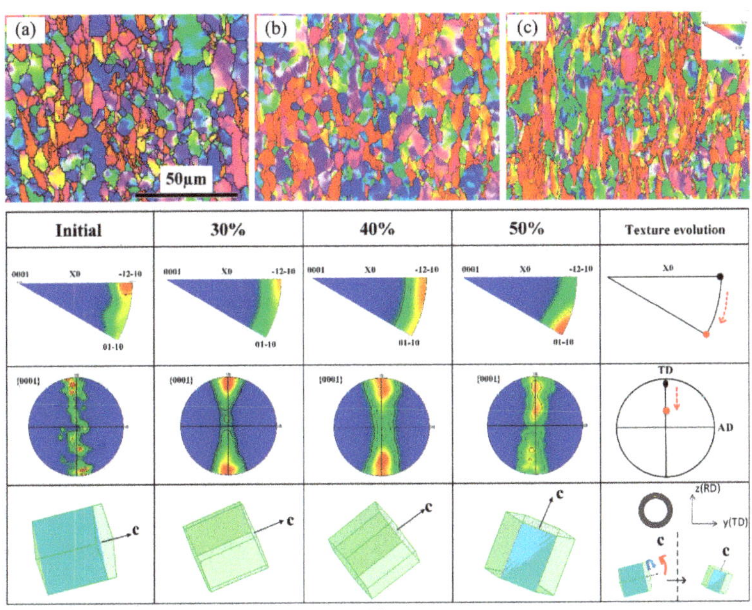

Figure 4. EBSD inverse pole figure (IPF) maps showing the microstructural developments for a Zr-4 tube at (**a**) 30%, (**b**) 40%, and (**c**) 50% pilgered condition. (**d**) The corresponding crystallographic texture evolution path explained with the help of inverse pole figure and crystal rotation schematic. (Reprinted with permission from ref. [7]. Copyright 2019, Elsevier).

2.5. Grain Fragmentation during Deformation

Zr undergoes heterogeneous deformation, leading to notable features such as the generation and decay of deformation twins and orientation-dependent fragmentation of grains. Grain fragmentation behavior has been reported for cold-rolled (CR) Zr sheet material undergoing plane strain deformation [19], unidirectional rolling [19], simple uniaxial deformation [6,91], and pilgering [10,12]. This heterogeneous microstructure, which develops during deformation, persists after thermal annealing [23] and significantly influences the microstructure-property relationships. The deformed microstructures are often multi-scale, as microstructural heterogeneities occur over a length scale from macro to meso and finally micro. Even a small imposed strain (only a few percent under plane strain compression (PSC)) can induce a significantly higher near-boundary mesoscopic shear strain, approximately an order of magnitude higher [20]. Strikingly, significant differences in plastic strain are locally more observable at specific microscopic features such as GBs and the second phase. These lead to SLs [92], which are essential contributors to grain fragmentation. Direct observations on plane strain-deformed microstructure show that mesoscopic strain distributions are responsible for residual stress (see Figure 5a) and orientation gradient development (see Figure 5b). Significant differences in mesoscopic strain further create local regions of high dislocation density near GBs, acting as precursors to grain fragmentation. Since grain fragmentation is orientation-sensitive, samples deformed along the rolling direction (*RD*) exhibit a different grain fragmentation behavior compared to samples deformed along the *ND*. The origin of the orientation-dependent deformation associated with grain fragmentation is linked to the activation of prismatic and pyramidal $\langle c+a \rangle$ slip [93]. With increasing temperature, the anisotropy between the CRSS of pyramidal $\langle c+a \rangle$ and prismatic $\langle a \rangle$ slip increases, similarly affecting grain fragmentation behavior.

2.6. Crystallographic Texture Evolution during Rolling and Annealing

Another intriguing aspect of the multi-scale heterogeneous deformed microstructure evolution is crystallographic texture development. Most texture measurements focus on the basal pole distribution, which is significant in terms of the anisotropy of HCP metals [70,84,94]. Little attention has been paid to the distribution of prism or pyramidal plane poles, as they are not as crucial for the mechanical behavior of the material as the basal pole distribution. However, they are strong indicators for the degree of annealing and should be quantified to determine the TMP effects on grain reorientation. The definitive orientation of the basal plane (0002) in all HCP metals under PSC deformation, such as cold-rolled semi-finished commodities (e.g., wire, sheet, or tubing), is parallel to the direction of elongation. Depending on individual properties, deviations from this desired orientation, i.e., tilts of the basal plane and their rotation around its pole, indicate distinctive variances in material deformation.

Variation in the deformation process (applied strain, deformation temperature) also produces a characteristic texture component in both deformed and annealed material for the same starting texture configuration [96]. These are summarized in Table 2 for cold-work stress-relieved (CWSR) and recrystallized (Rex) specimens of typical Zircaloy-4. A schematic showing the typical annealing and deformation textures for various deformation processes and different product types, such as tube, wire, and sheet specimens, is also illustrated in Figure 6a. For tube specimens, depending on the ratio of R_w (reduction in wall thickness) and R_D (reduction in tube diameter), significant mechanical anisotropy in HCP Zr alloys is observed during deformation. Understanding the textural changes that occur during TMP, particularly during DRx and static recrystallization (SRx), is of critical importance. Most reviews of hot deformation related to DRx focus on flow softening and the corresponding appearance of an inflection point in strain hardening behavior [14]. However, the origin of DRx grains in hot deformed microstructures is related to mechanisms such as continuous DRx (CDRx), geometric DRx, and flow softening leading to standard discontinuous DRx (DDRx). The DRx can be associated with flow saturation in specific orientations; no indication of work softening is noted. This can be evident from the variation in deformation flow behavior in Zr systems and its effect on grain recrystallization [88,97], as shown in Figure 6b. Since initial orientation is critical for deformed microstructure development, it naturally leads to orientation-dependent DRx behavior. As the angle between the deformation axis and c-axes increases, the DRx effect is weakened, thus reducing the fraction of recrystallized grains. It is suggested that with c-axes moving away from the deformation direction, the easier-to-activate prismatic slip dominates. Hence, the practical stacking faults and stored energy of grains during deformation are ultimately reduced, unlike the case of pyramidal $\langle c + a \rangle$ slip deformation. Regarding the operating mode of DRx, microstructural and misorientation characteristics show a variety of possible mechanisms, including CDRx [97,98], geometric DRx [99], and DDRx [91,100].

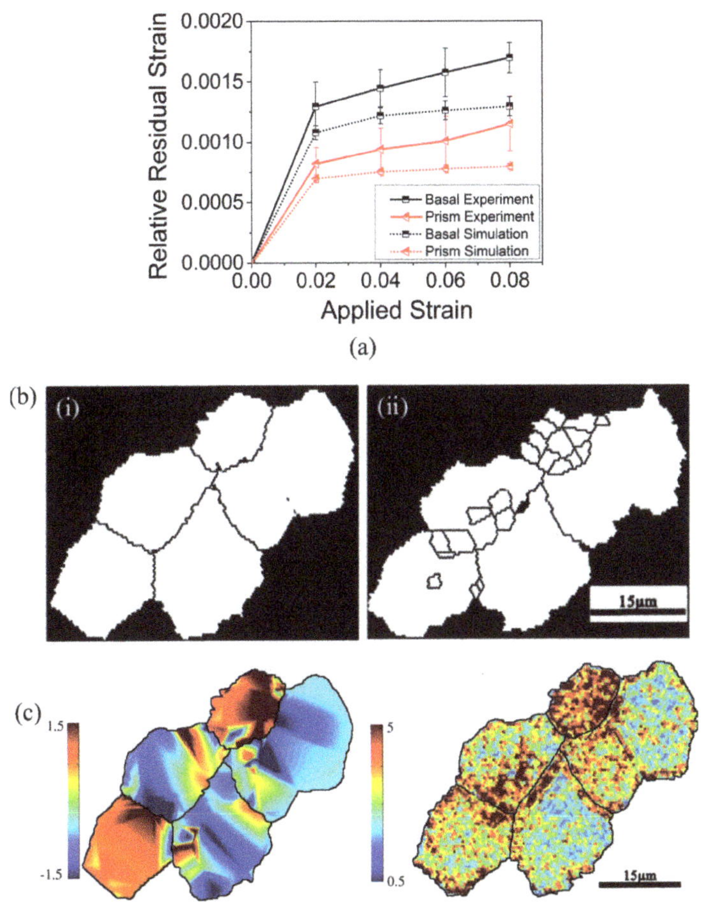

Figure 5. (**a**) Residual strain measurements on Zr-4 sample deformed in an interrupted manner using a tensile loading setup for a final strain of 8%. Specific grains with basal and non-basal crystallographic orientations are tracked during interrupted testing, and their residual strains are identified using laboratory scale micro-Laue technique, which shows a hierarchical development [95]. (**b**) Progressive meso-structure evolution during PSC. Changes in grain shape (or mesoscopic strain) are visible [20] with grain fragmentation after (**i**) 4% and (**ii**) 10% PSC. (**c**) Developments in near-boundary mesoscopic shear (**left**) and in-grain misorientations (**right**) shown after 4% PSC. (Reprinted with permission from refs. [20,95]. Copyrights 2014 and 2020, Elsevier).

Table 2. Ideal orientations in Euler Plots for CWSR and Rex Zircaloy-4 [70].

	CWSR	Recrystallized
φ	Ideal Orientations ($\{hkil\}[hkil]$)	Ideal orientations ($\{hkil\}[hkil]$)
0°	$\{0002\}$ $\{\bar{1}011\}[1\bar{2}10]$ $\{1011\}[1\bar{2}10]$	$\{\bar{1}016\}[5\bar{4}11]$ $\{\bar{1}012\}[2\bar{3}10]$ $\{\bar{1}012\}[2\bar{3}10]$ $\{\bar{1}010\}[1\bar{2}12]$
10°	$\{0002\}[1\bar{3}20]$ $\{1012\}[1\bar{2}10]$	$\{\bar{5}149\}[1\bar{2}10]$
20°	$\{\bar{1}011\}[1\bar{3}10]$ $\{\bar{3}125\}[1\bar{4}30]$	$\{\bar{3}128\}[1\bar{5}40]$
30°	$\{\bar{2}118\}[0\bar{1}10]$ $\{\bar{2}112\}[0\bar{1}10]$	$\{\bar{2}117\}[0\bar{1}10]$
40°	$\{\bar{1}1011\}[0\bar{1}10]$ $\{\bar{3}218\}[0\bar{1}10]$	$\{\bar{3}219\}[0\bar{1}10]$
50°	$\{1\bar{1}04\}[1\bar{4}50]$	$(\bar{1}120)\ [1560]$

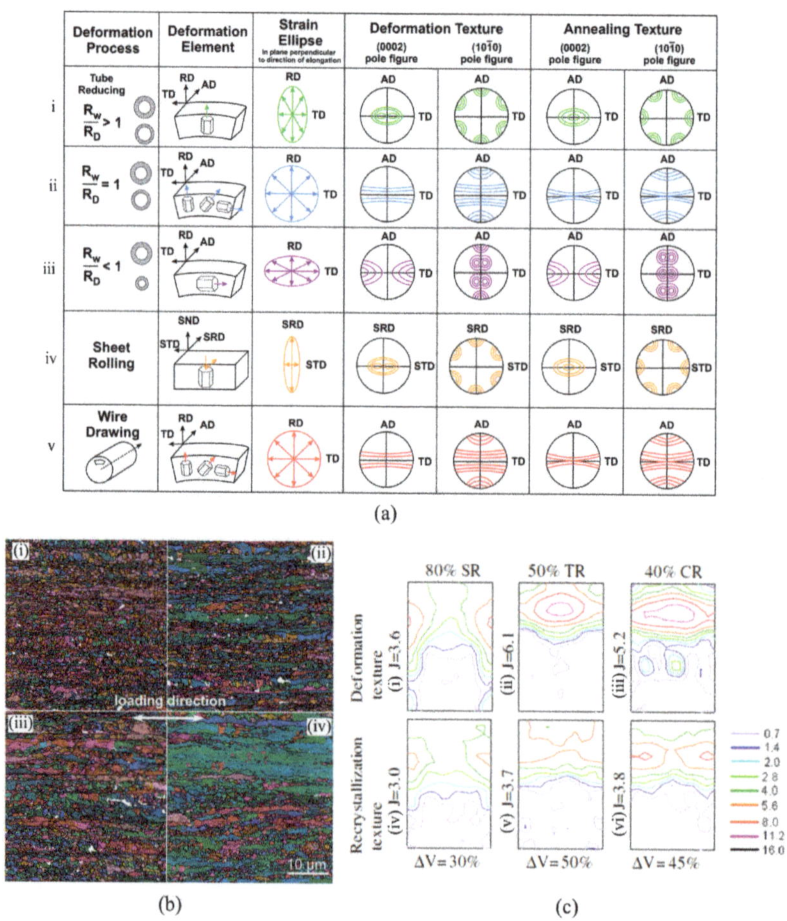

Figure 6. (**a**) Classical representation of deformation and annealing textures associated with different products and deformation modes for Zr. (**i–iii**) represents the tube deformation with varying R_w/R_D

ratio while sheet rolling and wire drawing are shown in (**iv**) and (**v**), respectively. Deformation directions are shown with the deformation element and the strain ellipse, data from [72]. (**b**) DRx evolution as a function of sample orientation at 923 K. The samples are cut such that the compression axis formed (**i**) 0° (**ii**) 30° (**iii**) 60° and (**iv**) 90° tilt angle with the ND direction of the parent sheet. No clear inflection point was observed in the stress-strain response during compression [97]. (**c**) Texture index J and the reoriented volume fraction (ΔV) are presented for each of the three typical conditions (simple [CR], transverse [TR], and cross-rolled) following (**i–iii**) deformation and (**iv–vi**) primary recrystallization for a low-alloyed Zr [101]. (Reprinted with permission from refs. [97,101]. Copyrights 2013 and 2009, Elsevier).

We further discuss the stability of deformation components as a function of deformation strain during typical rolling operations involving 'simple rolling' (SR) (with rolling along the RD), transverse rolling (TR) (rolling along the transverse direction (TD)) and cross-rolling (CR) (with alternate passes along the RD and TD), as shown in Figure 6c. After significant deformation of 80% SR, the final texture remains unchanged after recrystallization, with just a slight reduction in the texture index throughout primary recrystallization [101]. The recrystallization texture in the case of 50% TR material remains substantially weaker than the deformation texture, with a shift in maxima towards orientations with $\langle 10\overline{1}0\rangle \,||\, RD$, culminating in a weak maximum of about $\{0°, 0° - 20°, 0°\}$. During primary recrystallization, the 40% CR specimen also undergoes a considerable textural change: the maximum of the orientation distribution function (ODF) shifts from $\{0°, 20°, 30°\}$ to $\{0°, 20°, 0°\}$, representing a reoriented volume percentage of around 45% [101]. Thus, the deformation direction and the loading mode significantly affect the nature of SRx.

3. Deformed Microstructure and Texture in Titanium

3.1. Microstructure and Texture Evolution during Rolling

Several authors have investigated the microstructure and texture evolution during the CR of Ti [44–47,102]. Gurao et al. [45] investigated the microstructure and texture evolution during unidirectional rolling (UDR) of commercially pure titanium (CP-Ti) with a starting hot-rolled sheet as starting material as characterized by a basal radial texture, i.e., (0001) nearly parallel to ND, as shown in Figure 7a. The microstructure formed after UDR is characterized by elongated grains with a high aspect ratio with the longer axis along the rolling direction. Chun et al. [48] investigated the evolution of microstructure and texture during rolling, observing high twinning activity during the initial stages of deformation (up to about 40%). The evolution of twins is clearly illustrated using EBSD, as shown in Figure 7a–c. The higher fraction of compression twinning and a lower fraction of extension twinning in the early stages of deformation and their absence at higher rolling reductions are apparent from Figure 7d,e.

Though this initial profuse twinning and subsequent suppression is demonstrated by Nourbakhsh and O'Brien [49] using transmission electron microscopy (TEM) (Figure 7f,g), the characteristics of these twinning mechanisms have been understood much later. Rolling of Ti at ambient temperature generally results in a characteristic split-transverse direction (split-TD) type of texture [37,44,46,47,50,103]. This well-known split-TD texture in Ti is characterized by several essential texture components, listed in Table 3, that can be present in varying proportions in the deformed material. However, it should be noted that this type of texture is often characterized by predominantly two components (i.e., E and B), with other components in minor fractions. On the pole figure, several of these components share the same position. Hence, the texture evolution should be examined using the ODF sections to highlight the major changes.

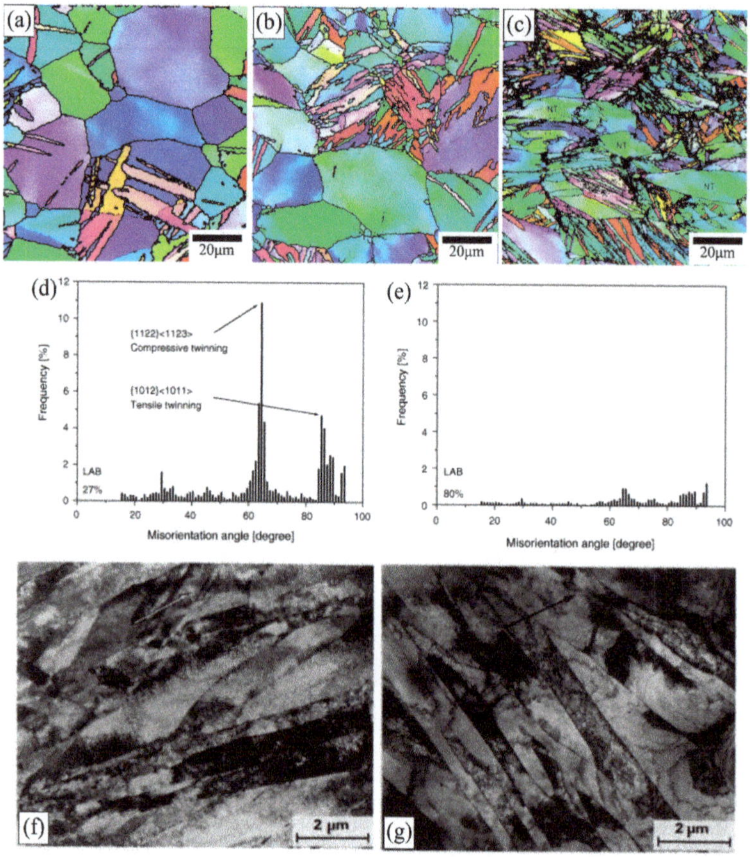

Figure 7. (**a**–**c**) EBSD-IPF maps of CP-Ti specimens at 10%, 20% and 30% rolling reduction, respectively, (**d**,**e**) misorientation angle distribution showing the fraction of twins in 10% and 40%, respectively [48], (**f**) TEM image showing profuse primary and secondary twinning in 40% CR material, (**g**) TEM image showing the absence of secondary twins after short-term annealing [49]. (Reprinted with permission from refs. [48,49]. Copyright 2005 and 1988, Elsevier).

Table 3. Important texture components in the rolling texture of Ti [44].

Texture Components	$\{\varphi_1,\varphi,\varphi_2\}$ Degrees	Description
E	$\{0°, 40°, 30°\}$	$\{0001\}\langle 11\bar{2}0\rangle$ tilted from ND toward TD
B	$\{0°, 40°, 0°\}$	$\{0001\}\langle 10\bar{1}0\rangle$ tilted from ND toward TD
M	$\{50°, 90°, 30°\}$	$\{01\bar{1}0\}\langle 2\bar{1}\bar{1}2\rangle$
D	$\{30°, 0°, 0°\}$	$\{0001\}\langle 11\bar{2}0\rangle$
A	$\{0°, 0°, 0°\}$	$\{0001\}\langle 10\bar{1}0\rangle$
C	$\{0°, 90°, 0°\}$	$\{11\bar{2}\bar{2}\}\langle 10\bar{1}0\rangle$

Gurao et al. [45] reported that a starting basal radial texture evolved into the well-known split-TD type texture with a higher fraction of the E component, as shown in Figure 8a, which displays important texture components in titanium. Conversely, a strong recrystallization texture with high intensity near the E component also developed into a split-TD type texture with a pronounced B component [46]. This characteristic split-TD texture with a strong B component was similarly observed by Ghosh [104], Atasi et al. [44], and Zhong et al. [50] (see Figure 8b,c). However, despite the similar starting texture, Chun et al. [48] noted E component strengthening (Figure 8d,e). Additionally, a weak basal fiber texture and infrequent weak $\langle 10\bar{1}0\rangle$ and $\langle 11\bar{2}0\rangle$ fibers were also observed during the rolling of α-Ti [44,49,51,53,54,103]. In summary, regardless of the starting texture, a split-TD type texture is the most commonly observed cold rolling texture in Ti. The mechanism of the formation of this texture will be explained in the following section.

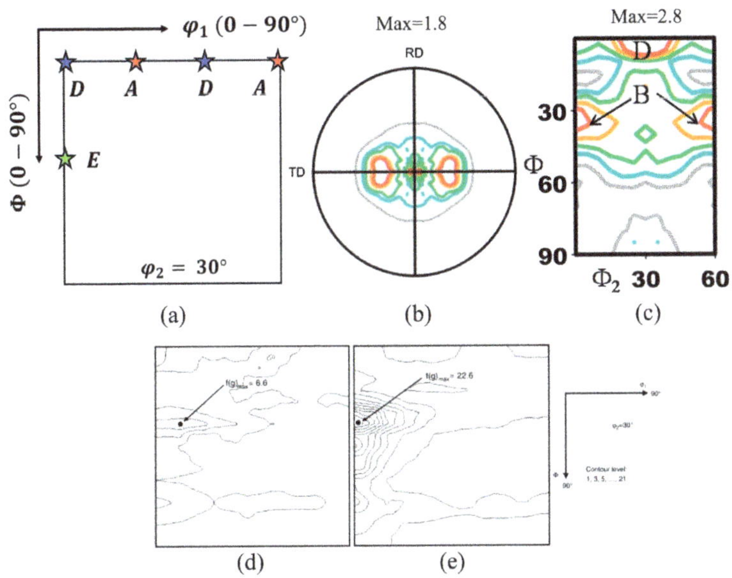

Figure 8. (**a**) $\Phi_2 = 30°$ ODF section showing the position of ideal texture components in the rolling texture of titanium. (**b**) (0001) pole figure showing texture after UDR [104]. (**c**) Cold rolling texture represented using $\Phi_1 = 0°$ ODF section showing texture after UDR [104]. (**d**,**e**) $\Phi_2 = 30°$ ODF sections showing texture after 10% and 80% cold rolling, respectively [48]. (Reprinted with permission from refs. [48,104]. Copyright 2005 and 2019, Elsevier).

3.1.1. Mechanism of Split-TD Texture Formation

Nourbakhsh and O'Brien [49] observed that at 20% deformation, the material exhibited a split rolling direction (SRD-basal poles deviated from ND by ±0–30° towards RD) type of texture. This texture undergoes a rapid transition to the commonly observed split-TD texture (STD-basal poles concentrated at 20° from ND towards TD) before the deformation reaches 40%. It is also observed that twinning was active only during the initial stages of deformation, i.e., up to 40%. These observations are explained by the following sequence of events. Initially, primary twinning occurs in favorably oriented grains, resulting in a strong SRD texture. Secondary twinning in these primary twins converts the SRD type of texture to an STD type. The presence of primary twins at 20% deformation and secondary twins at 40% is demonstrated using optical and TEM micrographs, see Figure 7f. The sudden orientation transition observed between 20% and 40% could not be attributed to slip, as such rapid orientation change cannot be generated by slip. Short-term annealing of 40% deformed specimens reveals that the secondary twins completely disappear due to partial recrystallization (see Figure 7g), and the texture measurement of this annealed specimen reveals SRD texture. This corroborates the earlier proposition that secondary twins are responsible for the STD texture development from SRD texture.

Twinning is suppressed at higher strains due to the extreme grain refinement caused by initial twining. Further deformation above 40% is carried by slip deformation and only results in slow rotation of individual grains around the basal pole and a rotation towards the TD till about 30–40° from ND, where the well-known split-TD texture stabilizes. Using EBSD, Zhong et al. [50] elaborate that primary compression twinning $(11\bar{2}2)\langle 11\bar{2}\bar{3}\rangle$ followed by secondary extension twinning $(10\bar{1}2)\langle 10\bar{1}\bar{1}\rangle$ results in the split-TD type texture formation with the metastable split RD texture at 20% strain [44,48,50]. This sequence of primary compression twinning (CT) followed by secondary extension twinning (ET) is also observed to be active in cryo-rolled titanium, as illustrated in Figure 9 [105]. It is further understood that grains not favorably oriented for twinning are termed 'white grains' (i.e., grains with (0002) making an angle of 30–70° with ND in ND − TD plane). Under the combined action of basal and prismatic slip, these grains slowly reorient towards the [0, 40, 0] position (B component) [44,103]. A more detailed sequence of events has also been proposed to explain the texture evolution. Primary compression twinning takes the B component to the M component, which undergoes secondary extension twinning, taking it to the D component [50].

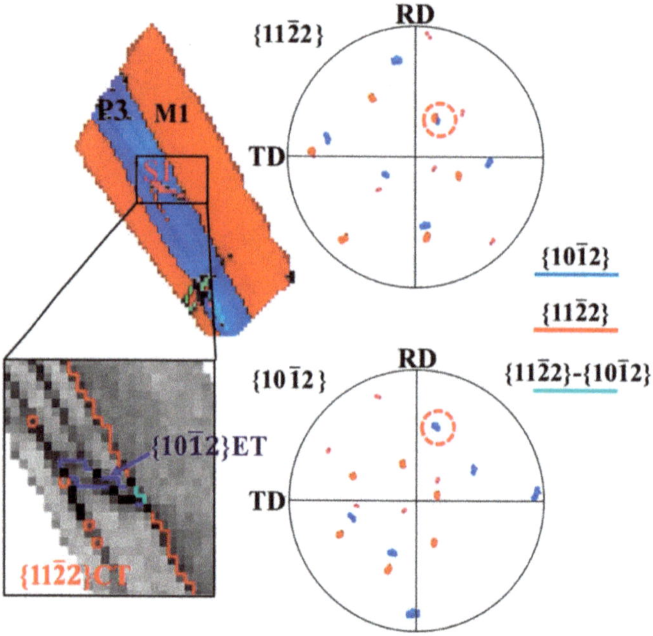

Figure 9. EBSD-IPF map and pole figures showing primary (compression) and secondary (extension) twinning in the rolled CP-Ti [105]. Note-The red circle in the $\{11\bar{2}2\}$ pole figure indicates the coincidence point between the parent grain and the primary compression twin. The red circle in the $\{10\bar{1}2\}$ pole figure indicates the coincident point between the primary compression and the secondary extension twin. (Reprinted with permission from ref. [105]. Copyright 2023, Elsevier).

Similarly, twinning events take the E component to the M′ and then to the A component. However, it is observed that the E component twinning occurs with a much lower probability. The gradual strengthening of the B component occurs due to the rotation of E towards B due to dislocation slip. A reverse mechanism of B reorienting towards E could strengthen the E component, as observed by several others [45,48]. This could be due to a higher probability of E component grains twinning than B due to local incompatibility or grain size effects [36]. Nevertheless, in either case, the characteristic split-TD texture evolves. Hence, the deformation texture evolution in Ti can be categorized into two distinct regimes: (i) the low to intermediate strain level, where texture evolution is predominantly due to compression and extension twinning, and (ii) the high strain regime, where texture evolution is relatively sluggish and is caused solely by slip.

3.1.2. Influence of Strain Path

Two-step cross-rolling (TSCR), multi-step cross-rolling (MSCR), and reverse rolling (RR) are other types of commonly studied strain paths in rolling [44,45]. It is observed that Kernel Average Misorientation (KAM) is low for the cross-rolled specimens. Also, viscoplastic self-consistent simulations (VPSC) show that the average number of active slip systems in the cross-rolled specimens is higher than in the UDR and RR specimens [51,53]. This observation has been used to substantiate microstructural observations. When more slip systems are active, it is well known that the lattice rotation caused would be lower, resulting in lower KAM as seen in cross-rolling. The lower KAM in the cross-rolled specimens is also attributed to the stacking fault energy (SFE) difference between the basal and prismatic planes. Higher basal slip system activity in the cross-rolled specimens can result in a higher amount of cross slip from the high SFE basal plane to the low SFE prism plane, resulting in a lower KAM in cross-rolled samples. Furthermore, the texture post-deformation is also different between the UDR and MSCR samples (see Figure 10a,b, respectively). While the intensity of the E component is higher for the UDR samples, the fraction of the M and C components is higher for the cross-rolled samples. A very strong $\langle 10\bar{1}0 \rangle$ and $\langle 11\bar{2}0 \rangle$ fiber is also present in the cross-rolled samples. The strength of these fibers is relatively lower for the UDR samples. The difference in texture between the UDR and MSCR samples is attributed to the difference in slip activities.

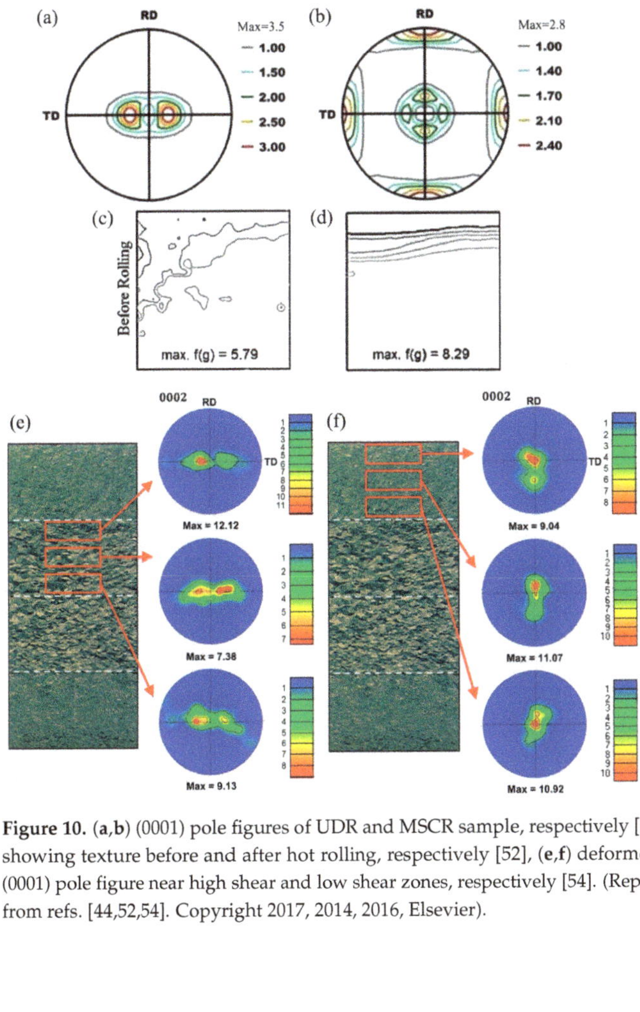

Figure 10. (**a**,**b**) (0001) pole figures of UDR and MSCR sample, respectively [44], (**c**,**d**) ODF sections showing texture before and after hot rolling, respectively [52], (**e**,**f**) deformed microstructure and (0001) pole figure near high shear and low shear zones, respectively [54]. (Reprinted with permission from refs. [44,52,54]. Copyright 2017, 2014, 2016, Elsevier).

3.1.3. Mechanism of Texture Formation during Cross Rolling

Due to the 90° alternation in the rolling direction, the activity of deformation mechanisms is altered during MSCR [44,51]. The slip or twin systems that are not active while rolling along one direction can get activated while rolling along the orthogonal direction. Due to the activation of these systems, latent hardening of the prismatic slip system would be higher. Hence, it would result in the reduction of prismatic slip and a simultaneous increase in basal slip activity [50]. A sequence of events can also be envisaged to explain the texture formation. Initially, grains favorable for twinning undergo compression twinning, resulting in a split-RD type of texture. However, due to the change in rolling direction, secondary extension twinning will not get activated, and hence, the transition to split TD texture will not happen in these grains. This is corroborated by the relative intensities of the contraction twin component (M) and extension twin component (D) in the UDR and MSCR samples [44]. Hence, the split-RD texture stabilizes at higher strains. Additionally, the 'white grains' undergo contraction twinning due to rolling along TD and form the split-TD type of texture, which is further stabilized when rolling along RD by activating basal slip. Due to the simultaneous development of split-TD and split-RD components, an overall weak texture with a split along TD and RD is developed. Twinning in white grains is also responsible for the overall weakening and the difference in texture from UDR [44].

3.1.4. Influence of Temperature

At high rolling temperatures around 873 K or 1073 K, a strong basal fiber texture ($\langle 0001 \rangle \parallel ND$) is commonly observed, see Figure 10c,d [47,52]. To understand the temperature effect, CP-Ti with an initial random texture is hot rolled until 90% under three different strain paths: UDR, MSCR, and RR. The microstructure remained essentially equiaxed for all the deformation paths and did not elongate like in the CR condition. A higher fraction of twins is observed in the MSCR sample, which can be attributed to the change in rolling direction during every alternate pass. Decreased low-angle GB (LAGB) fraction at higher deformation levels indicates DRx. The texture post-deformation is a homogeneous basal fiber ($(0001) \parallel ND$), independent of the strain paths. The volume fraction of all components except D and A, which are essentially part of the basal fiber, is very low. This kind of texture is also observed by Bahl et al. [47] in CP-Ti after hot rolling at 873 K and 1073 K. This texture differs entirely from the split-TD type texture that it develops after cold rolling. This difference is often rationalized by the higher activity of basal slip at elevated temperatures. It is well known that the CRSS for basal slip is considerably lowered at elevated temperatures [106,107]. Once the basal slip gets activated, rolling deformation, which essentially involves compression along ND, rotates the basal planes to align perpendicular to ND, i.e., $(0001) \parallel ND$ [108].

3.1.5. Influence of Shear

Rolling is generally a plane strain deformation process. However, under certain conditions, shear strain components can also be induced during the rolling process. Milner et al. [54] observed a sheared microstructure in the regions near the surface of a CR sheet due to higher shear strain components near the surface. This deformation state is also observed in conditions where friction between the rolls and the sheet is not negligible or in processes like asymmetric rolling where the roll speeds are different. The microstructure and texture have also been distinct between the low shear zone (central regions) and the high shear zones (near the surface); see Figure 10e,f. The grains appear elongated, highly deformed, and inclined by approximately 50° from the normal axis due to intense shearing. A strong basal fiber texture is observed in the high shear zone, while a characteristic split-TD texture prevails in the no-shear zone. Yang et al. [53] studied the influence of shear components of strain on the texture evolution in rolling. It is shown that during pure plane strain deformation, the E component and a heterogeneous $\langle 10\bar{1}0 \rangle \parallel RD$ are formed. However, with the increase in the shear component, a strong basal fiber, $\langle 0001 \rangle \parallel ND$, and a partial $\langle 11\bar{2}0 \rangle \parallel RD$ formed. This variation is attributed to a higher basal and pyramidal slip activity, which has been confirmed by VPSC simulations. This also corroborated the observations made by Milner et al. [54].

3.1.6. Influence of Grain Size

As discussed in the previous section, the role of twinning in the deformation texture and microstructure evolution is very pronounced in Ti. To understand the evolution when twin mechanisms are inactive, Gurao and Suwas [51] performed unidirectional rolling of submicron grain (SMG) size material generated using equal channel angular extrusion (ECAE) and cryo-rolling. The annealed specimen after these operations had a grain size of 500 nm. This SMG material was subjected to unidirectional rolling at RT to a 90% reduction in thickness. The starting texture was a weak shear-type texture generally observed in ECAE-processed Ti [109]. The characteristic texture formed was a split-TD type texture like microcrystalline Ti. However, a close inspection of the ODF revealed that the fraction of the B component was relatively lower compared to the microcrystalline sample. However, a high fraction of the E component was also observed. Furthermore, the overall texture intensity was low for the SMG sample. An interesting observation was the low fraction of M and D components in the deformation texture, pointing towards a lower activity of twins, unlike microcrystalline material. The absence of twins was confirmed using TEM, which also provided evidence of slip activity at very high strain levels. The suppression of twinning with a decrease in grain size is a well-known phenomenon attributed to the increase in twin nucleation stress [110,111]. This increase in twin nucleation stress is attributed to the hindrance of the coordinated movement of $\langle 10\bar{1}0 \rangle$ and $\langle 11\bar{2}3 \rangle$ zonal partial dislocations. At the sub-micron length scale, the adjacent slip planes are weakly coupled by threading partial dislocations, which hinders the movement of partials on the adjacent planes, thus hindering twinning. Consequently, a sluggish texture evolution, which is solely due to the activity of basal, prismatic, and pyramidal $\langle c + a \rangle$ slip systems, occurs.

3.2. *Microstructure and Texture Evolution during Extrusion and Drawing*

Extrusion of CP-Ti produces a fiber texture symmetric about the extrusion axis [55]. Here, the fiber is characterized by basal poles deviated by ~90° from the extrusion axis and spread along the TD line. The extrusion texture does not change significantly, even up to about 623 K or 723 K extrusion temperatures. The microstructure revealed the beginning of recrystallization at 623 K and 723 K. The RT extruded samples showed a large scatter. However, the samples extruded at high temperatures showed a fiber texture symmetric about the extrusion axis. During extrusion, the deformation state is indirect compression, where the specimen experiences a compressive load normal to the extrusion axis. Due to this, the slip planes tend to rotate until they are normal to the compression axis. Since the active deformation mode is basal slip at high temperatures, the basal planes tend to align normally to the compression axis. In axisymmetric products like extruded rods, this manifests as a fiber texture symmetric about the extrusion axis. Sabat et al. [56] investigated the cold drawing of CP-Ti wires and reported a fiber texture symmetric about the drawing axis. The deformation state in the drawing is also an indirect compression normal to the drawing axis, and the fiber-like texture would only develop if basal slip is activated. However, with the prismatic active slip system at RT, the rationale used for hot extrusion may not apply. However, since the reduction per step and the strain rate are high, the difference in strain rate sensitivity between the basal and prismatic slip systems might cause a difference in their relative activities, resulting in the fiber texture [56].

3.3. *Microstructure and Texture during Annealing*

CP-Ti is found to have the widest application in heat exchangers and storage tanks, where the starting material is always a rolled and recrystallized sheet. The most commonly used annealing treatment is performed at 973 K for 2 h. Several reports discuss the so-called "recrystallization texture" of Ti after annealing in this temperature range. Wagner et al. [103] conducted a systematic study to understand the microstructure and texture evolution mechanism. The evolution of recrystallization texture and microstructure in low alloyed Ti was studied during the annealing of 80% CR sheets. The primary intention was to understand the primary recrystallization stage.

3.3.1. Microstructure and Texture Evolution during Primary Recrystallization

Primary recrystallization in CP-Ti (T40) was divided into two stages. The first stage corresponds to a bimodal microstructure with the appearance of new grains in about 80% of the material volume, with about 25% changing their orientation. The second stage was sluggish and characterized by the disappearance of the so-called white grains, which correspond to [0, 45, 0] orientation, which did not twin during deformation. Most of the recrystallization experiments carried out in the study were done at 773 K for various durations. The samples annealed at 773 K for 1 min showed the beginning of dislocation rearrangement into cell structures. Annealing for a slightly longer time evidenced the polygonization process due to recovery processes; see Figure 11a. Squared and hexagonal dislocation networks forming low-angle boundaries were observed even after longer annealing. The cells formed were depleted of dislocations in the interior, and the recovery process was found to be active homogeneously throughout the material. The recrystallization process was highly heterogeneous as certain deformed grains resisted recrystallization even after long-term annealing (Region 1 in Figure 11b). EBSD analysis confirmed that these grains had an orientation very close to the white grains, see Figure 11b. The transition of cells and sub-grains into recrystallized grains separated by high-angle GBs was also vividly demonstrated using TEM [103].

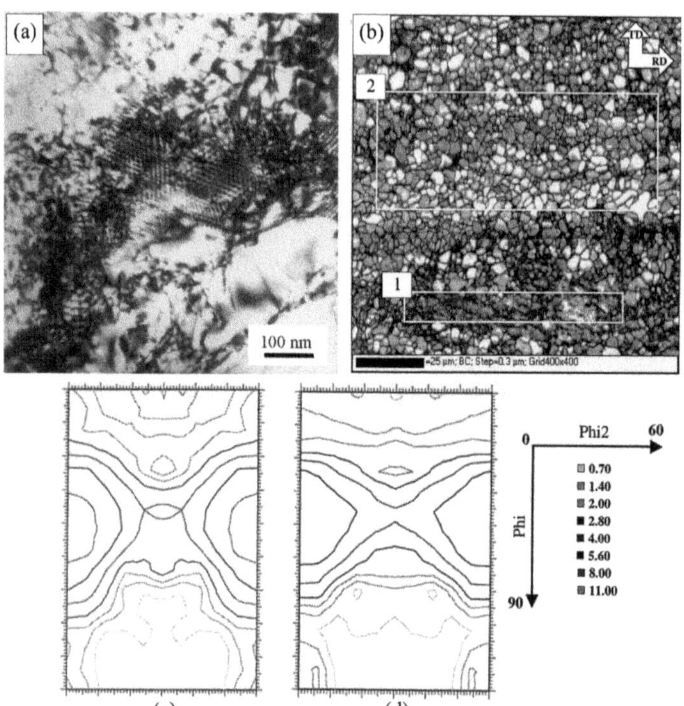

Figure 11. (**a**) TEM image showing the polygonization process during the recovery stage, (**b**) EBSD image quality map of the sample annealed at 773 K for 40 min, (**c**,**d**) ODF of the sample after the first stage and at the end of primary recrystallization, respectively [103]. (Reprinted with permission from ref. [103]. Copyright 2002, Elsevier).

Microbands and elongated lamellar boundaries characterized the microstructure before annealing. As observed by several authors, twinning dominates in Ti up to about 30–40% strain. However, twinning was not observed in orientations close to that of the white grains, which also agrees with Bozzolo et al. [46] observations. The deformation texture was characterized by a main peak at {0°, 45°, 0°}, i.e., $(\bar{1}2\bar{1}3)\langle 10\bar{1}0\rangle$ with a large spread around it, and a weak peak around {0°, 35°, 30°}, i.e., $(10\bar{1}3)\langle 1\bar{2}10\rangle$, which are close to the B and E components common in Ti rolling texture. The texture post-annealing at 973 K for one hour was characterized by the {0°, 32°, 30°}, i.e., $(10\bar{1}3)\langle 1\bar{2}10\rangle$, and {0°, 30°, 0°}, i.e., $(\bar{1}2\bar{1}5)\langle \bar{1}0\bar{1}0\rangle$ and was very similar to the so-called "recrystallization texture" in Ti. These components were also present in the deformed material, but a large spread around these ideal orientations is often seen, which has vanished. By comparing the ODF sections of fully recrystallized, i.e., at 973 K annealed and 773 K annealed samples that correspond to the end of primary crystallization, it was concluded that there was only a moderate variation in the texture during primary recrystallization.

Only minor changes were observed by comparing the ODF sections of the CR primary recrystallized samples; see Figure 11c,d. It was observed that the orientation with low and high ϕ angles disappeared, and more orientations appeared around {0°, 32°, 30°}. Additionally, only 25% of the grains changed orientation, with most of this occurring within the first 60 min of annealing at 773 K. The first stage of primary recrystallization is completed fairly rapidly, with about 80% of the material recrystallizing within the first 40 to 80 min at 773 K. Similar trends of recrystallization kinetics were also reported by Shankar et al. [57]. The second stage, which involves consuming the remaining material that resists recrystallization, is sluggish and completed in about 300 min. It was observed that the grains that changed their orientation at the end of the first stage (about 25%) were essentially those that underwent twinning during the rolling deformation and changed orientation towards {0°, 30°, 30°} during the first stage of primary recrystallization. Continuous recrystallization was identified as the mechanism of recrystallization in these grains, which was reinforced by the extent of recovery in this material (Figure 11a). However, unlike cubic materials, where the recovery process triggers the nucleation of new grains that subsequently grow with large misorientation with the matrix, Ti behaves differently as the texture change is observed to be very moderate [57]. A mechanism leading to recrystallized grains by forming sub-grains through dislocation rearrangement and their coarsening was proposed. The remaining material, other than the white grains, also undergoes early continuous recrystallization and a texture change due to the competition during the growth of these grains occurs. Additionally, the observation of new grains near HAGBs, though very scarce, was attributed to DDRx involving classical nucleation and growth.

The white grains were observed to be deformed with a much more homogeneous dislocation structure, resisting discontinuous recrystallization and not giving rise to significant nucleation. The stability of this particular orientation was attributed to the zero-rotation field around this orientation and the absence of twinning. These grains, which primarily deform by a combination of prism and pyramidal slip activity, recrystallize only in the second stage of primary recrystallization. The mechanism by which these grains recrystallize was understood to be a continuous recrystallization process or extended recovery. In the second stage of primary recrystallization, no significant change in texture occurred. The fact that primary recrystallization did not cause any significant changes to the deformation texture emphasizes that the so-called 'recrystallization textures' generally reported in Ti annealed above 773 K are not exactly recrystallization textures. They are textures that evolve during the secondary recrystallization process or grain growth.

3.3.2. Microstructure and Texture Evolution during Secondary Recrystallization

Bozzolo et al. [58] investigated the secondary recrystallization process by keeping the primary recrystallized material as the initial material. During secondary recrystallization, significant grain growth occurs, following the common trend concerning time and temperature (Figure 12a). When primary recrystallization was complete, the maximum was near

{0°, 35°, 0°}. During secondary recrystallization or grain growth, a broad peak centered around {0°, 35°, 30°} developed. It was also observed that the orientations that disappeared during grain growth were highly misoriented concerning the {0°, 35°, 30°} component and had the smallest grain size. The starting microstructure consisted of regions with small grain sizes and regions with larger grain sizes, likely due to differences in local primary recrystallization kinetics. It was observed that all the small-grained regions disappeared during grain coarsening due to their higher growth rates. The starting material had a texture very close to the deformed state, with a major volume fraction of the {0°, 35°, 0°} component (Figure 12b). The texture after grain growth was characterized by orientations located around {0°, 35°, 30°} with a large spread (Figure 12c). The intermediate stages of grain growth showed these components in varying proportions. Analyzing the differences in the ODF between the starting material and the material after the secondary recrystallization process, the major difference was that {0°, 35°, 30°} increased in volume fraction, whereas the initially major {0°, 35°, 0°} component did not change significantly (Figure 12d). The disappearing orientations were closely located at ϕ_2 near 0°/60°, and ϕ below 20° or above 40°, (Figure 12d). The {0°, 35°, 0°} component also increased, but to a much lesser extent compared to the {0°, 35°, 30°} component. The analysis of the ODF of the smallest and the largest grains separately showed that the global texture obtained after extended grain growth was very similar to that of the largest grains. The texture of the smallest grains was close to that of the initial global texture.

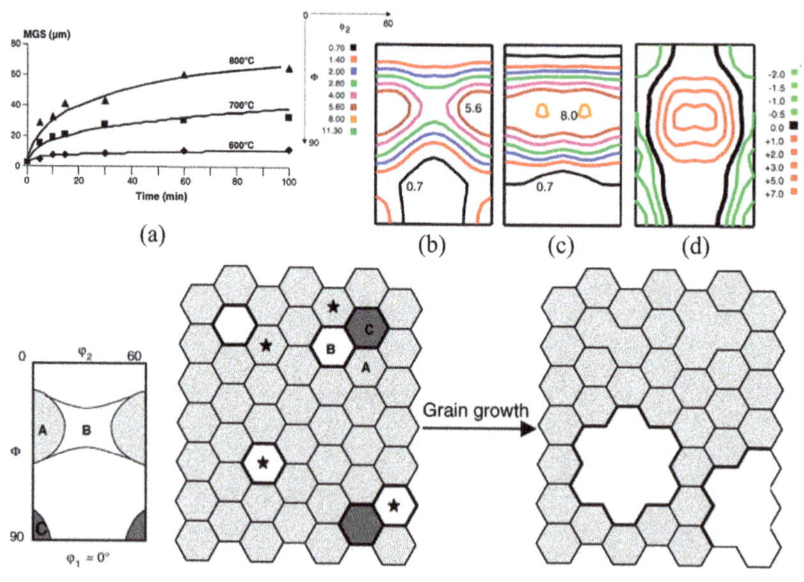

Figure 12. (**a**) Evolution of grain size as a function of annealing time and temperature (**b**) Texture of starting material (after primary recrystallization), (**c**) texture after annealing at 873 K for 30 min, (**d**) ODF difference between c and b, and (**e**) schematic showing the mechanism of texture evolution during secondary recrystallization [58]. (Reprinted with permission from ref. [58]. Copyright 2005 Elsevier).

The development of certain orientations and the disappearance of others was explained using the correlation between grain size, orientation, and the nature of the GBs according to the energy and mobility criteria. It is well known that mobility and energy are high for highly misoriented GBs. The disappearing grains were observed to have a misorientation greater than 30° with the ideal orientation {0°, 35°, 30°}. Hence, these boundaries were considered to have high mobility. The mechanism responsible for texture evolution predominantly involves boundaries with misorientations greater than 30° with higher mobility. Figure 12e shows grains of A, B, and C types, where A and B correspond to {0°, 35°, 0°} and {0°, 35°, 30°}, respectively.

The C-type grains are widely dispersed in the orientation space and have much smaller grain sizes than the A and B grains. The C-type grains also have a misorientation of more than 30° with the growing {0°, 35°, 30°} component. Their disappearance is due to a combined effect of their small grain size and highly misoriented GBs. The initial major texture component, the {0°, 35°, 0°} (A type) grain, grows much less than the grains close to {0°, 35°, 30°}. The grain size distribution of the A and B-type grains is quite similar. This lower growth rate of A-type grains was attributed to a higher proportion of them being present in the initial material, which causes lower GB mobility due to the lower misorientation angle. The A-type or B-type grains can grow to consume each other; however, due to the orientation pinning effect, the B-type grains consuming the A-type grains are more favored. Thus, B-type grains continue to grow, and the corresponding texture component {0°, 35°, 30°} keeps increasing until it becomes predominant. A different texture evolution was observed when the starting CR texture was dominated by the {0°, 35°, 30°} component. At annealing temperatures up to 873 K, the mechanism as described earlier holds good to an extent; however, at higher annealing temperatures, the component around {0°, 35°, 0°} became the major component, and the {0°, 35°, 30°} component almost vanished [44]. A reverse mechanism like the previous one can be envisaged to explain this.

Despite all these observations and proposed mechanisms, it is important to note that the recrystallization texture of Ti is largely dependent upon the deformation texture and the recrystallization temperature. The annealing texture that develops between 773 K and 973 K consists of $(0001)\langle 1\bar{2}10\rangle$ and $(0001)\langle 10\bar{1}0\rangle$, $\pm 30°$ to the TD. When the temperature range is above 973 K, substantial grain growth occurs, and the recrystallization texture consists of either $(10\bar{1}3)\langle 1\bar{2}10\rangle$ or $(20\bar{2}5)\langle 1\bar{2}10\rangle$ components, which manifest as $\pm 30°$ across the TD line on the (0001) pole figure. At temperatures below 773 K, the recrystallization texture is similar to the cold rolling texture. The cross-rolling textures are also different from the unidirectionally rolled textures, and the textures post-recrystallization are also different [44]. The mechanism of recrystallization is also dependent upon the deformed state. Shankar et al. [57] reported sub-grain coarsening in the deformed lath region and random nucleation in the highly deformed regions. 50% CR specimens upon recrystallization showed substantial occurrence of discontinuous sub-grain coarsening and sub-grain coalescence, i.e., oriented nucleation as the recrystallization mechanism [57]. Hence, a unified theory to explain the evolution of recrystallization texture in α-Ti is still absent.

4. Microstructure and Texture Evolution in Magnesium Alloys

4.1. Microstructure and Texture during Annealing

Mg-based alloys are deformed by various processing routes, such as rolling, extrusion, and equal channel angular processing (ECAP). Various rolling processes are performed over Mg alloys to activate the slip systems and modify the grain orientations [112,113]. After hot extrusion, Mg alloy viz. Mg-Y-Zn alloy displays a bimodal microstructure comprising lamellar 14H and block 18R phases. This dual-phase structure enhances the mechanical strength of the Mg alloy by combining the strength of block phases with the deformability of lamellar phases [67]. Hot rolling leads to refined microstructures with 78% and 22% of LAGBs and HAGBs with equiaxed grains. DRx and deformed grains were 68% and 32%, respectively, see Figure 13 [66]. Similarly, hot-rolling annealed samples have HAGBs and LAGBs of 78% and 22%, respectively, with precipitate distribution along the GBs. In the case of Mg-Sn alloys, complete recrystallization occurs at an annealing temperature of 488 K [66]. In the case of ultralight BCC Mg alloys such as Mg-Li-Al alloys, anisothermal aging results in a phase transformation characterized by spinodal decomposition, forming Al-rich zones followed by the nucleation of rod-like θ phase. Subsequent coarsening of the θ phase (Mg_3Al) leads to its transformation into AlLi with a core-shell structure, thereby affecting hardness [68].

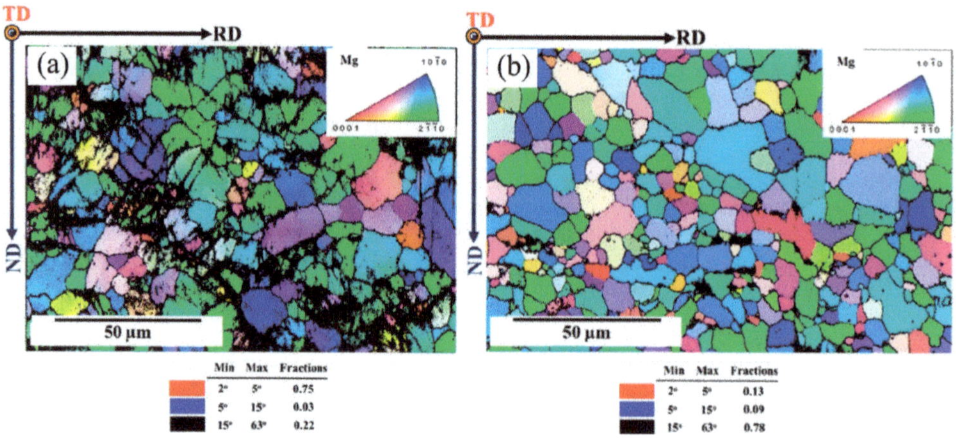

Figure 13. EBSD analysis showing typical inverse pole figure maps of Mg alloy. (**a**) Under hot rolled condition; (**b**) after hot rolling followed by annealing [66]. (Reprinted with permission from ref. [66]. Copyright 2021, Elsevier).

In the case of Mg-Re alloy hot rolling at an initial 10% reduction, the parental grains (PG) exhibit large spreads of off-basal orientations that activate $(10\overline{1}2)$ twins. Twins occupy significant portions of the PG, and these twins can be identified as extension twins, as shown in Figure 14b,c. They exhibit an elliptical or irregular shape and appear thick. The inherent high mobility of $(10\overline{1}2)$ twins enable them to swiftly occupy and reorient the surrounding matrix, typically by an angle of approximately 86.3°. This, in turn, promotes the facilitation of dislocation slip within the twinned region. However, dislocation slips do not dominate due to the even distributions of intragranular misorientation (IGM) in PG. Still, basal $\langle a \rangle$ slip is significantly activated in the twins [114,115]. The basal $\langle a \rangle$ slip is notably activated within the twins due to the presence of IGM. Despite a high Schmid factor (SF), prismatic $\langle a \rangle$ slip tends to exhibit low activity within the matrix of these PG. This can be attributed to the relatively high CRSS required for prismatic $\langle a \rangle$ slip, especially when compared to the CRSS values for $(10\overline{1}2)$ twins and basal $\langle a \rangle$ slip at the rolling temperature of 713 K. At 20% reduction $\{10\overline{1}2\}$ twins activated in the twinned regions and uniform IGM can synergistically activate various dislocation systems within the material. As a result, even though PG with different initial orientations may exhibit a variety of deformation mechanisms and activities, they collectively align the c-axes of most PG parallel to the *ND* [116].

This alignment contributes to the overall development of the basal texture in the material. At 20% reduction, the prevalence of twins within PG significantly increases, as evident in Figure 14d–f. For instance, twin lamellae within 'Grain 2-A' can be observed. In these twin lamellae, twin pairs located at GBs of the PGs are indicated by the dotted arrows in Figure 14d. Moreover, secondary twins can be detected within these twin lamellae, as shown by the solid arrows in Figure 14e,f. Meanwhile, neighboring grains like 'Grain 2-B' and 'Grain 2-C' appear less favorable for twinning activation [118,119]. This highlights the significant role played by the initial orientations of the PG in influencing the twinning response. In the case of Mg-Y-Zn alloy, CR leads to the uniform distribution of SB composed of fine micro-bands. These micro-bands are interspersed with high fractions of compression twins that contain a high density of dislocations.

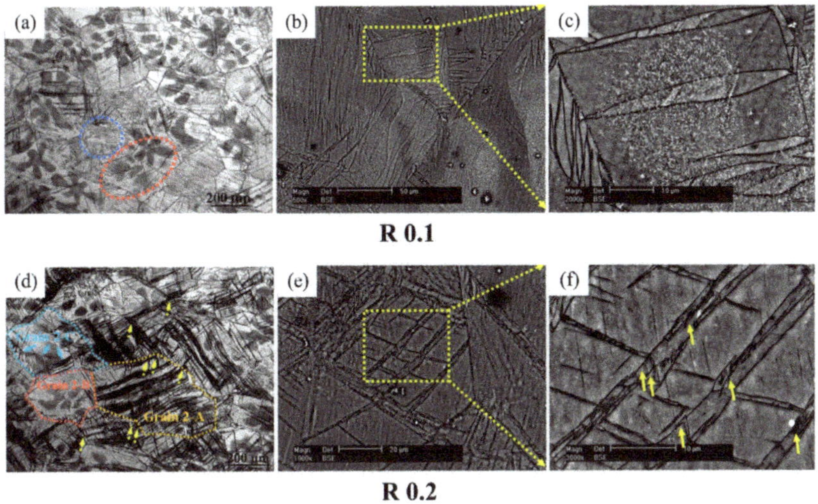

Figure 14. SEM characterization of Mg alloy subjected to hot rolling. Sub-figure (**a–c**) corresponds to 10% reduction along the RD–ND section; sub-figure (**d–f**) corresponds to a 20% reduction. In sub-figure (**d**), dotted color lines indicate cross-grain boundary twin pairs and arrows in sub-figure (**f**) highlight secondary twins [117]. Note—Orange, red and blue colors are marked to grains 2-A, 2-B, and 2-C, respectively. (Reprinted with permission from ref. [117]. Copyright 2022, Elsevier).

In general, the twin shear stress was found to have a close relationship with the axial ratio (c/a) of the close-packed hexagonal lattice. In the case of Mg alloys, the $(10\bar{1}2)$ twin exhibited the lowest shear stress, making it the most likely twin type to occur during rolling, as shown in Figure 15. This could be attributed to the random orientation of grains in the sintered Mg alloy [120]. In the early stages of deformation, grains often need continuous adjustments to facilitate slip occurrence during deformation. Consequently, a substantial quantity of twins tends to form in the Mg matrix during the initial stages of deformation as the microstructure transitions from the sintered state to the deformed state. Furthermore, when a significant rolling reduction is applied, the grains experience intense plastic deformation. Given the limited number of slip systems in Mg alloys, dislocations readily accumulate rapidly, reaching the dislocation density necessary for recrystallization. This situation is more conducive to DRx [121]. While a significant quantity of $(10\bar{1}2)$ twins was observed in the specimen with a 20% rolling reduction, it's important to note that Mg alloys can exhibit various types of twins. These typically include $(10\bar{1}2)$ tensile twins, $(10\bar{1}1)$ compression twins, and $(10\bar{1}1) - (10\bar{1}2)$ secondary twins. In the case of asymmetric rolling $(101\bar{1}) - (101\bar{2})$ double twins, $(101\bar{2})$ tensile twins, $(101\bar{1})$ compression twins appear, and DRx occurs during rolling.

Asymmetric rolling with a 14% reduction rate induces twinning and CDRx phenomena in Mg alloys. Notably, the CDRx grains are smaller and more abundant than those in the sample subjected to symmetric rolling. Additionally, the twins observed in the asymmetrically rolled sample were primarily of the tensile twin variety [63,123]. The prismatic slip typically requires a higher CRSS than basal slip under uniaxial tension conditions; however, the prismatic slip system exhibits significantly higher activity during shear deformation in processes like ECAP. In ECAP, hydrostatic back pressure is applied to materials with HCP structure, introducing an additional stress component to potential slip planes. This added stress compensates for the necessary stress levels to activate non-basal slip systems. Consequently, the imposition of hydrostatic back pressure leads to a notable reduction in yield anisotropy (the CRSS ratio of non-basal to basal slip) and results in the increased activity of non-basal slip, as opposed to basal slip. Furthermore, when added to Mg alloys, rare earth (RE) elements like yttrium can reduce the SFE by altering the dislocation core structure, facilitating the activation of non-basal slip systems [124].

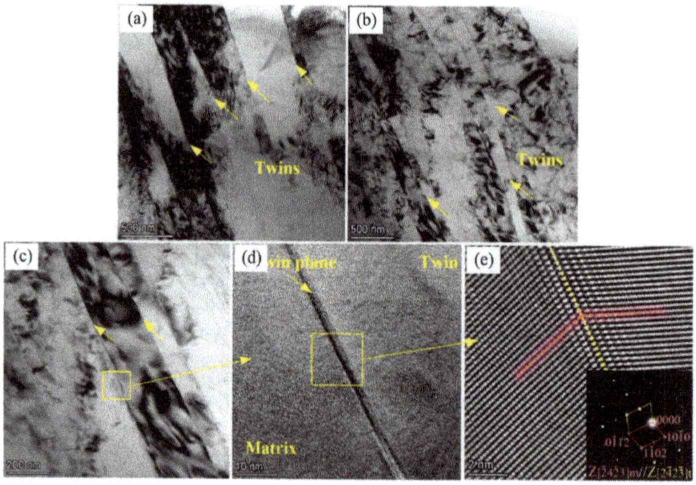

Figure 15. TEM micrographs of 20% rolled Mg alloy. (**a–c**) Bright-field micrographs highlighting twins; (**d**,**e**) high-resolution images detailing $\{10\bar{1}2\}$ tensile twins [122]. Note—Arrows indicating twins. (Reprinted with permission from ref. [122]. Copyright 2022 Elsevier).

4.2. Texture Evolution in Plastically Deformed and Heat-Treated Materials

In the case of Mg and its alloys, the common slip systems are (i) Basal Slip System: This system is the most dominant at RT deformation because it has the lowest CRSS. It enables the movement of dislocations along the basal plane. (ii) Prismatic Slip System: The prismatic slip system activates when the deformation temperature surpasses 523 K, allowing dislocations to move in the prismatic plane of the crystal. (iii) Pyramidal-I and Pyramidal-II Slip Systems: These systems become significantly active at deformation temperatures exceeding 623 K, contributing to increased formability [125–127]. They facilitate dislocation motion in different pyramidal planes. For hot rolling, the resulting texture is primarily basal (0002), similar to the conventional texture observed in rolled Mg. However, the intensity of the basal pole extends from the *ND* to the *RD* due to the activation of non-basal slip systems. The presence of solid solution in alloys plays a key role in this effect [128]. The crystallographic texture remains similar, but there is a reduction in intensity in the case of the hot rolled annealed (HRA) sample; see Figure 16. Furthermore, for the HRA sample, there is a more pronounced spread of the basal texture along the *TD*. This spread is approximately 30° and 35° from *ND* to *RD* and *TD*, respectively, for the HRA sample.

In the context of $(10\bar{1}1) - (10\bar{1}2)$ double twins, it's important to note that only the variants corresponding to their primary $(10\bar{1}1)$ twins were considered for SF calculations. Among the primary $(10\bar{1}1)$ twin variants, one particular variant (oriented at approximately 37.5° around the $\langle 10\bar{1}2\rangle$ axis concerning the matrix) is significantly favored over the others due to its minimal compatibility strain. Both primary $(10\bar{1}2)$ twins and primary $(10\bar{1}1)$ twins, adhere well to the SF law under the current rolling conditions. These twins are significantly activated in PG with off-basal orientations, typically in PGs where the *c*-axes are tilted by 60–90° away from the *ND*. Based on SF law, twin variants with higher SF exhibit greater activity compared to those with lower SFs. Similarly, the $(10\bar{1}2)$ twin variants with the highest SFs for typical PG led to twinned regions tilted by 30° from the *ND*. For PG with off-basal orientations, all these $(10\bar{1}2)$ twin variants cause the matrix orientations to rotate within 30° away from the *ND*. This rotation significantly contributes to the formation of the primary basal texture component, which aligns with findings in other Mg alloys subjected to uniaxial compression after a 10% deformation [129]. As for the secondary intensities tilted by more than 30° from the ND toward the TD, these orientations can be attributed to contributions from either $(10\bar{1}2)$ twins with different variants. At higher levels of reduction, such as 50% reduction, there will be a noticeable increase and subsequent dramatic decrease in the activity of shear bands and double twins, see Figure 17. As a result, SBs and double twins may have a more prominent role in contributing to the formation of the 'RD-split' texture as rolling reduction increases. This 'RD-split' texture remains stable even after further rolling with accumulated reductions of 0.6 and 0.85, coupled with intermediate annealing. This stability can primarily be attributed to the contribution of SBs, which exhibit higher activity at larger rolling reductions [130].

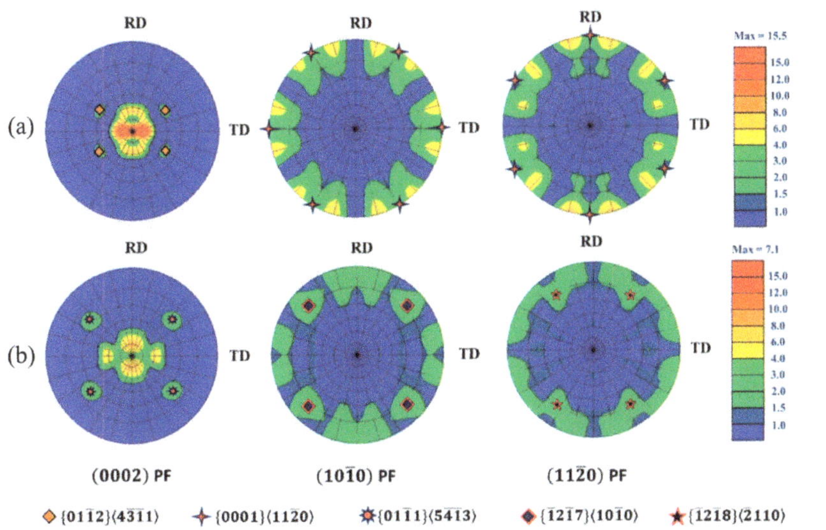

Figure 16. Crystallographic texture pole figures depicting Mg alloy. (**a**) Under hot rolled condition; (**b**) after hot rolling followed by annealing [66]. (Reprinted with permission from ref. [66]. Copyright 2021, Elsevier).

In the case of ultrasonic surface rolling processing, Mg alloy reveals double twins (DT) in this region of $(10\bar{1}2)\langle 10\bar{1}1\rangle$ extension twin type. Compression twins and double twins are not formed in this process because the CRSS required to form extension twins is much lower than that needed for other twin types. The c-axis of the $(10\bar{1}2)\langle 10\bar{1}1\rangle$ extension twins will be perpendicular to the applied stress direction. The misorientation angle between the matrix and the extension twins is approximately 86°. Due to multiple extension twins, the GB content of 86° $(10\bar{1}2)\langle 10\bar{1}1\rangle$ in this region is significantly higher than that of other HAGB [131]. During ultrasonic surface rolling, there are gradient changes in twin density within the material. As the depth of the deformation layer increases, there is a corresponding decrease in twin density [132]. Concurrently, the grain size of the Mg alloy increases with the depth of the deformation layer. Statistical analysis of the grain sizes at different layer depths indicates that near the treated surface, $(101\bar{1})$ and $(101\bar{3})$ compression twins are readily generated. This phenomenon is attributed to the high plastic strain experienced by the Mg alloy near the surface, which meets the CRSS requirements for compression twin formation. As the depth of the deformation layer increases, the DTs transition to $(101\bar{2})$ extension twins. This change can be explained by varying levels of plastic strain experienced by the Mg alloy during ultrasonic surface rolling processing [133]. When the deformation layer depth is shallow, the Mg alloy undergoes significant plastic strain, providing the necessary CRSS for compression twin formation. However, as the plastic strain decreases with increasing deformation layer depth, compression twins are no longer favored. Instead, $(10\bar{1}\bar{2})$ extension twins become prominent because their CRSS requirements are much lower than those of compression twins, leading to the generation of numerous $(10\bar{1}\bar{2})$ extension twins in the Mg alloy. After rolling, the texture exhibits enhancement in the presence of $(1\bar{1}\bar{2}0)$ | | TD texture. This texture variation leads to distinct mechanical performance differences between the TD and RD. The increased $(112\bar{3})$ | | ND texture promotes slip possibilities, potentially resulting in increased elongation. The SF of $(0001)\langle 11\bar{2}0\rangle$ in RD for Mg alloy samples was greater than in the TD. Additionally, the SF of $(0001)\langle 11\bar{2}0\rangle$ exceeded 0.35. These SF values provide evidence of anisotropy in the sample, indicating that initiating cylindrical slip is challenging. This anisotropy may enhance the material's strength while reducing its elongation, contributing to its mechanical behavior [134]. The ECAP Mg sample reveals the presence of all three slip systems, and both twinning mechanisms play a role in accommodating the strain generated. The increased intensity of the peaks associated with the prismatic and pyramidal planes, along with the centralized pole figures observed in the ECAP-processed sample compared to the homogenized sample, indicates the activation of non-basal slip systems during the ECAP process. After this process, the dominant basal component disappeared, giving way to weak basal and prismatic fibers [135].

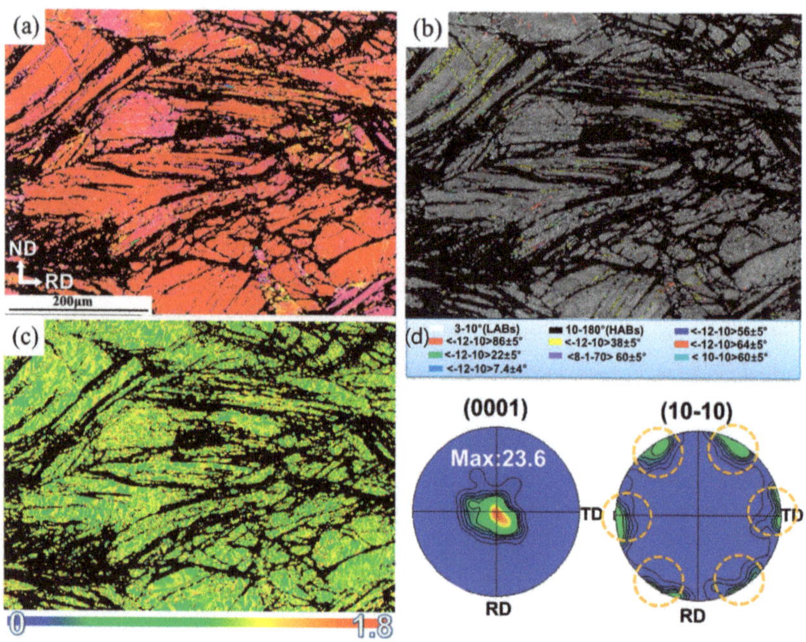

Figure 17. Electron backscattered diffraction of rolled Mg alloy at 50% reduction (**a**) the inverse pole figure mapping, (**b**) boundary mapping, (**c**) local strain mapping, (**d**) (0001) and $(10\bar{1}0)$ pole figures [117]. (Reprinted with permission from ref. [117]. Copyright 2022, Elsevier).

4.3. Influence Strain Rate on Plastic Deformation and Texture Development

Strain rate sensitivity in Mg alloys manifests in various deformation mechanisms such as basal, prismatic, pyramidal, and extension twinning, each exhibiting unique sensitivities to different strain rates [64,136]. Higher strain rates typically increase stress, especially in compression along the ND. Texture development is also influenced by strain rate, with initial textures evolving distinctly under varying strain rates (see Figure 18).

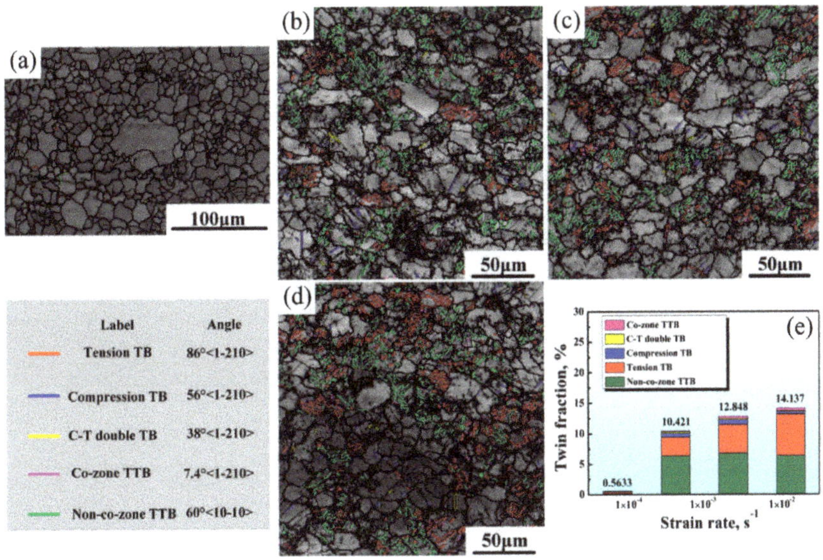

Figure 18. Band contrast maps of extruded-annealed AZ91 alloy before and after tensile deformation under different strain rates: (**a**) initial, (**b**) 1×10^{-4} s^{-1}, (**c**) 1×10^{-3} s^{-1}, (**d**) 1×10^{-2} s^{-1}, and (**e**) volume fraction of twins [64]. (Reprinted with permission from ref. [64]. Copyright 2023, Elsevier).

For instance, studies using the elasto-viscoplastic self-consistent model have shown that in-plane compression at different strain rates results in different texture evolutions, which are crucial for understanding the mechanical behavior of Mg alloy [137–139]. Additionally, higher strain rates in extruded-annealed AZ91 Mg alloy enhance strength but reduce elongation and promote rapid dislocation accumulation. The strain rate also affects the twinning behavior, which is crucial for yield strength and deformation mechanisms in AZ31 Mg alloys [59,60].

4.4. *Effect of Temperature on Plastic Deformation and Texture Development*

Deformation mechanisms at lower temperatures, such as 373 K, are less effective than at higher temperatures, like 473 K, leading to more inhomogeneous deformation, increased SBs, and cracking [140]. Temperature influences texture development, with initial textures such as basal or prismatic playing a crucial role in the resulting deformation texture. Likewise, the Kink Band mechanism occurs at 473 K, which is related to microstructural refinement and dislocation pile-up [141]. Higher temperatures generally enhance ductility due to increased thermal activation of dislocations and reduced resistance to slip [137,140]. At 523 K, the slip band mechanism is prominent in Mg alloy, where prismatic slip predominates due to increased activation energy. In cases of plastic deformation in Mg-Gd-Nd(-Zn)-Zr alloys, shear bands form at 673 K, leading to micro-cracks and shear fractures. Additionally, reticular shear bands, which are prevalent at 673 K, cause obstructions that contribute to deformation by accommodating strain through twin propagation and dynamic recrystallization [69]. In Figure 19a, the intensity differences along the perimeter suggest varying amounts of twinning at RT strain rates. Figure 19b shows nearly no change in texture during simple shear at 423 K, especially at low strain rates, indicating possible GBs sliding [142]. During low-strain conditions, a noticeable basal texture occurs with c-axes perpendicular to ND. At moderate strains, basal planes tilt toward transverse directions, activating basal and limited non-basal slip systems. High strains induce significant basal texture rotation, reducing intensity and enhancing non-basal slip and twinning [143,144]. Yield strength and flow stress decrease with increasing temperature, which is attributed to the activation of different slip systems and twinning mechanisms. Studies have shown that textures evolve differently under various temperatures; for example, the basal texture intensity decreases while prismatic texture intensity increases as temperature rises [145]. The temperature sensitivity of deformation mechanisms such as basal slip and twinning significantly impacts the mechanical properties and behavior of Mg alloys [146,147].

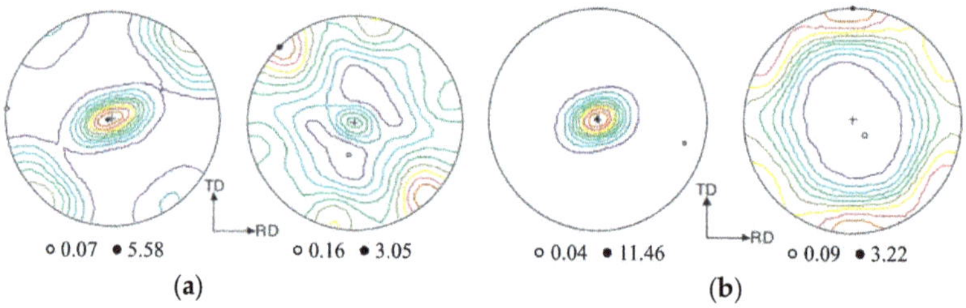

Figure 19. Pole figures at a strain rate of 10^{-4} s^{-1} in *RD* (**a**) at RT and (**b**) 423 K [148]. (Reprinted with permission from ref. [148]. Copyright 2011, Elsevier).

5. Summary

Understanding the interplay between microstructure and texture evolution during plastic deformation and recrystallization is pivotal for HCP alloys, tailoring their properties to specific application demands. Factors such as processing routes, temperature, and crystallographic orientation strongly influence the resulting microstructure and textures. These are summarized for Zr, Ti, and Mg as follows:

1. In Zr, the anisotropy in plastic deformation leads to significant heterogeneities in strain accommodation. For example, during RT and elevated temperature plastic deformation, Zr undergoes grain fragmentation for specific crystallographic orientations, and a significant difference in imposed shear strain is observed near the GBs. Due to the deformation anisotropy in Zr, which is both orientation and temperature-sensitive, strong crystallographic textures often lead to hierarchical differences in deformation texture evolution and residual stresses. The activation of different slip and twin systems is highly dependent on the starting crystallographic texture and working temperatures. Anomalous behavior of basal slip is also observed at higher working temperatures in some cases, which is currently attributed to the oxygen content in the samples. Stored energy and dislocation development in Zr strongly depend on the crystallographic texture and deformation configuration, i.e., strain path and grain morphology. On the other hand, annealing textures are highly dependent on starting crystallographic orientations.
2. In CP-Ti, the most common rolling texture is a split-TD type texture, which forms because of primary compression twinning followed by secondary extension twinning. Activation of basal slip systems changes the final texture to basal fiber texture ($\langle 0001 \rangle$ | | ND) due to activation of basal slip systems. Variation in strain path causes lower deformation heterogeneities and variation in crystallographic texture due to a relatively higher number of active slip systems. The presence of high shear components leads to higher activity of basal and pyramidal slip systems and results in a strong basal fiber and a partial $\langle 11\bar{2}0 \rangle$ | | RD texture. The recrystallization texture of Ti is largely dependent upon the deformation texture and the recrystallization temperature. The annealing texture that develops between 773 K and 973 K consists of $(0001)\langle 1\bar{2}10 \rangle$ and $(0001)\langle 10\bar{1}0 \rangle$, $\pm 30°$ to the TD. When the temperature range is above 973 K, substantial grain growth occurs, and the recrystallization texture consists of either $(10\bar{1}3)\langle 1\bar{2}10 \rangle$ or $(20\bar{2}5)\langle 1\bar{2}10 \rangle$ components manifest as $\pm 30°$ across the TD line on the (0001) pole figure.

3. In Mg alloys, minor variations in the strain path lead to variations in deformation magnitude and have been shown to cause asymmetry in texture evolution. Various types of twins evolve in Mg alloys during deformation, including $(10\bar{1}2)$ tensile twins, $(10\bar{1}1)$ compression twins, and $(10\bar{1}1) - (10\bar{1}2)$ secondary twins. Deformation mechanisms at lower temperatures lead to inhomogeneous deformation, increased SBs, and cracking. Higher temperatures generally enhance the ductility of Mg alloy due to increased thermal activation of dislocations and reduced resistance to slip. Yield strength and flow stress decrease with increasing temperature, attributed to the activation of different slip systems and twinning mechanisms. The strain rate also affects the twinning behavior, which is crucial for yield strength and deformation mechanisms in AZ31 Mg alloys. Higher strain rates in extruded-annealed AZ91 Mg alloy enhance strength but reduce elongation and promote rapid dislocation accumulation.

Author Contributions: K.T.: Investigation, writing—original draft preparation, review, and editing; K.U.Y.: writing—Ti section; V.K.: writing—Mg section; A.G.: writing: review and editing; S.-H.C.: conceptualization, writing—review and editing, supervision, funding. All authors have read and agreed to the published version of the manuscript.

Funding: This work was supported by the Technology Innovation Program (or Industrial Strategic Technology Development Program) (20015158, development of processing and fastening technology to minimize damage to the plating layer of highly corrosion-resistant, highly-formed plated steel materials for plant farms) funded By the Ministry of Trade, Industry & Energy (MOTIE, Republic of Korea). Also, this research was supported by "Regional Innovation Strategy (RIS)" through the National Research Foundation of Korea (NRF) funded by the Ministry of Education (MOE) (2021RIS-002).

Data Availability Statement: Not applicable.

Conflicts of Interest: The authors declare no conflicts of interest.

Abbreviations

BCC: body-centered cubic; CDRx: continuous dynamic recrystallization; CR: cold rolled; CP-Ti: commercially pure titanium; CR: Cross- rolling; CRSS: critical resolved shear stress; CWSR: cold work stress relieved; DRx: dynamic recrystallization; DDRx: discontinuous dynamic recrystallization; DT: double twins; $\bar{\varepsilon}$: effective strains; EBSD: electron backscattered diffraction; ECAP: equal channel angular processing; ECAE: equal channel angular extrusion; FCC: face-centered cubic; GBS: grain boundary sliding; HAGBs: high angle grain boundaries; HCP: hexagonal close-packed; HRA: hot rolled annealed; IGM: intragranular misorientation; IP: in-plane; IPF: inverse pole figure; KAM: kernel average misorientation; LN: liquid nitrogen; LAGB: low angle grain boundary; Mg: Magnesium; MSCR: multi-step cross rolling; ND: normal direction; ODF: orientation distribution function; PG: parental grains; PSC: plain stain compression; RD: Rolling direction; RT: room temperature; RR: reverse rolling; RE: rare earth; Rex: recrystallized; R_W: reduction in wall thickness; R_D: reduction in tube diameter; SB: Shear Bands; S: shear stress; SFE: stacking fault energy; SF: Schmid factor; SLs: strain localizations; SRD: split rolling direction; SMG: submicron grain; STD: split-TD; SRx: static recrystallization; SR: simple rolling; TEM: transmission electron microscopy; Ti: Titanium; TMP: thermo-mechanical processing; TT: through-thickness; TSCR: two-step cross rolling; TR: transverse rolling; TD: transverse direction; UDR: unidirectional rolling; VPSC: viscoplastic self-consistent simulations; Zr: Zirconium.

References

1. Bishoyi, B.D.; Sabat, R.K.; Sahu, J.; Sahoo, S.K. Effect of Temperature on Microstructure and Texture Evolution during Uniaxial Tension of Commercially Pure Titanium. *Mater. Sci. Eng. A* **2018**, *703*, 399–412. [CrossRef]
2. Beer, A.G.; Barnett, M.R. Microstructural Development during Hot Working of Mg-3Al-1Zn. Metall. *Mater. Trans. A* **2007**, *38*, 1856–1867. [CrossRef]
3. Guo, L.; Fujita, F. Effect of Deformation Mode, Dynamic Recrystallization and Twinning on Rolling Texture Evolution of AZ31 Magnesium Alloys. *Trans. Nonferrous Met. Soc. China* **2018**, *28*, 1094–1102. [CrossRef]

4. Lutjering, G.; Williams, J.; Titanium. *Engineering Materials and Processes*, 2nd ed.; Springer: Berlin/Heidelberg, Germany, 2007.
5. Banerjee, D.; Williams, J.C. Perspectives on Titanium Science and Technology. *Acta Mater.* **2013**, *61*, 844–879. [CrossRef]
6. McCabe, R.J.; Cerreta, E.K.; Misra, A.; Kaschner, G.C.; Tomé, C.N. Effects of Texture, Temperature and Strain on the Deformation Modes of Zirconium. *Philos. Mag.* **2006**, *86*, 3595–3611. [CrossRef]
7. Deng, S.; Song, H.; Zheng, C.; Zhao, T.; Zhang, S.; Nielsen, K.B. Selection of Deformation Modes and Related Texture Evolution in Zircaloy-4 during One Pass Cold Pilgering. *Mater. Sci. Eng. A* **2019**, *764*, 138280. [CrossRef]
8. Azevedo, C.R.F. Selection of Fuel Cladding Material for Nuclear Fission Reactors. *Eng. Fail. Anal.* **2011**, *18*, 1943–1962. [CrossRef]
9. Hanawa, T.; Nomura, N. Zirconium and Zirconium Alloys for Biomedical Use. In *Advances in Materials Science Research: Volume 52*; Nova Science Publishers, Inc.: New York, NY, USA, 2022; pp. 107–144.
10. Singh, J.; Mahesh, S.; Kumar, G.; Pant, P.; Srivastava, D.; Dey, G.K.; Saibaba, N.; Samajdar, I. Texture Development and Plastic Deformation in a Pilgered Zircaloy-4 Tube. *Metall. Mater. Trans. A* **2015**, *46*, 1927–1947. [CrossRef]
11. Abe, H.; Iwamoto, T.; Yamamoto, Y.; Nishida, S.; Komatsu, R. Dimensional Accuracy of Tubes in Cold Pilgering. *J. Mater. Process. Technol.* **2016**, *231*, 277–287. [CrossRef]
12. Gurao, N.P.; Akhiani, H.; Szpunar, J.A. Pilgering of Zircaloy-4: Experiments and Simulations. *J. Nucl. Mater.* **2014**, *453*, 158–168. [CrossRef]
13. Krishna, K.V.M.; Sahoo, S.K.; Samajdar, I.; Neogy, S.; Tewari, R.; Srivastava, D.; Dey, G.K.; Das, G.H.; Saibaba, N.; Banarjee, S. Microstructural and Textural Developments during Zircaloy-4 Fuel Tube Fabrication. *J. Nucl. Mater.* **2008**, *383*, 78–85. [CrossRef]
14. Verlinden, B.; Cahn, R.W. *Thermo-Mechanical Processing of Metallic Materials*, 1st ed.; Pergamon Materials Series; Pergamon materials series; Elsevier Science: Amsterdam, The Netherlands, 2007; ISBN 978-0-08-044497-0.
15. Saibaba, N. Fabrication of Seamless Calandria Tubes by Cold Pilgering Route Using 3-Pass and 2-Pass Schedules. *J. Nucl. Mater.* **2008**, *383*, 63–70. [CrossRef]
16. Beyerlein, I.J.; Tomé, C.N. A Dislocation-Based Constitutive Law for Pure Zr Including Temperature Effects. *Int. J. Plast.* **2008**, *24*, 867–895. [CrossRef]
17. Bharat Reddy, G.; Sarkar, A.; Kapoor, R.; Kanjarla, A.K. Effect of Temperature on the Selection of Deformation Modes in Zircaloy-4. Mater. *Sci. Eng. A* **2018**, *734*, 210–223. [CrossRef]
18. Knezevic, M.; Zecevic, M.; Beyerlein, I.J.; Bingert, J.F.; McCabe, R.J. Strain Rate and Temperature Effects on the Selection of Primary and Secondary Slip and Twinning Systems in HCP Zr. *Acta Mater.* **2015**, *88*, 55–73. [CrossRef]
19. Sahoo, S.K.; Hiwarkar, V.D.; Samajdar, I.; Dey, G.K.; Srivastav, D.; Tiwari, R.; Banerjee, S. Heterogeneous Deformation in Single-Phase Zircaloy 2. *Scr. Mater.* **2007**, *56*, 963–966. [CrossRef]
20. Keskar, N.; Mukherjee, S.; Mani Krishna, K.V.; Srivastava, D.; Dey, G.K.; Pant, P.; Doherty, R.D.; Samajdar, I. Quantifying the Mesoscopic Shear Strains in Plane Strain Compressed Polycrystalline Zirconium. *Acta Mater.* **2014**, *69*, 265–274. [CrossRef]
21. Sahoo, S.K.; Hiwarkar, V.D.; Mani Krishna, K.V.; Samajdar, I.; Pant, P.; Pujari, P.K.; Dey, G.K.; Srivastav, D.; Tiwari, R.; Banerjee, S. Grain Fragmentation and Twinning in Deformed Zircaloy 2: Response to Positron Lifetime Measurements. *Mater. Sci. Eng. A* **2010**, *527*, 1427–1435. [CrossRef]
22. Hiwarkar, V.D.; Sahoo, S.K.; Samajdar, I.; Satpathy, A.; Krishna, K.V.M.; Dey, G.K.; Srivastav, D.; Tewari, R.; Banarjee, S. Defining Recrystallization in Pilgered Zircaloy-4: From Preferred Nucleation to Growth Inhibition. *J. Nucl. Mater.* **2011**, *412*, 287–293. [CrossRef]
23. Kumar, G.; Singh, R.; Singh, J.; Srivastava, D.; Dey, G.K.; Samajdar, I. Defining the Stages of Annealing in a Moderately Deformed Commercial Zirconium Alloy. *J. Nucl. Mater.* **2015**, *466*, 243–252. [CrossRef]
24. Adamson, R.B.; Coleman, C.E.; Griffiths, M. Irradiation Creep and Growth of Zirconium Alloys: A Critical Review. *J. Nucl. Mater.* **2019**, *521*, 167–244. [CrossRef]
25. Rodriguez, P. Nuclear Reactor Materials: Irradiation Effects. In *Encyclopedia of Materials: Science and Technology*; Buschow, K.H.J., Cahn, R.W., Flemings, M.C., Ilschner, B., Kramer, E.J., Mahajan, S., Veyssière, P., Eds.; Elsevier: Oxford, UK, 2001; pp. 6349–6361. ISBN 978-0-08-043152-9.
26. Kumar, G.; Kanjarla, A.K.; Lodh, A.; Singh, J.; Singh, R.; Srivastava, D.; Dey, G.K.; Saibaba, N.; Doherty, R.D.; Samajdar, I. Burst Ductility of Zirconium Clads: The Defining Role of Residual Stress. *Metall. Mater. Trans. A* **2016**, *47*, 3882–3896. [CrossRef]
27. Nagase, F.; Fuketa, T. Investigation of Hydride Rim Effect on Failure of Zircaloy-4 Cladding with Tube Burst Test. *J. Nucl. Sci. Technol.* **2005**, *42*, 58–65. [CrossRef]
28. Gupta, A.; Khatirkar, R.K.; Kumar, A.; Thool, K.S.; Bhibhanshu, N.; Suwas, S. Texture Development During Cold Rolling of a β-Ti Alloy: Experiments and Simulations. *Metall. Mater. Trans. A* **2021**, *52*, 1031–1043. [CrossRef]
29. Gupta, A.; Khatirkar, R.K.; Kumar, A.; Parihar, M.S. Investigations on the Effect of Heating Temperature and Cooling Rate on Evolution of Microstructure in an α + β Titanium Alloy. *J. Mater. Res.* **2018**, *33*, 946–957. [CrossRef]
30. Jawed, S.F.; Liu, Y.J.; Wang, J.C.; Rabadia, C.D.; Wang, L.Q.; Li, Y.H.; Zhang, X.H.; Zhang, L.C. Tailoring Deformation and Superelastic Behaviors of Beta-Type Ti-Nb-Mn-Sn Alloys. *J. Mech. Behav. Biomed. Mater.* **2020**, *110*, 103867. [CrossRef] [PubMed]
31. Singh, A.K.; Kaushik, L.; Pawar, S.; Singh, J.; Das, H.; Mondal, M.; Hong, S.-T.; Choi, S.-H. Unraveling the Heterogeneous Evolution of the Microstructure and Texture in the Thermomechanically Affected Zone of Commercially Pure Titanium during Friction Stir Processing. *Int. J. Mech. Sci.* **2023**, *239*, 107894. [CrossRef]
32. Gupta, A.; Khatirkar, R.K.; Kumar, A.; Thool, K.; Bibhanshu, N.; Suwas, S. Microstructure and Texture Development in Ti-15V-3Cr-3Sn-3Al Alloy – Possible Role of Strain Path. *Mater. Charact.* **2019**, *156*, 109884. [CrossRef]

33. Wang, Y.N.; Huang, J.C. Texture Analysis in Hexagonal Materials. *Mater. Chem. Phys.* **2003**, *81*, 11–26. [CrossRef]
34. Salem, A.A.; Kalidindi, S.R.; Doherty, R.D.; Semiatin, S.L. Strain Hardening Due to Deformation Twinning in α-Titanium: Mechanisms. *Metall. Mater. Trans. A* **2006**, *37*, 259–268. [CrossRef]
35. Rabadia, C.D.; Jawed, S.F.; Wang, J.; Siddhpura, M.; Siddhpura, A. Revised Semiempirical Approach to Predict the Occurrence of Twinning in Titanium Alloys. *ACS Omega* **2021**, *6*, 34056–34064. [CrossRef] [PubMed]
36. Yazar, K.U.; Shamitha, M.; Suwas, S. Texture-Dependent Dwell Fatigue Response of Titanium. *Philos. Mag.* **2021**, *101*, 1443–1470. [CrossRef]
37. Yazar, K.U.; Mishra, S.; Kumar, L.; Bahl, S.; Kumar, T.K.; Suwas, S. Texture Induced Planar Anisotropy of Dwell Fatigue Response in Titanium: Insights from Experiments and Crystal Plasticity Simulations. *Int. J. Plast.* **2022**, *152*, 103140. [CrossRef]
38. Shin, E.J.; Jung, A.; Choi, S.-H.; Rollett, A.D.; Park, S.S. A Theoretical Prediction of Twin Variants in Extruded AZ31 Mg Allbys Using the Microstructure Based Crystal Plasticity Finite Element Method. *Mater. Sci. Eng. A* **2012**, *538*, 190–201. [CrossRef]
39. Song, J.; Chen, J.; Xiong, X.; Peng, X.; Chen, D.; Pan, F. Research Advances of Magnesium and Magnesium Alloys Worldwide in 2021. *J. Magnes. Alloy* **2022**, *10*, 863–898. [CrossRef]
40. Choi, S.-H.; Shin, E.J.; Seong, B.S. Simulation of Deformation Twins and Deformation Texture in an AZ31 Mg Alloy under Uniaxial Compression. *Acta Mater.* **2007**, *55*, 4181–4192. [CrossRef]
41. Pandey, A.; Kabirian, F.; Hwang, J.-H.; Choi, S.-H.; Khan, A.S. Mechanical Responses and Deformation Mechanisms of an AZ31 Mg Alloy Sheet under Dynamic and Simple Shear Deformations. *Int. J. Plast.* **2015**, *68*, 111–131. [CrossRef]
42. Bhattacharyya, J.J.; Wang, F.; Stanford, N.; Agnew, S.R. Slip Mode Dependency of Dislocation Shearing and Looping of Precipitates in Mg Alloy WE43. *Acta Mater.* **2018**, *146*, 55–62. [CrossRef]
43. Singh, J.; Mahesh, S.; Kumar, G. Deformation Twinning in Zirconium: Direct Experimental Observations and Polycrystal Plasticity Predictions. *Metall. Mater. Trans. A* **2015**, *46*, 5058–5071. [CrossRef]
44. Ghosh, A.; Singh, A.; Gurao, N.P. Effect of Rolling Mode and Annealing Temperature on Microstructure and Texture of Commercially Pure-Titanium. *Mater. Charact.* **2017**, *125*, 83–93. [CrossRef]
45. Gurao, N.P.; Sethuraman, S.; Suwas, S. Evolution of Texture and Microstructure in Commercially Pure Titanium with Change in Strain Path during Rolling. *Metall. Mater. Trans. A Phys. Metall. Mater. Sci.* **2013**, *44*, 1497–1507. [CrossRef]
46. Bozzolo, N.; Dewobroto, N.; Wenk, H.R.; Wagner, F. Microstructure and Microtexture of Highly Cold-Rolled Commercially Pure Titanium. *J. Mater. Sc.* **2007**, *42*, 2405–2416. [CrossRef]
47. Bahl, S.; Suwas, S.; Chatterjee, K. The Importance of Crystallographic Texture in the Use of Titanium as an Orthopedic Biomaterial. *RSC Adv.* **2014**, *4*, 38078–38087. [CrossRef]
48. Chun, Y.B.; Yu, S.H.; Semiatin, S.L.; Hwang, S.K. Effect of Deformation Twinning on Microstructure and Texture Evolution during Cold Rolling of CP-Titanium. *Mater. Sci. Eng. A* **2005**, *398*, 209–219. [CrossRef]
49. Shankar, S.; O'Brien, T.D. Texture Formation and Transition in Cold-Rolled Titanium. *Mater. Sci. Eng. A* **1988**, *100*, 109–114. [CrossRef]
50. Zhong, Y.; Yin, F.; Nagai, K. Role of Deformation Twin on Texture Evolution in Cold-Rolled Commercial-Purity Ti. *J. Mater. Res.* **2008**, *23*, 2954–2966. [CrossRef]
51. Gurao, N.P.; Suwas, S. Evolution of Crystallographic Texture during Deformation of Submicron Grain Size Titanium. *J. Mater. Res.* **2011**, *26*, 523–532. [CrossRef]
52. Sahoo, S.K.; Sabat, R.K.; Sahni, S.; Suwas, S. Texture and Microstructure Evolution of Commercially Pure Titanium during Hot Rolling: Role of Strain-Paths. *Mater. Des.* **2016**, *91*, 58–71. [CrossRef]
53. Yang, H.; Wang, Y. Effect of Shear Deformation on Texture Evolution in Rolled Pure Titanium. *MATEC Web Conf.* **2018**, *186*, 02002. [CrossRef]
54. Milner, J.L.; Abu-farha, F.; Kurfess, T.; Hammond, V.H. Effects of Induced Shear Deformation on Microstructure and Texture Evolution in CP-Ti Rolled Sheets. *Mater. Sci. Eng. A* **2014**, *619*, 12–25. [CrossRef]
55. Zherebtsov, S.; Lojkowski, W.; Mazur, A.; Salishchev, G. Structure and Properties of Hydrostatically Extruded Commercially Pure Titanium. *Mater. Sci. Eng. A* **2010**, *527*, 5596–5603. [CrossRef]
56. Bishoyi, B.; Vinjamuri, R.; Sabat, R.K.; Patro, S.K.; Suwas, S.; Sahoo, S.K. Cold Drawing of Commercially Pure Titanium and Its Effect on Microstructure and Texture Evolution. *Metall. Mat. Trans. A* **2022**, *53*, 1845–1858. [CrossRef]
57. Shankar, G.; Raguraman, S.; Barrales-mora, L.A.; Suwas, S. Development of Recrystallization Texture in Commercially Pure Titanium: Experiments and Simulation. *JOM* **2020**, *72*, 4559–4573. [CrossRef]
58. Bozzolo, N.; Dewobroto, N.; Grosdidier, T.; Wagner, F. Texture Evolution during Grain Growth in Recrystallized Commercially Pure Titanium. *Mater. Sci. Eng. A* **2005**, *397*, 346–355. [CrossRef]
59. Lee, J.U.; Kim, Y.J.; Kim, S.-H.; Lee, J.H.; Kim, M.-S.; Choi, S.-H.; Moon, B.G.; Kim, Y.M.; Park, S.H. Texture Tailoring and Bendability Improvement of Rolled AZ31 Alloy Using {10–12} Twinning: The Effect of Precompression Levels. *J. Magnes. Alloy* **2019**, *7*, 648–660. [CrossRef]
60. Zhao, L.; Zhu, W.; Chen, W.; Zhao, X.; Yan, C.; Hong, R.; Jin, Z. An Insight into Mechanical Response and Twinning Behavior of Bimodal Textured AZ31 Magnesium Alloy under Quasi-Static and High Strain Rate Compression. *J. Mater. Res. Tech.* **2023**, *27*, 4692–4705. [CrossRef]
61. Mo, Y.; Jiang, F.; Xu, H.; Tang, J.; Fu, D.; Zhang, H.; Teng, J. Deformation Mechanisms and Microstructural Characteristics of AZ61 Magnesium Alloys Processed by a Continuous Expansion Extrusion Approach. *J. Magnes. Alloy*, 2022; *in press*. [CrossRef]

62. Qin, D.H.; Wang, M.J.; Sun, C.Y.; Su, Z.X.; Qian, L.Y.; Sun, Z.H. Interaction between Texture Evolution and Dynamic Recrystallization of Extruded AZ80 Magnesium Alloy during Hot Deformation. *Mater. Sci. Eng. A* **2020**, *788*, 139537. [CrossRef]
63. Wang, D.; Jing, Y.; Lin, B.; Li, J.; Shi, Y.; Misra, R.D.K. On the Structure, Mechanical Behavior, and Deformation Mechanism of AZ91 Magnesium Alloy Processed by Symmetric and Asymmetric Rolling. *Mater. Charact.* **2022**, *194*, 112444. [CrossRef]
64. Chai, F.; Ma, Z.; Han, X.; Hu, X.; Chang, Z.; Zhou, J. Effect of Strain Rates on Mechanical Behavior, Microstructure Evolution and Failure Mechanism of Extruded-Annealed AZ91 Magnesium Alloy under Room-Temperature Tension. *J. Mat. Res. Tech.* **2023**, *27*, 4644–4656. [CrossRef]
65. Cui, C.; Zhang, W.; Chen, W.; He, J.; Chen, X.; Hou, J. Microstructure, Texture Evolution and Yield Strength Symmetry Improvement of as-Extruded ZK60 Mg Alloy via Multi-Directional Impact Forging. *J. Magnes. Alloy* **2022**, *10*, 2745–2760. [CrossRef]
66. Verma, K.K.; Kumar, S.; Suwas, S. Evolution of Microstructure and Texture during Hot Rolling and Subsequent Annealing of the TZ73 Magnesium Alloy and Its Influence on Tensile Properties. *Mater. Sci. Eng. A* **2021**, *821*, 141480. [CrossRef]
67. Liu, W.; Zhao, Y.; Zhang, Y.; Shuai, C.; Chen, L.; Huang, Z.; Hou, H. Deformation-Induced Dynamic Precipitation of 14H-LPSO Structure and Its Effect on Dynamic Recrystallization in Hot-Extruded Mg-Y-Zn Alloys. *Int. J. Plast.* **2023**, *164*, 103573. [CrossRef]
68. Xin, T.; Tang, S.; Ji, F.; Cui, L.; He, B.; Lin, X.; Tian, X.; Hou, H.; Zhao, Y.; Ferry, M. Phase Transformations in an Ultralight BCC Mg Alloy during Anisothermal Ageing. *Acta Mater.* **2022**, *239*, 118248. [CrossRef]
69. Zhang, Q.; Chen, Z.; Li, Q.; Chen, X.; Zhao, J.; Bao, J. Deformation Behavior Characterized by Reticular Shear Bands and Long Chain Twins in Mg-Gd-Nd(-Zn)-Zr Alloys. *J. Mat. Res. Tech.* **2021**, *15*, 5326–5342. [CrossRef]
70. Murty, K.L.; Charit, I. Texture Development and Anisotropic Deformation of Zircaloys. *Prog. Nucl. Ener.* **2006**, *48*, 325–359. [CrossRef]
71. MacEwen, S.R.; Christodoulou, N.; Salinas-Rodríguez, A. Residual Grain-Interaction Stresses in Zirconium Alloys. *Metall. Trans. A* **1990**, *21*, 1083–1095. [CrossRef]
72. Tenckhoff, E. *Deformation Mechanisms, Texture, and Anisotropy in Zirconium And Zircaloy*; ASTM Special Technical Publication 966; American Society for Testing and Materials: Philadelphia, PA, USA, 1988; ISBN 978-0-8031-5039-3.
73. Kalidindi, S.R. Modeling Anisotropic Strain Hardening and Deformation Textures in Low Stacking Fault Energy Fcc Metals. *Int. J. Plast.* **2001**, *17*, 837–860. [CrossRef]
74. Cockeram, B.V.; Hollenbeck, J.L. The Role of Stress-State on the Deformation and Fracture Mechanism of Hydrided and Non-Hydrided Zircaloy-4. *J. Nucl. Mater.* **2015**, *467*, 9–31. [CrossRef]
75. Withers, P.J.; Bhadeshia, H.K.D.H. Residual Stress. Part 2—Nature and Origins. *Mater. Sc. Tech.* **2001**, *17*, 366–375. [CrossRef]
76. Kumar, G.; Lodh, A.; Singh, J.; Singh, R.; Srivastava, D.; Dey, G.; Samajdar, I. Experimental Characterization and Finite Element Modeling of through Thickness Deformation Gradient in a Cold Rolled Zirconium Sheet. *CIRP J. Man. Sc. Tech.* **2017**, *19*, 176–190. [CrossRef]
77. Li, Y.; Po, G.; Cui, Y.; Ghoniem, N. Prismatic-to-Basal Plastic Slip Transition in Zirconium. *Acta Mater.* **2023**, *242*, 118451. [CrossRef]
78. Gong, J.; Benjamin Britton, T.; Cuddihy, M.A.; Dunne, F.P.E.; Wilkinson, A.J. <a> Prismatic, <a> Basal, and <c+a> Slip Strengths of Commercially Pure Zr by Micro-Cantilever Tests. *Acta Mater.* **2015**, *96*, 249–257. [CrossRef]
79. Wu, Z.; Curtin, W.A. Mechanism and Energetics of <c+a> Dislocation Cross-Slip in Hcp Metals. *Proc. Natl. Acad. Sci. USA* **2016**, *113*, 11137–11142. [CrossRef] [PubMed]
80. Soo, P.; Higgins, G.T. The Deformation of Zirconium-Oxygen Single Crystals. *Acta Mater.* **1968**, *16*, 177–186. [CrossRef]
81. Akhtar, A.; Teghtsoonian, A. Plastic Deformation of Zirconium Single Crystals. *Acta Mater.* **1971**, *19*, 655–663. [CrossRef]
82. Akhtar, A. Schmid's Law and Prismatic Slip of Zirconium. *Scr. Mater.* **1975**, *9*, 859–861. [CrossRef]
83. Akhtar, A. Compression of Zirconium Single Crystals Parallel to the C-Axis. *J. Nucl. Mater.* **1973**, *47*, 79–86. [CrossRef]
84. Xu, F.; Holt, R.A.; Daymond, M.R. Modeling Texture Evolution during Uni-Axial Deformation of Zircaloy-2. *J. Nucl. Mater.* **2009**, *394*, 9–19. [CrossRef]
85. Tenckhoff, E. Review of Deformation Mechanisms, Texture, and Mechanical Anisotropy in Zirconium and Zirconium Base Alloys. *J. ASTM Int.* **2005**, *2*, 1–26. [CrossRef]
86. Chai, L.; Luan, B.; Xiao, D.; Zhang, M.; Murty, K.L.; Liu, Q. Microstructural and Textural Evolution of Commercially Pure Zr Sheet Rolled at Room and Liquid Nitrogen Temperatures. *Mater. Des.* **2015**, *85*, 296–308. [CrossRef]
87. McCabe, R.J.; Proust, G.; Cerreta, E.K.; Misra, A. Quantitative Analysis of Deformation Twinning in Zirconium. *Int. J. Plast.* **2009**, *25*, 454–472. [CrossRef]
88. He, W.; Chapuis, A.; Chen, X.; Liu, Q. Effect of Loading Direction on the Deformation and Annealing Behavior of a Zirconium Alloy. *Mater. Sci. Eng. A* **2018**, *734*, 364–373. [CrossRef]
89. Vaibhaw, K.; Rao, S.V.R.; Jha, S.K.; Saibaba, N.; Jayaraj, R.N. Texture and Hydride Orientation Relationship of Zircaloy-4 Fuel Clad Tube during Its Fabrication for Pressurized Heavy Water Reactors. *J. Nucl. Mater.* **2008**, *383*, 71–77. [CrossRef]
90. Liu, H.; Song, H.-W.; Deng, S.-Y.; Chen, S.-F.; Zhang, S.-H. Study on Formability Improvement of Zr-4 Sheets Based on Texture Optimization. *Metals* **2024**, *14*, 725. [CrossRef]
91. Thool, K.S.; Mani Krishna, K.V.; Srivastava, D.; Patra, A.; Doherty, R.D.; Samajdar, I. Confirmation of Dynamically Recrystallized Grains in Hexagonal Zirconium through Local Internal Friction Measurements. *Metall. Mat. Trans. A* **2019**, *50*, 5000–5014. [CrossRef]

92. Erinosho, T.O.; Dunne, F.P.E. Strain Localization and Failure in Irradiated Zircaloy with Crystal Plasticity. *Int. J. Plast.* **2015**, *71*, 170–194. [CrossRef]
93. Honniball, P.D.; Preuss, M.; Rugg, D.; Quinta da Fonseca, J. Grain Breakup During Elevated Temperature Deformation of an HCP Metal. *Metall. Mat. Trans. A* **2015**, *46*, 2143–2156. [CrossRef]
94. Guo, W.; Li, G.; Yuan, F.; Han, F.; Zhang, Y.; Ali, M.; Ren, J.; Yuan, G. Texture Development and Mechanical Behavior of Zircaloy-4 Alloy Plates Fabricated by Cold Rolling and Annealing. *Mater. Sci. Eng. A* **2021**, *807*, 140846. [CrossRef]
95. Thool, K.; Patra, A.; Fullwood, D.; Krishna, K.V.M.; Srivastava, D.; Samajdar, I. The Role of Crystallographic Orientations on Heterogeneous Deformation in a Zirconium Alloy: A Combined Experimental and Modeling Study. *Int. J. Plast.* **2020**, 102785. [CrossRef]
96. Kumar, G.; Balo, S.; Dhoble, A.; Singh, J.; Singh, R.; Srivastava, D.; Dey, G.K.; Samajdar, I. Through-Thickness Deformation Gradient in a Part-Pilgered Zirconium Tube: Experimental Measurements and Numerical Validation. *Metall. Mat. Trans. A* **2017**, *48*, 2844–2857. [CrossRef]
97. Luan, B.; Gao, S.; Chai, L.; Li, X.; Chapuis, A.; Liu, Q. Compression Deformation Behavior of Zr-1Sn-0.3Nb Alloy with Different Initial Orientations at 650 °C. *Mat. Des.* **2013**, *52*, 1065–1070. [CrossRef]
98. Chauvy, C.; Barbéris, P.; Montheillet, F. Dynamic Recrystallization of Zircaloy-4 during Working within the Upper α-Range. *Mat. Sc. For.* **2004**, *467–470*, 1151–1156. [CrossRef]
99. Pérez-Prado, M.T.; Barrabes, S.R.; Kassner, M.E.; Evangelista, E. Dynamic Restoration Mechanisms in α-Zirconium at Elevated Temperatures. *Acta Mater.* **2005**, *53*, 581–591. [CrossRef]
100. Chakravartty, J.K.; Kapoor, R.; Sarkar, A.; Banerjee, S.; Barberis, P.; Dean, S.W. Dynamic Recrystallization in Zirconium Alloys. *J. ASTM Int.* **2010**, *7*, 1–14. [CrossRef]
101. Gerspach, F.; Bozzolo, N.; Wagner, F. About Texture Stability during Primary Recrystallization of Cold-Rolled Low Alloyed Zirconium. *Scr. Mat.* **2009**, *60*, 203–206. [CrossRef]
102. Rawles, J.; Fialkova, S.; Hubbard, K.; Xu, Z.; Hale, C.; Sankar, J. Optimizing the Rolling Process of Lightweight Materials. *Crystals* **2024**, *14*, 582. [CrossRef]
103. Wagner, F.; Bozzolo, N.; Van Landuyt, O.; Grosdidier, T. Evolution of Recrystallisation Texture and Microstructure in Low Alloyed Titanium Sheets. *Acta Mater.* **2002**, *50*, 1245–1259. [CrossRef]
104. Ghosh, A. Anisotropic Tensile and Ratcheting Behavior of Commercially Pure Titanium Processed via Cross Rolling and Annealing. *Int. J. Fatigue* **2019**, *120*, 12–22. [CrossRef]
105. Chen, C.; Han, D.; Song, Y.; Wang, M.; Li, Y.; Xu, S. Thermal Stability of Deformation Twins in Cryogenic Rolled CP-Ti. *Mater. Charact.* **2023**, *196*, 112587. [CrossRef]
106. Williams, J.C.; Baggerly, R.G.; Paton, N.E. Deformation Behavior of HCP Ti-Al Alloy Single Crystals. *Metall. Mat. Trans. A Phys. Metall. Mater. Sci.* **2002**, *33*, 837–850. [CrossRef]
107. Yazar, K.U.; Mishra, S.; Karmakar, A.; Bhattacharjee, A.; Suwas, S. On the Temperature Sensitivity of Dwell Fatigue of a Near Alpha Titanium Alloy: Role of Strain Hardening and Strain Rate Sensitivity. *Metall. Mat. Trans. A* **2020**, *51*, 5036–5042. [CrossRef]
108. Gurao, N.P.; Kapoor, R.; Suwas, S. Deformation Behaviour of Commercially Pure Titanium at Extreme Strain Rates. *Acta Mater.* **2011**, *59*, 3431–3446. [CrossRef]
109. Fundenberger, J.; Gottstein, G. Texture Evolution in Commercially Pure Titanium after Warm Equal Channel Angular Extrusion. *Acta Mater.* **2011**, *59*, 1121–1133. [CrossRef]
110. Meyers, M.A.; Vöhringer, O.; Lubarda, V.A. The Onset of Twinning in Metals: A Constitutive Description. *Acta Mater.* **2001**, *49*, 4025–4039. [CrossRef]
111. Rabadia, C.D.; Liu, Y.J.; Zhao, C.H.; Wang, J.C.; Jawed, S.F.; Wang, L.Q.; Chen, L.Y.; Sun, H.; Zhang, L.C. Improved Trade-off between Strength and Plasticity in Titanium Based Metastable Beta Type Ti-Zr-Fe-Sn Alloys. *Mater. Sci. Eng. A* **2019**, *766*, 138340. [CrossRef]
112. Panchal, M.; Kaushik, L.; Kottan Renganayagalu, R.; Choi, S.-H.; Singh, J. Microstructure and Texture Evolution in AZX311 Mg Alloy during In-Plane Shear Deformation. *J. Alloys Compd.* **2024**, *999*, 174977. [CrossRef]
113. Su, C.J.; Xu, T.T.; Zhang, K.; Zhang, K.; Lou, S.M.; Wang, Q. Plastic Deformation of Magnesium Alloy with Different Forming Parameters during Ultrasonic Vibration-Assisted Single-Point Incremental Forming. *Rare Met.* **2022**, *41*, 3878–3886. [CrossRef]
114. Chen, W.; He, W.; Jiang, B.; Pan, F. Simultaneous Enhancement of Strength and Ductility for AZ31 Magnesium Alloy by Pre-Twinning Induced Heterostructure. *Int. J. Plast.* **2022**, *159*, 103483. [CrossRef]
115. Li, L.; Suh, B.-C.; Suh, J.-S.; Kim, C.; Go, Y.; Kim, Y.M. Static Recrystallization Behavior of the Cold-Rolled Mg-1Al-1Zn-0.1Ca-0.2Y Magnesium Alloy Sheet. *J. Alloys Compd.* **2023**, *938*, 168508. [CrossRef]
116. Li, N.; Yang, L.; Wang, C.; Monclús, M.A.; Shi, D.; Molina-Aldareguía, J.M. Deformation Mechanisms of Basal Slip, Twinning and Non-Basal Slips in Mg–Y Alloy by Micropillar Compression. *Mater. Sci. Eng. A* **2021**, *819*, 141408. [CrossRef]
117. Shi, B.Q.; Wang, Y.Z.; Shang, X.L.; Zhao, L.Y.; Li, C.Q.; Chen, D.C.; Nie, B.H.; Chen, R.S.; Ke, W. Microstructure Evolution of Twinning-Induced Shear Bands and Correlation with 'RD-Split'Texture during Hot Rolling in a Mg-1.1 Zn-0.76 Y-0.56 Zr Alloy. *Mater. Charact.* **2022**, *187*, 111853. [CrossRef]
118. Mu, S.; Jonas, J.J.; Gottstein, G. Variant Selection of Primary, Secondary and Tertiary Twins in a Deformed Mg Alloy. *Acta Mater.* **2012**, *60*, 2043–2053. [CrossRef]

119. Hong, S.-G.; Park, S.H.; Lee, C.S. Role of {10–12} Twinning Characteristics in the Deformation Behavior of a Polycrystalline Magnesium Alloy. *Acta Mater.* **2010**, *58*, 5873–5885. [CrossRef]
120. Chino, Y.; Kimura, K.; Mabuchi, M. Twinning Behavior and Deformation Mechanisms of Extruded AZ31 Mg Alloy. *Mater. Sci. Eng. A* **2008**, *486*, 481–488. [CrossRef]
121. Liu, Y.; Cai, S.; Dai, L. A New Method for Grain Refinement in Magnesium Alloy: High Speed Extrusion Machining. *Mater. Sci. Eng. A* **2016**, *651*, 878–885. [CrossRef]
122. Wang, T.; Yang, L.; Tang, Z.; Wu, L.; Yan, H.; Liu, C.; Ma, Y.; Liu, W.; Yu, Z. Microstructure, Mechanical Properties and Deformation Mechanism of Powder Metallurgy AZ31 Magnesium Alloy during Rolling. *Mater. Sci. Eng. A* **2022**, *844*, 143042. [CrossRef]
123. Peng, J.; Zhang, Z.; Guo, P.; Zhou, W.; Wu, Y. The Effect of Grain Size on Texture Evolution and Mechanical Properties of an AZ31 Magnesium Alloy during Cold-Rolling Process. *J. Alloys Compd.* **2020**, *817*, 153302. [CrossRef]
124. Daghigh, M.; Mohri, M.; Ghanbari, H.; Nili-ahmadabadi, M. Enhanced Deformability of WE43 Magnesium Alloy by Activation of Non-Basal Slip and Twinning Deformation Modes in the Equal Channel Angular Pressing at Room Temperature. *Mater. Let.* **2023**, *351*, 135047. [CrossRef]
125. Kandalam, S.; Sabat, R.K.; Bibhanshu, N.; Avadhani, G.S.; Kumar, S.; Suwas, S. Superplasticity in High Temperature Magnesium Alloy WE43. Mater. *Sci. Eng. A* **2017**, *687*, 85–92. [CrossRef]
126. Dharmendra, C.; Rao, K.P.; Prasad, Y.; Hort, N.; Kainer, K.U. Hot Working Mechanisms and Texture Development in Mg-3Sn-2Ca-0.4 Al Alloy. *Mat. Chem. Phy.* **2012**, *136*, 1081–1091. [CrossRef]
127. Panchal, M.; Kaushik, L.; Kim, M.-S.; Kottan Renganayagalu, R.; Choi, S.-H.; Singh, J. Effect of Pre-Twinning and Heat Treatment on Formability of AZX311 Mg Alloy. *J. Magnes. Alloy* **2024**, *12*, 1154–1169. [CrossRef]
128. Koundinya, N.; Raman, L.; Chawake, N.; Kottada, R.S. Hot Deformation Behaviour of Mg-3Al-3Sn and Mg-3Al-3Sn-1 Zn Alloys: Role of Zn. *Materialia* **2018**, *3*, 274–287. [CrossRef]
129. Jiang, M.G.; Yan, H.; Chen, R.S. Twinning, Recrystallization and Texture Development during Multi-Directional Impact Forging in an AZ61 Mg Alloy. *J. Alloys Compd.* **2015**, *650*, 399–409. [CrossRef]
130. Lentz, M.; Risse, M.; Schaefer, N.; Reimers, W.; Beyerlein, I.J. Strength and Ductility with {1011}—{1012} Double Twinning in a Magnesium Alloy. *Nat. Commun.* **2016**, *7*, 11068. [CrossRef] [PubMed]
131. Shi, X.Y.; Liu, Y.; Li, D.J.; Chen, B.; Zeng, X.Q.; Lu, J.; Ding, W.J. Microstructure Evolution and Mechanical Properties of an Mg–Gd Alloy Subjected to Surface Mechanical Attrition Treatment. *Mater. Sci. Eng. A* **2015**, *630*, 146–154. [CrossRef]
132. Zhu, S.Q.; Ringer, S.P. On the Role of Twinning and Stacking Faults on the Crystal Plasticity and Grain Refinement in Magnesium Alloys. *Acta Mater.* **2018**, *144*, 365–375. [CrossRef]
133. Tang, J.; Fan, H.; Wei, D.; Jiang, W.; Wang, Q.; Tian, X.; Zhang, X. Interaction between a {101¯2} Twin Boundary and Grain Boundaries in Magnesium. *Int. J. Plast.* **2020**, *126*, 102613. [CrossRef]
134. Rajput, S.K.; Chaudhari, G.P.; Nath, S.K. Characterization of Hot Deformation Behavior of a Low Carbon Steel Using Processing Maps, Constitutive Equations and Zener-Hollomon Parameter. *J. Mater. Proc. Tech.* **2016**, *237*, 113–125. [CrossRef]
135. Sandlöbes, S.; Friák, M.; Zaefferer, S.; Dick, A.; Yi, S.; Letzig, D.; Pei, Z.; Zhu, L.-F.; Neugebauer, J.; Raabe, D. The Relation between Ductility and Stacking Fault Energies in Mg and Mg–Y Alloys. *Acta Mater.* **2012**, *60*, 3011–3021. [CrossRef]
136. Xu, J.; Guan, B.; Zhao, X.; Fu, R.; Hu, Q.; Liu, C. Strain Rate Dependence of Twinning Behavior in AZ31 Mg Alloys. *Metals* **2023**, *13*, 1882. [CrossRef]
137. Wang, H.; Wu, P.; Kurukuri, S.; Worswick, M.J.; Peng, Y.; Tang, D.; Li, D. Strain Rate Sensitivities of Deformation Mechanisms in Magnesium Alloys. *Int. J. Plast.* **2018**, *107*, 207–222. [CrossRef]
138. Ardeljan, M.; Beyerlein, I.J.; McWilliams, B.A.; Knezevic, M. Strain Rate and Temperature Sensitive Multi-Level Crystal Plasticity Model for Large Plastic Deformation Behavior: Application to AZ31 Magnesium Alloy. *Int. J. Plast.* **2016**, *83*, 90–109. [CrossRef]
139. Singh, J.; Kim, M.-S.; Kaushik, L.; Kang, J.-H.; Kim, D.; Martin, E.; Choi, S.-H. Twinning-Detwinning Behavior of E-Form Mg Alloy Sheets during in-Plane Reverse Loading. *Int. J. Plast.* **2020**, *127*, 102637. [CrossRef]
140. Gehrmann, R.; Frommert, M.M.; Gottstein, G. Texture Effects on Plastic Deformation of Magnesium. *Mater. Sci. Eng. A* **2005**, *395*, 338–349. [CrossRef]
141. Wang, Y.-Y.; Jia, C.; Xu, M.; Kaseem, M.; Tayebi, M. Microstructural Changes Caused by the Creep Test in ZK60 Alloy Reinforced by SiCp at Intermediate Temperature after KOBO Extrusion and Aging. *Materials* **2023**, *16*, 3885. [CrossRef]
142. Wang, Z.; Cao, G.; Wang, F.; Zhou, L.; Mao, P.; Jiang, X.; Liu, Z. Investigation of the Microstructure and Properties of Extrusion-Shear Deformed ZC61 Magnesium Alloy under High Strain Rate Deformation. *Mater. Charact.* **2021**, *172*, 110839. [CrossRef]
143. Wang, L.; Li, Y.; Zhang, H.; Zhang, Z.; Yang, Q.; Zhang, Q.; Wang, H.; Cheng, W.; Shin, K.S.; Vedani, M. Review: Achieving Enhanced Plasticity of Magnesium Alloys below Recrystallization Temperature through Various Texture Control Methods. *J. Mat. Res. Tech.* **2020**, *9*, 12604–12625. [CrossRef]
144. Wang, Y.-Y.; Jia, C.; Tayebi, M.; Hamawandi, B. Microstructural Evolution during Accelerated Tensile Creep Test of ZK60/SiCp Composite after KoBo Extrusion. *Materials* **2022**, *15*, 6428. [CrossRef]
145. Miao, X.; Li, Z.; Liu, S.; Hou, K.; Wang, J.; Yang, S. Ionic Bridging Strengthened MXene/GO Nanocomposite Films with Extraordinary Mechanical and Tribological Properties. *Appl. Surf. Sc.* **2023**, *625*, 157181. [CrossRef]
146. Li, X.; Guo, F.; Ma, Y.; Jiang, L.; Lai, H.; Liu, H.; Zhang, D.; Pei, R. Rolling Texture Development in a Dual-Phase Mg-Li Alloy: The Role of Temperature. *J. Magnes. Alloy* **2023**, *11*, 2980–2990. [CrossRef]

147. Watanabe, H.; Ishikawa, K. Effect of Texture on High Temperature Deformation Behavior at High Strain Rates in a Mg-3Al-1Zn Alloy. *Mater. Sci. Eng. A* **2009**, *523*, 304–311. [CrossRef]
148. Khan, A.S.; Pandey, A.; Gnäupel-Herold, T.; Mishra, R.K. Mechanical Response and Texture Evolution of AZ31 Alloy at Large Strains for Different Strain Rates and Temperatures. *Int. J. Plast.* **2011**, *27*, 688–706. [CrossRef]

MDPI AG
Grosspeteranlage 5
4052 Basel
Switzerland
Tel.: +41 61 683 77 34

Crystals Editorial Office
E-mail: crystals@mdpi.com
www.mdpi.com/journal/crystals

www.ingramcontent.com/pod-product-compliance
Lightning Source LLC
Chambersburg PA
CBHW061839210926
51850CB00006B/927
* 9 7 8 3 7 2 5 8 6 0 2 7 2 *